Neuronal Acetylcholine Receptors

Neuronal Acetylcholine Receptors

V. I. Skok,
A. A. Selyanko, and
V. A. Derkach

Bogomolets Institute of Physiology
Academy of Sciences of the USSR
Kiev, USSR

Translated from Russian by
the authors

CONSULTANTS BUREAU
NEW YORK AND LONDON

Library of Congress Cataloging in Publication Data

Skok, Vladimir I. (Vladimir Ivanovich)
[Neĭronal'nye kholinoret͡septory. English]
Neuronal acetylcholine receptors / V. I. Skok, A. A. Selyanko, and V. A. Derkach: translated from Russian by the authors.
p. cm.
Translation of: Neĭronal'nye kholinoret͡septory.
Includes bibliographies and index.
ISBN-13: 978-1-4684-1670-1 e-ISBN-13: 978-1-4684-1668-8
DOI: 10.1007/ 978-1-4684-1668-8
1. Acetylcholine—Receptors. 2. Neural receptors. 3. Muscarinic receptors. 4. Nicotinic receptors. I. Selyanko, A. A. II. Derkach, V. A. III. Title.
[DNLM: 1. Receptors, Cholinergic—physiology. WL 102.8 S628n]
QP364.7.S5813 1989
591.1'88—dc19
DNLM/DLC 88-36687
for Library of Congress CIP

This volume is published under an agreement with the Copyright Agency of the USSR (VAAP)

Softcover reprint of the hardcover 1st edition 1989
A Division of Plenum Publishing Corporation
233 Spring Street, New York, N.Y. 10013

PREFACE

Among the different types of receptors for neurotransmitters, nicotinic acetylcholine receptors were the first to be studied systematically; at present they are very well characterized. This is due to the discovery of two very convenient objects that are endowed with nicotinic acetylcholine receptors – the skeletal muscle and the electric organ. The large size of skeletal muscle fibers, which simplifies the intracellular recording of transmembrane potentials and currents, played a crucial role in obtaining the first quantitative estimates of the activity of acetylcholine receptors and the kinetics of their interaction with ligands. On the other hand, the extremely high content of receptor protein in the electric organ tissue – two orders higher than in muscle tissue – rendered it highly suitable for studying the biochemistry of receptors. The combination of pharmacological, electrophysiological, and biochemical approaches resulted in rapid progress in the investigation of acetylcholine receptors.

Nicotinic acetylcholine receptors are also present in the neurons of autonomic ganglia, in the central nervous system of vertebrates, and in the ganglion neurons of invertebrates. Although each of these three types of receptors has its own pharmacological specificity, some of their properties are common and differ from those in the acetylcholine receptors of skeletal muscle and electric organ. One of these differences is that neuronal nicotinic receptors usually coexist in the same nerve cell with other receptors, e.g., muscarinic, serotoninergic, or peptidergic. Another difference is the relatively low density of receptor distribution on the postsynaptic membrane and the slow kinetics of each single receptor–channel complex activity compared with those in skeletal muscle fiber. The fact that these properties are unique for neuronal nicotinic receptors implies that they are related to some specific neuronal functions, e.g., to integration of nerve activity achieved through interaction of several presynaptic inputs at the same postsynaptic neuron.

Neuronal acetylcholine receptors, both nicotinic and muscarinic, are involved in numerous important physiological processes which include visceral functions, body movements, and behavior. However, in spite of their great importance in physiology, neuronal acetylcholine receptors until recently remained little studied compared to those in skeletal muscle and in electric organ, with the exception of classical pharmacological studies of nicotinic receptors in autonomic ganglia which were mostly aimed at creating hypotensive drugs.

This peculiar situation has begun to change rapidly in recent years. First, modern electrophysiological techniques, in particular single-channel current recording, have been applied to autonomic ganglion neurons and have given surprising results that suggested new molecular mechanisms. These results provide a promising approach to the study of how other neuronal receptors operate, including those still little studied in the central nervous system. Second, the ionic properties and pharmacology of central muscarinic receptors have been studied in detail using slice preparation, cell culture, and histochemical techniques. The great importance of these recent findings in neurobiology made necessary their systematization and analysis, which is the main goal of this book.

Insofar as this book is a first attempt at describing neuronal acetylcholine receptors as a separate group of receptors, the description starts with their localization in the nervous system and in a nerve cell, their physiological functions, and their general molecular structure (Chapter 1). This is followed by an analysis of the molecular mechanisms involved in the activation of neuronal nicotinic receptors (Chapters 2 and 3) and in their pharmacological blockade (Chapters 4–7). Then, some common properties of neuronal muscarinic receptors are analyzed (Chapters 8 and 9). Finally, other mechanisms related to the functions of nicotinic and muscarinic receptors in the biology of nerve cells are discussed (Chapters 10 and 11).

V. I. Skok
A. A. Selyanko
V. A. Derkach

CONTENTS

Chapter 1. General Characteristics of Neuronal Acetylcholine Receptors

Chapter 2. Recognition Center of the Neuronal Nicotinic Acetylcholine Receptor

Chapter 3. Ionic Channel of the Nicotinic Acetylcholine Receptor

Chapter 4. General Characteristics of the Effects Produced by Blocking Agents in Neuronal Acetylcholine Receptors

Chapter 6. Localization and Possible Functional Significance of the Sites in the Neuronal Nicotinic Acetylcholine Receptor Interacting with Blocking Agents

Chapter 7. The Relation of Nicotinic Receptor Open-Channel Blockade to Blockade of Synaptic Transmission

Chapter 8. Recognition Center of Muscarinic Acetylcholine Receptor

Chapter 9. Mechanisms Underlying Muscarinic Postsynaptic Response

Chapter 10. Mechanisms of Muscarinic Regulation of Neuronal Electrical Activity

Chapter 1

GENERAL CHARACTERISTICS OF NEURONAL ACETYLCHOLINE RECEPTORS

1.1. STRUCTURE AND FUNCTIONS OF INTERNEURONAL CHOLINERGIC SYNAPSES

1.1.1. Autonomic Ganglia of Vertebrates

The autonomic ganglia and plexuses are accumulations of nerve cells located outside the central nervous system and connected with the central nervous system through preganglionic fibers. The cell bodies of preganglionic neurons lie within the brain stem or within the spinal cord, and their long processes, preganglionic fibers, make cholinergic synapses with the neurons in autonomic ganglia or plexuses. There are two systems of autonomic ganglia: sympathetic, which obtain preganglionic fibers from the thoracic and lumbar regions of the spinal cord, and parasympathetic, which obtain them from the brain stem and from the sacral region of the spinal cord. The long processes of neurons in the ganglia, or postganglionic fibers, connect the ganglia with peripheral organs (smooth muscles and glands) where they form synaptic contacts. For a more detailed description of autonomic ganglia, see reviews by Skok (1973) and Gabella (1976). The ganglia located within the walls of the stomach and gut, although they formally belong to the parasympathetic system, are much more complicated in their neuronal organization than sympathetic ganglia or the parasympathetic ganglia of the head, and for this reason they are sometimes classified as a separate division of the autonomic nervous system.

The principal scheme of pathways in sympathetic ganglion is shown in Fig. 1.1. The preganglionic fibers and the postganglionic neurons that they innervate constitute centrifugal pathways. These pathways are involved in the central regulation of peripheral target organs. In addition, there are peripheral reflex pathways in the ganglion that are formed by afferent fibers of peripheral origin making synaptic contacts with postganglionic neurons. Through peripheral reflex pathways some

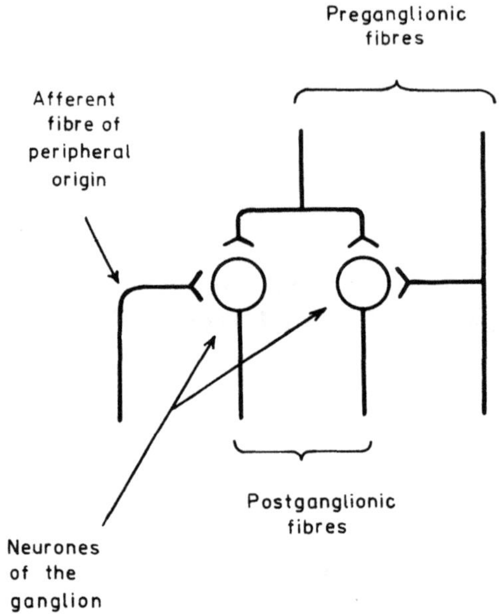

Fig. 1.1. Principal scheme of pathways in extramural autonomic ganglion (Skok, 1980).

peripheral organs may influence themselves or other peripheral organs independently of the central nervous system.

Preganglionic presynaptic terminals in the ganglia release acetylcholine which interacts with either nicotinic or muscarinic receptors of postganglionic neurons. Nicotinic transmission triggers a postsynaptic spike, while muscarinic transmission plays mostly, if not exclusively, a modulatory role. More detailed analysis of pathways in autonomic ganglia is given in special reviews (Bulygin, 1964, 1983; Skok, 1973, 1980; Nishi, 1974; Gabella, 1976; Kuba and Koketsu, 1978). Nicotinic synapses in centrifugal pathways, which have been better studied than those in other pathways, are described here in detail. Each preganglionic fiber, after it enters the ganglion, ramifies and then forms synaptic contacts with many neurons (multiplication). On the other hand, each neuron in mammalian ganglia receives a presynaptic supply from more than one preganglionic fiber (convergence). By contrast, in birds and in amphibia large groups of neurons are singly innervated, i.e., each neuron receives only one preganglionic fiber. The neurons that receive multiple innervation are multipolar, the synaptic contacts are located mainly on their dendrites, and, to a lesser extent, on their somata. The neurons with single innervation are unipolar (see Gabella, 1976).

Recent work has revealed that one of the converging preganglionic fibers usually is much more potent in its postsynaptic excitatory effect that other fibers converging upon the same neuron, and a presynaptic spike in this fiber always evokes a postsynaptic spike (dominant presynaptic input; see Skok and Ivanov, 1983). Other

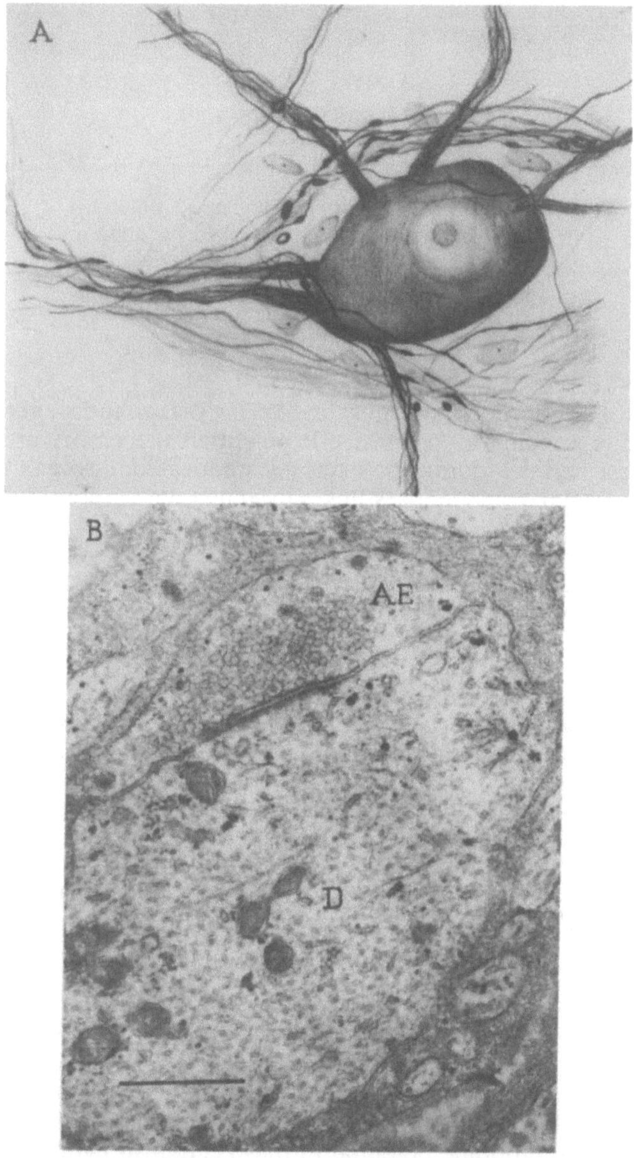

Fig. 1.2. Structure of interneuronal synapse in cat superior cervical ganglion neuron. A) Synaptic contacts between preganglionic fibers and neuronal dendrites and soma. Golgi–Deineke method (Babmindra and Dyachkova, 1968). B) Axo-dendritic contact. The axon ending (AE) forms the synapse with the large dendrite (D). The axon contains rounded vesicles. Dense areas are seen in the sites of contact between pre- and postsynaptic membranes. Solid bar measures 1 μm (L. M. Koval, unpublished data).

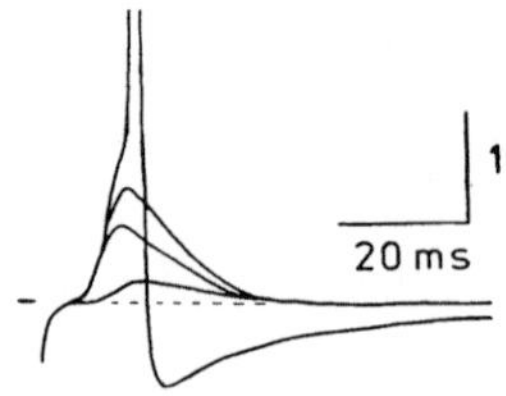

Fig. 1.3. Excitatory postsynaptic potentials and action potential evoked in rabbit superior cervical ganglion neuron by a single orthodromic stimulus of variable strength. The top of the action potential is not shown (Skok and Ivanov, 1983).

converging preganglionic fibers may evoke a postsynaptic spike only through summation of their excitatory postsynaptic potentials (accessory presynaptic input). It has been suggested that dominant synapses are located at soma close to where an axon initiates, similarly to synapses in singly innervated amphibian sympathetic B cells.

The ultrastructure of the cholinergic synapse in nerve cell is very similar to that in other types of cells, e.g., in muscle fibers. Presynaptic terminals contain numerous synaptic vesicles where acetylcholine is stored (Matthews, 1983). Acetylcholine can also be stored in extravesicular space (Birks and Fitch, 1974). In the synaptic area, the axoplasm of the presynaptic terminal is separated from the axoplasm in the postsynaptic nerve cell by a distance of 150-300 Å. The increased electron density areas are localized in both pre- and postsynaptic sites (Fig. 1.2).

In contrast to nicotinic transmission, where even a single presynaptic spike may trigger a postsynaptic spike (Fig. 1.3), in muscarinic transmission repetitive presynaptic stimulation is usually necessary to evoke a postsynaptic discharge. It is therefore not clear whether under normal conditions muscarinic transmission may result in a postsynaptic spike or whether its effect on the postsynaptic cell is only modulatory (Skok, 1980, 1981).

Nicotinic and muscarinic receptors have been found on the cells of the carotid body that obtain preganglionic innervation and modulate the chemical sensitivity of carotid receptors similarly to SIF cells in sympathetic ganglia described below (see Shepherd, 1979; Eyzaguirre and Monti-Bloch, 1982). Schwann cells also possess nicotinic receptors located in peripheral nerves; they may be responsible for long-lasting changes in membrane potential in these cells following repetitive firing of adjacent nerve fibers (Villegas, 1975).

In addition to acetylcholine receptors, the neurons of autonomic ganglia have receptors for dopamine, serotonin, and various peptides (see Skok, 1981). Sympathetic ganglia contain small fluorescent cells (SIF cells) that are thought to release dopamine or noradrenaline in response to activation of their muscarinic receptors with acetylcholine released by preganglionic fibers. With respect to peptides, they are considered as either transmitters or cotransmitters in the ganglia, including vasoactive intestinal polypeptide, substance P, somatostatin, enkefalins, gastrin, and cholecistokinin. Such multiple chemical sensitivity, found mostly in enteric neurons, is probably a common feature of many other types of neurons, including central neurons that are less accessible to pharmacological investigation.

The time interval between the presynaptic action potential and the onset of a change in membrane potential in the postsynaptic cell evoked by this action potential (synaptic delay) in nicotinic interneuronal transmission ranges from less than a millisecond to a few milliseconds, while in all other types of transmission in peripheral neurons (muscarinic, peptidergic, etc.) it may be as long as hundreds of milliseconds. There is, therefore, fast and slow synaptic transmission. This difference reflects a fundamental difference between nicotinic and other receptors of peripheral neurons: in the former, the receptor and ionic channel are coupled together in the same oligomeric protein molecule, while in the latter they are separate molecules linked through a long chain of intracellular chemical messengers, e.g., cyclic nucleotides or Ca^{2+}. In many central neurons fast synaptic transmission occurs through glutamate receptors instead of nicotinic receptors in peripheral neurons.

1.1.2. Central Nervous System of Vertebrates

In accordance with Dale's principle, acetylcholine receptors are found in Renshaw cells, the neurons in the spinal cord that obtain presynaptic input from axon collaterals of cholinergic motoneurons of the spinal cord. Renshaw cells evoke recurrent inhibition in the motoneurons. The receptors on Renshaw cells are of two types, nicotinic and muscarinic. The synaptic transmitters for other spinal interneurons are amino acids (see Shepherd, 1979).

Acetylcholine receptors have been found in some retinal nerve cells. Horizontal cells that generate only slow membrane potential changes without spikes have muscarinic receptors (Gerschenfeld and Piccolino, 1977). By contrast, amacrine cells, which generate spikes, obtain activator influences from bipolar cells through nicotinic receptors (Daniels and Vogal, 1980).

Cholinergic transmission is believed to occur in many parts of the brain. The main evidence for this is histochemical localization in the brain of acetylcholine, choline acetyltransferase (ChAc), the enzyme involved in the synthesis of acetylcholine, and acetylcholinesterase (AChE), the enzyme that destroys acetylcholine after it is released from presynaptic terminals. The highest content of ChAc and AChE is observed in the striatum, one of the basal ganglia in the brain (see Tuček, 1978). This correlates with muscarinic transmission into GABA-ergic striatal neurons (Ariano, 1985) that form an inhibitory pathway to other basal ganglia (see Shepherd, 1979). Muscarinic receptors are also located in hippocampal pyramidal neurons (Salmoiraghi and Stefanis, 1965; Bird and Aghajanian, 1975; Benardo and Prince, 1982a). Muscarinic transmission from nerve fibers that originate from the septum evokes slow depolarization in hippocampal pyramidal neurons followed by their firing (Cole and Nicoll, 1984a). This conclusion is supported by the finding that repetitive stimulation of the septal area results in the appearance in the hippocampal neurons of slow inward transmembrane currents which can be blocked by atropine (Gahwiler and Brown, 1985).

Evidence for the existence of nicotinic receptors in the brain is also provided by the specific binding of labeled nicotine and α-bungarotoxin in brain tissue homogenate (Yoshida and Imura, 1979; Romano and Goldstein, 1980; Clarke et al., 1984). This binding has been observed in the hypothalamus, hippocampus, and thalamus. Using the DNA hybridization technique, Boulter et al. (1986) have found nicotinic receptors in the hippocampus, in the dentate gyrus, and in the medial habenula of the striatum. Both nicotinic and muscarinic receptors exist in cells cultured from the brain stem of the rat embryo (Saji and Miura, 1982). It should be noted that nicotinic receptors revealed in the brain using autoradiographic or histochemical techniques may not be involved in synaptic transmission. This suggestion is based on the fact that intensive binding of α-bungarotoxin can be observed in sensory ganglia and in the areas of the hippocampus devoid of synaptic contacts (Hunt and Schmidt, 1979; Polz-Tejera et al., 1980). The functions of nicotinic receptors in these areas remain unclear. One more line of evidence that acetylcholine receptors may play an important role in brain activity is the strong effect the agonists and antagonists of acetylcholine exert upon learning and memory in animals and in man (Flood et al., 1981; Rinne et al., 1985).

Central nicotinic receptors are more sensitive to ganglion-blocking agents, e.g., hexamethonium, mecamylamine, and pempidine, than receptors in the endplate (see Morley, 1981) and are not homologous to nicotinic receptors from electric organ (Morley et al., 1983). Thus, in their pharmacological specificity nicotinic receptors of central neurons are more similar to those in the neurons of autonomic ganglia than to those in skeletal muscle and electric organ.

1.1.3. The Ganglia of Invertebrates

Neuronal acetylcholine receptors are present in the ganglia of molluscs (Kehoe, 1972; Gerschenfeld, 1973), annelids (Kerkut and Walker, 1967), and arthropods, particularly in crustacea and insects (Pitman, 1971).

Most detailed pharmacological characteristics have been obtained for molluscan receptors. Three different types of acetylcholine receptors have been found. Two nicotinic types cause depolarization of neuronal membrane, i.e., excitatory effect. The third type is muscarinic and causes hyperpolarization of neuronal membrane, i.e., inhibitory effect (Kehoe, 1972; Ger et al., 1979b). However, muscarinic receptors in molluscan neurons differ from muscarinic receptors in vertebrates in that they are insensitive to atropine and to some muscarinic agonists (see Chapter 4).

1.2. LOCALIZATION OF ACETYLCHOLINE RECEPTORS ON THE NEURONAL SURFACE

1.2.1. Binding of Labeled Antagonists

1.2.1.1. Nicotinic Receptors

To study the localization of nicotinic receptors in the muscle, the labeled α-neurotoxins from snake venom, which have a highly specific and irreversible blocking effect on synaptic transmission, have traditionally been used (Chang and Lee, 1963; Changeux et al., 1970; Lee et al., 1972a, b). This approach is also used in the investigation of neuronal nicotinic receptors, although the blocking effect of α-neurotoxins on nicotinic interneuronal synaptic transmission is not always observed (see Chapter 5).

Autoradiographic experiments reveal binding of α-neurotoxins in many regions of the central and peripheral nervous system (Greene et al., 1973; Fumagalli et al., 1976; McQuarrie et al., 1976; Wilson and Kirshner, 1977). However, only in two objects, in frog prevertebral sympathetic ganglion and in rat superior cervical ganglion, are the binding sites localized in the synaptic areas of a neuron (Marshall, 1981; Smolen, 1983). In frog sympathetic ganglion the binding sites for α-bungarotoxin are concentrated in the postsynaptic membrane under active zones of presynaptic nerve terminals. Only in these areas do the antibodies directed against acetylcholine receptors of electric organ attach to the postsynaptic membrane. These results correlate with the blocking effect of α-bungarotoxin on synaptic transmission in frog sympathetic ganglion (Marshall, 1981), indicating that α-bungarotoxin binds those nicotinic receptors that are involved in synaptic transmission. They are located at the postsynaptic soma membrane adjacent to the axon hillock (Weitsen and Weight, 1977).

In rat sympathetic ganglion α-bungarotoxin likewise binds to the subsynaptic areas of the neuronal membrane (Smolen, 1983). The neurons here, in contrast to those in amphibian ganglia, are multipolar and their subsynaptic areas are located mostly at their dendrites.

The mean number of toxin-binding sites per 1 μm^2 is 575 in the subsynaptic membrane and only 35.6 in the extrasynaptic membrane of the neuron. The latter value is close to that found for the extrasynaptic area in skeletal muscle fiber, while the former one is one to two orders lower than in the muscle end-plate (Fertuck and Salpeter, 1976; Matthews-Bellinger and Salpeter, 1983). Recent electrophysiological results have confirmed that α-neurotoxins interact with nicotonic receptors in the neurons of mammalian sympathetic ganglia (Quik and Lamarca, 1982; Selyanko, 1983), which implies that the locations of α-toxin-binding sites and receptors should be identical.

In chick ciliary ganglion the binding sites for α-bungarotoxin are predominantly located on neuronal somata (Gangitano et al., 1978), both in the subsynaptic and extrasynaptic membranes (Cocchia and Fumagalli, 1981; Jacob and Berg, 1983; Messing and Gonatas, 1983; Loring et al., 1985). They are distributed homogeneously at rather low density, 100-200 sites per 1 μm^2 (Loring et al., 1985). This density is intermediate between the above densities for toxin-binding sites for subsynaptic and extrasynaptic membrane areas in rat sympathetic ganglion neuron.

The sites that bind α-toxins in the ciliary ganglion of birds and mammals participate in synaptic transmission, which is evidenced by the blocking effect produced by some α-toxins on the transmission through the ganglion (Chiappinelli and

Zigmond, 1978; Conti-Tronconi et al., 1979; Chiappinelli et al., 1981). Therefore, the ciliary ganglion is a unique ganglion in that its nicotinic receptors are homogeneously distributed over the neuronal surface rather than being concentrated in subsynaptic membrane areas as in sympathetic ganglia and muscles. The physiological significance of this difference is not yet clear.

Each neuron in rat sympathetic ganglion contains, on the average, $9.2 \cdot 10^5$ binding sites for α-bungarotoxin (Fumagalli et al., 1976). This value is higher in avian ciliary ganglion: $3 \cdot 10^6$ (Chiappinelli and Giacobini, 1978) or even $9 \cdot 10^6$ sites (Gangitano et al., 1978). The difference between these ganglia, at least partially, may be due to the difference in the size of neurons, which implies that the mean density of nicotinic receptors is similar in different types of neurons (see Loring et al., 1985).

Although most of the toxin-binding sites in the neuron of autonomic ganglion are located in the neuronal membrane, some of them, rather stable in different measurements (about 20%), are usually revealed in the cytoplasm where they are concentrated in multivesicular bodies and lysosomes (Smolen, 1983; Loring et al., 1985). The appearance of labeled α-bungarotoxin in the cytoplasm can be prevented by tubocurarine, suggesting that this toxin is accumulated via activation of nicotinic receptors (Loring et al., 1985).

This suggestion is supported by the finding that the number of synaptic contacts and the sensitivity of a neuron to acetylcholine correlate with the number of toxin-binding sites. For example, in the course of development of autonomic ganglion the toxin-binding sites appear simultaneously with synaptic transmission (Gangitano et al., 1978). In cultured ganglion neurons the number of toxin-binding sites increases fivefold during the first seven days, and then ceases to increase (Messing and Kim, 1981). This process correlates with an increase in the number of functional synapses in the ganglion (Hirst and McLachlan, 1984). Postganglionic axotomy is followed by a decrease in the number of synaptic contacts in ganglion neurons (Matthews and Nelson, 1975) and by a decrease in the number of sites that bind α-bungarotoxin (Fumagalli and De Renzis, 1980). A similar loss in the number of synaptic contacts in the ganglion is observed after preganglionic denervation (Raisman et al., 1974). An intriguing finding is that denervation may not affect the binding of α-bungarotoxin to ganglion neurons (Fumagalli et al., 1976; Fumagalli and De Renzis, 1980) and their sensitivity to acetylcholine (Brown, 1969; see, however, Dun et al., 1976).

The neurons of the central nervous system, unlike the neurons of autonomic ganglia, have been little studied with respect to the distribution of their nicotinic receptors on the neuronal surface. The regional distribution of α-bungarotoxin-binding sites in rat brain is not identical to that of acetylcholine, which implies that the toxin-binding sites normally do not interact with acetylcholine (Schwartz et al., 1982). The number of α-bungarotoxin-binding sites in rat brain is two times that of binding sites to dendrotoxin which, unlike α-bungarotoxin, blocks nicotinic transmission in the central nervous system. This differs from what has been found in fish electric organ where the numbers of binding sites for α-bungarotoxin and for

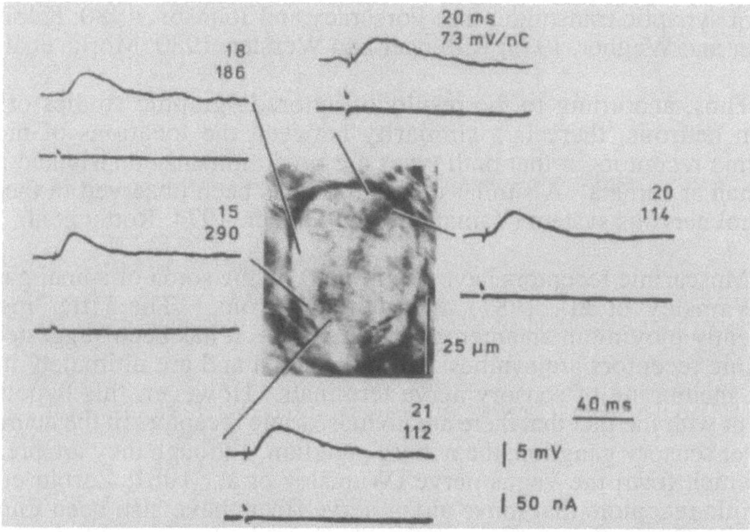

Fig. 1.4. Distribution of acetylcholine sensitivity over the surface of rabbit superior cervical ganglion neuron. Acetylcholine was applied to the places indicated by arrows. Membrane depolarizations (upper records) were elicited by 2-ms-long iontophoretic current pulses (lower records) of variable strength. Near each record the values of rise time of a response (in ms, above) and acetylcholine sensitivity (in mV/nC, below) are indicated (Selyanko, 1979).

dendrotoxin, which both block synaptic transmission, are identical. These results suggest that although α-bungarotoxin and dendrotoxin bind to nicotinic receptors in central neurons, binding of α-bungarotoxin does not prevent interaction of central nicotinic receptors with acetylcholine. Other evidence that supports this suggestion has been mentioned above.

1.2.1.2. Muscarinic Receptors

After incubation of rabbit superior cervical ganglion with [^{3}H]-atropine ($1 \cdot 10^{-6}$ M), the highest density of labels is observed in the areas between the neuronal somata, i.e., in the neuropile (Koval et al., 1982). Comparison of the autoradiograms with histological preparations stained using different methods indicates that the neuronal components that are most labeled are dendrites. In 62% of neurons the soma is completely devoid of labeling, while in the rest the labeling found in the soma is only about 12% of that in the dendrites.

A similar predominant distribution of muscarinic receptors on neuronal dendrites has been found in cultured enteric neurons (Buckley and Burnstock, 1984). The sites that bind labeled propylbenzilylcholine, a muscarinic antagonist, are concentrated on the neuronal processes, while only 10-20% of the label appears on the soma. The density of muscarinic receptors approaches 100 per 1 μm^2 (Buckley and Burnstock, 1986). Some muscarinic receptors are located at the terminal part of nerve fibers, suggesting that they may play a modulatory role in the release of synaptic transmitter (see Forsbracy and Johnson, 1980; Kilbinger, 1977; Kilbinger and Wagner, 1975; Kilbinger and Wessler, 1980; Morita et al., 1982).

Thus, according to the results of autoradiographic studies of autonomic ganglion neurons, there is a similarity between the locations of nicotinic and muscarinic receptors in that both types are predominantly distributed at dendrites rather than at somata. A similar distribution has been observed in the neurons of the central nervous system (Yamamura and Snyder, 1974; Rotter et al., 1979).

Muscarinic receptors have been found on the soma of spinal ganglion neurons (Wamsley et al., 1981) and in their axons. The latter receptors are permanently moving in an anterograde direction. It has been suggested that these muscarinic receptors are synthesized in the soma and are ultimately incorporated into the membrane of sensory nerve terminals. However, this hypothesis is not consistent with the fact that there are no muscarinic receptors in the neuronal somata in another sensory ganglion, the nodose ganglion, although they are present in their axons which form the vagus nerve (Wamsley et al., 1981; Zarbin et al., 1982). Muscarinic receptors that move along nerve fibers have also been found in many other nerve fibers of central or peripheral origin (Laduron, 1980; Wamsley, 1983). The functions of these receptors remain unclear.

1.2.2. Distribution of Sensitivity to Acetylcholine and to Acetylcholine Agonists on the Neuronal Surface

1.2.2.1. Nicotinic Receptors

The distribution of acetylcholine receptors on the neuronal surface obtained from autoradiographic experiments has further been clarified using another approach – intracellular recording of membrane potential changes (ACh potential) induced by local iontophoretic application of acetylcholine. Such a method performed under visual control through a micropipette with a tip diameter of about 0.1 µm allows one to activate acetylcholine receptors within an area of less than 4-µm radius (Roper, 1976).

With a soma as large as 40–50 µm in diameter, this approach allows one to test the sensitivity to acetylcholine in different areas of the soma surface. Most importantly, one can compare the sensitivity to acetylcholine in the membrane areas close to the presynaptic nerve terminals (which can be visualized through differential interference optics) with that in the extrasynaptic membrane areas. Experiments of this kind have been performed on singly innervated neurons of amphibian cardiac ganglia (Harris et al., 1971) and in multiply innervated neurons of mudpuppy cardiac ganglion (Roper, 1976) and rabbit superior cervical ganglion (Selyanko, 1979).

It has been found that the sensitivity to acetylcholine is two to three times higher in synaptic than in extrasynaptic membrane or cardiac neuron (Harris et al., 1971; Roper, 1976). This ratio is lower than in amphibian skeletal muscle fiber, where it reaches about 100 (Dreyer and Peper, 1974). On the other hand, the absolute value of the sensitivity to acetylcholine calculated as a ratio of the ACh

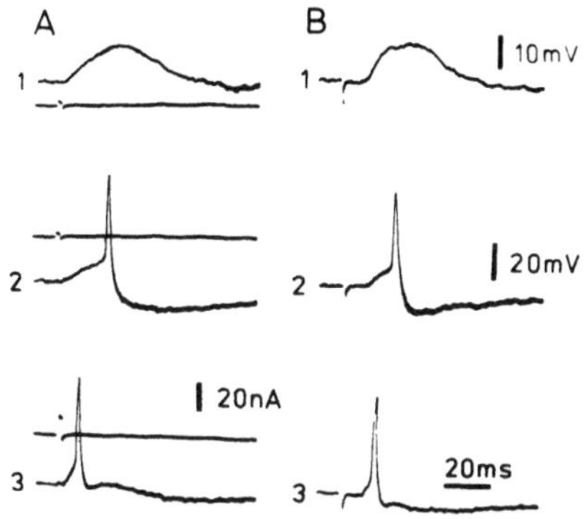

Fig. 1.5. Comparison between responses of the same rabbit superior cervical ganglion neuron to iontophoretic applications of acetylcholine (A) and to a single orthodromic stimulus (B). Iontophoretic and stimulating currents increase from upper to lower records. Iontophoretic current is shown in A. Voltage calibrations for 1 and 2–3 are shown separately (Selyanko and Skok, 1979).

potential to the iontophoretic charge (see Miledi, 1960) is several times lower in the neuron than in the muscle fiber. This difference correlates with the above difference between the neuron and the muscle fiber in the density of distribution of α-bungarotoxin-binding sites.

Figure 1.4 shows an example of the variability in acetylcholine sensitivity over the surface of a neuron. ACh potentials of equal amplitudes evoked by different doses of acetylcholine (indicated) are shown. The maximum difference in sensitivity observed in this experiment is fourfold. Maximum sensitivity correlates with the shortest rise time of the ACh potential. The variability in acetylcholine sensitivity is apparently due to the variable density in the distribution of acetylcholine receptors; the density is maximum in synaptic areas and minimum in extrasynaptic areas of the neuronal surface.

Application of acetylcholine to the site with maximum sensitivity is followed by the appearance of an ACh potential that is almost as long as the excitatory postsynaptic potential evoked by a single orthodromic stimulus (Fig. 1.5). The rise time of this ACh potential does not increase with increase in acetylcholine dose, indicating that the density of acetylcholine receptors is high and that they do not saturate (see Katz and Miledi, 1964; Dreyer and Peper, 1974).

The distribution of acetylcholine receptors on the membrane of amphibian cardiac ganglion neuron is strongly related to synaptic contacts. The changes in distribution of this sensitivity that follow denervation of the neuron are similar to those observed in muscle fiber: The sensitivity in extrasynaptic areas increases and finally reaches the level normally observed in synaptic areas (Kuffler et al., 1971;

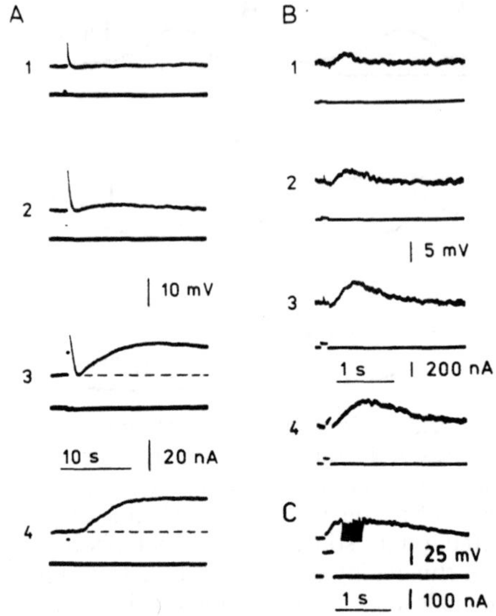

Fig. 1.6. Depolarizations elicited by local iontophoretic applications of acetylcholine and 5-methylfurmethide to the surface of rabbit superior cervical ganglion neuron. A) Monophasic responses (fast depolarizations) and biphasic responses (fast and slow depolarizations) caused by acetylcholine (iontophoretic current was increased from 1 to 3), while 5-methylfurmethide causes slow depolarization (4). B) Depolarization caused by increasing (from 1 to 4) doses of 5-methylfurmethide. C) Discharge caused by 5-methylfurmethide (the tops of the spikes are off the screen). In each pair of records the upper record indicates the membrane potential and the lower record indicates the iontophoretic current; durations of iontophoretic current pulses are 50, 100, and 200 ms for A, B, and C, respectively. Records A, B, and C were obtained from 3 different neurons; resting membrane potentials in these neurons were –45, –48, and –49 mV, respectively.

Roper, 1976). Reinnervation returns extrasynaptic sensitivity to its normal level (Dennis and Sargent, 1979).

In contrast to these results, no change in the distribution of acetylcholine sensitivity following denervation has been observed in amphibian sympathetic ganglion neurons (Dunn and Marshall, 1985). Thus, the effects of denervation may differ between sympathetic and parasympathetic ganglion. This finding is consistent with the fact that neither the total sensitivity to acetylcholine of each separate neuron nor the total number of binding sites to α-bungarotoxin reveals any change after denervation in mammalian sympathetic ganglion (Dun et al., 1976; Fumagalli et al., 1976; Fumagalli and De Renzis, 1980).

1.2.2.2. Muscarinic Receptors

The localization of muscarinic receptors on the neuron of autonomic ganglion has been studied with the use of local iontophoretic application of acetylcholine and of the muscarinic agonist 5-methylfurmethide (5-MF) (Koval et al., 1982). Experiments performed on rabbit superior cervical ganglion have shown that in about one-third of the neurons studied acetylcholine causes a biphasic response consisting of fast and slow depolarizations. Fast depolarization is illustrated in Figs. 1.4 and 1.5A. 5-MF evokes only slow depolarization, or MF-depolarization (Fig. 1.6A, B). The latency and duration of MF-depolarization are comparable to those in the slow excitatory postsynaptic potential evoked in ganglion neurons by stimulation of preganglionic fibers (Figs. 9.1 and 9.2).

The rise time and half-decay time of MF-depolarization increase with increasing dose of 5 MF, indicating that more distant receptors are activated with increased iontophoretic current. Usually a stronger iontophoretic current is needed to evoke an MF-depolarization of the same amplitude if 5-MF is applied to the soma than if it is applied to cell processes near the soma. Sometimes MF-depolarization is accompanied by action potentials (Fig. 1.6C).

These results suggest that the highest sensitivity to 5-MF is localized in cell processes. Another observation favors this suggestion. The rise time and half-decay time of MF-depolarization vary in different neurons from 150 ms to 10 s and from 150 ms to 21 s, respectively. This is apparently due to the fact that the MF-sensitive areas of the neuronal membrane are localized at greatly varying distances from the site of 5-MF application. Some neurons, in addition to MF-depolarization, respond to 5-MF by MF-hyperpolarization of much shorter latency. This kind of response is described in Chapter 9.

It can be concluded from the above results that muscarinic receptors are localized mostly on the dendrites of autonomic ganglion neurons and, to a lesser extent, on their somata, and are concentrated at the sites of synaptic contacts.

1.2.2.3. Acetylcholine Receptors in the Neurons of Invertebrate Ganglia

Our knowledge of the distribution of acetylcholine receptors in invertebrate ganglia is not as complete as in the case of vertebrate autonomic ganglia. Only molluscan ganglia have been studied in detail in this respect. As has been mentioned above, two types of acetylcholine receptors responsible for depolarization and hyperpolarization are present in molluscan ganglion neurons. For example, stimulation of the neuron L10 from the abdominal ganglion of *Aplysia* always causes excitation in the neuron R15 and inhibition in other neurons L2, L3, L4, and L6 (see Kandel, 1976). The total soma surface in these postsynaptic neurons is acetylcholine-sensitive. In other molluscs, only that part of the neuronal soma facing the neuropile, i.e., the internal medium of the ganglion, is sensitive to acetylcholine (Levitan and Tauc, 1972).

An interesting observation has been made in cultured neurons of the leech (Pellegrino and Simonneau, 1984). Immediately after isolation of the neurons from the ganglia, the acetylcholine sensitivity of their somatic surface is homogeneous. Later, after the cell processes appear, the areas adjacent to the soma become more sensitive to acetylcholine than other areas of the neuronal surface. The areas possessing different acetylcholine sensitivities are not otherwise pharmacologically different, which is evidenced by the fact that neuronal responses to application of acetylcholine are fully blocked by α-bungarotoxin and d-tubocurarine.

1.3. BIOCHEMICAL PROPERTIES, MOLECULAR STRUCTURE, AND FUNCTIONAL ORGANIZATION OF AN ACETYLCHOLINE RECEPTOR

The quaternary and primary structures are at present best studied for nicotinic receptors of electric organ and of skeletal muscle. Although, most recently, information of this kind has also been obtained for neuronal acetylcholine receptors – both nicotinic and muscarinic (see also Chapter 8.2) – this information remains incomplete. For this reason, some general properties of nicotinic receptors obtained mostly from an analysis of the electric organ tissue are briefly summarized in this chapter. A more detailed description is given in special reviews (Lindstrom, 1980; Kistler et al., 1982; Devillers-Thiery et al., 1983; Karlin et al., 1983; Raftery et al., 1983; Popot and Changeux, 1984).

The nicotinic receptor from fish electric organ is a glycoprotein that contains about 3–5% carbohydrates, including mannose, glucose, and N-glucosamine (Karlin et al., 1976; Raftery et al., 1976). The receptor contains more than 10 bound calcium ions per one acetylcholine-binding site (Eldefrawi et al., 1975). The receptor pentameric protein is composed of five subunits of four types with a resulting quaternary structure α_2, β, γ, δ. All four polypeptide chains are glycosylated, and their true molecular masses M_r obtained from their amino acid sequences are 50,200 (α), 53,700 (β), 56,300 (γ), and 57,600 (δ), yielding an M_r for the whole protein of 268,000 (see Fairclough et al., 1983; Young et al, 1985; Hucho, 1986).

Although the native nicotinic receptor is thought to form a dimer covalently linked together by a disulfide bridge between the subunits (Chang and Bock, 1977; Hamilton et al., 1979), the monomeric form is likewise functionally active: Its activation by acetylcholine is followed by ionic channel opening (Lindstrom, 1980).

A three-dimensional model of the nicotinic receptor developed from electron microscopy and x-ray analysis (see Kistler et al., 1982) is shown in Fig. 1.7A. In the membrane plane the receptor molecule is a roughly circular rosette of five subunits (Fig. 1.7B) with an overall diameter of 8–9 nm and diameter of each subunit 3–4 nm, and with an arrangement αβαγδ, as confirmed by later observations (Hamilton et al., 1985). The central pit is believed to be an ionic channel. In its axis normal to the membrane the receptor may have a dimension of 7–8 nm or more. The receptor spans the lipid bilayer, extending about 5.5 nm beyond the surface that faces the extracellular space and 1.5 nm beyond the other surface (see Klymkowsky and Stroud, 1979; Fambrough, 1979; Brisson and Unwin, 1985). In their amino acid sequences all four subunits are different but highly homologous

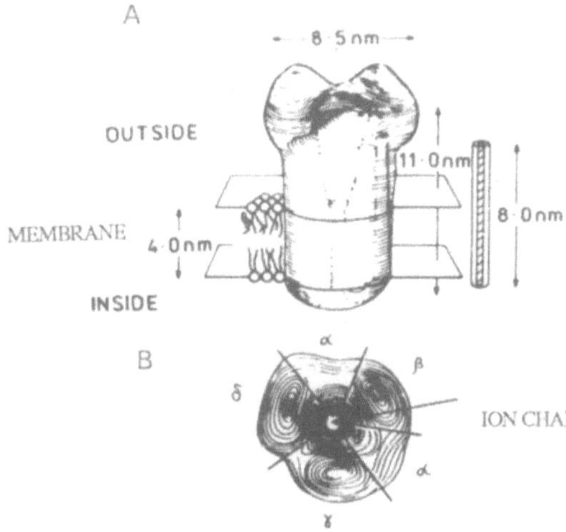

Fig. 1.7. Schematic three-dimensional representations of nicotinic acetylcholine receptor in a lipid bilayer. The protein topography was deduced from both electron-microscopy and x-ray diffraction. A) Side view; B) top view; subunits α, β, γ, and δ are indicated (McNamee and Ochoa, 1982).

polypeptides with an average homology of about 40% (see Gullick and Lindstrom, 1982). This indicates that the genes for the four polypeptide chains may have one common ancestor gene. This ancestor gene may still be preserved in some insects whose nicotinic receptor contains only one polypeptide chain (see Hucho, 1986). By cloning and sequencing the complementary DNAs encoding the respective subunit precursors, the amino acid sequences of the four subunits have been determined for fish electric organ nicotinic receptor (Noda et al., 1982, 1983b, c) and for the skeletal muscle nicotinic receptor of man and calf (Noda et al., 1983a; Takai et al., 1984; Tanabe et al., 1984; Shibahara et al., 1985). In particular, it has been shown that there is a high degree of sequence conservation between the mouse muscle γ subunit and human and calf γ subunits, while chicken and *Torpedo* γ subunits are more distant (Yu et al., 1986).

The amino acid sequence of the α subunit of a neuronal nicotinic receptor has very recently been obtained (Boulter et al., 1986). The rat phaeochromocytoma cell line PC12 expresses a neuronal acetylcholine receptor that is pharmacologically similar to that found on the neurons of sympathetic ganglion. The cDNA clone coding for the α subunit of mouse muscle nicotinic receptor has been used to detect, by low-stringency DNA/DNA hybridization, related sequences expressed in the PC12 cell line. One of the cDNA clones thus obtained, PCA38, has been used to deduce the amino acid sequence of the putative α subunit of a neuronal nicotinic receptor. This sequence reveals 47% homology with that in the α subunit of mouse muscle nicotinic receptor. Figures 2.8 and 3.20 illustrate two functionally important regions of the amino acid sequence in the neuronal α subunit compared with those in mouse muscle and in fish electric organ.

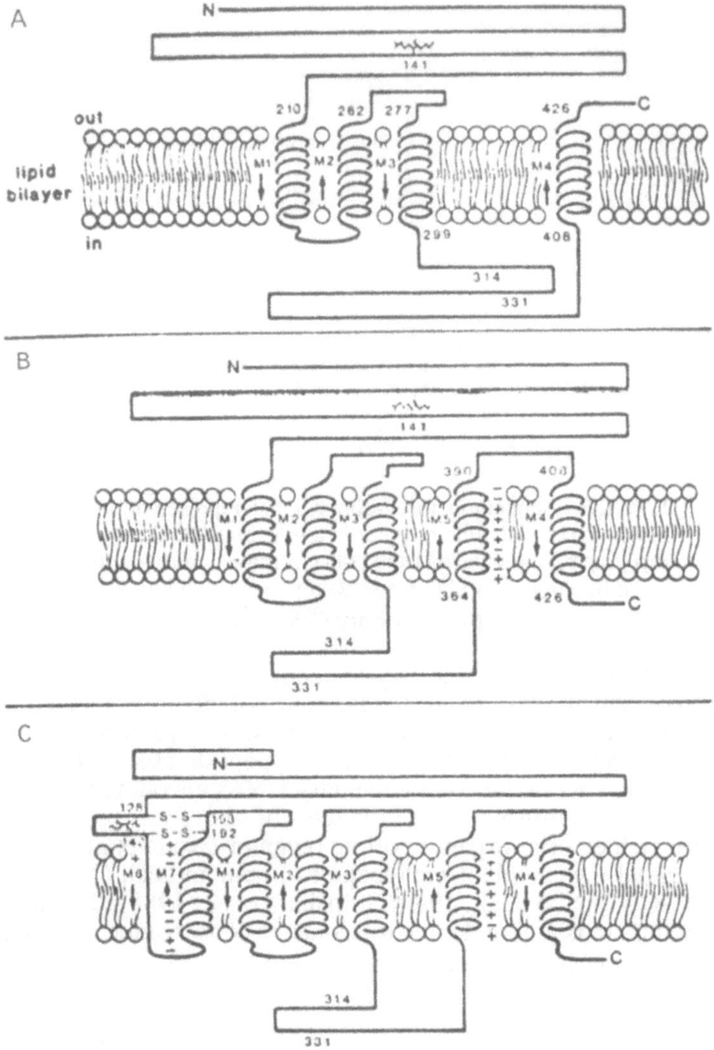

Fig. 1.8. Models of polypeptide chain organization in nicotinic acetylcholine receptor subunits. A) The four hydrophobic transmembranous domain model proposed by Claudio et al. (1983) and Devillers-Thiery et al. (1983) on the basis of an analysis of the distribution of hydrophobic residues in the sequences of receptor subunits. B) A model with amphipathic transmembrane domain proposed by Finer-Moore and Stroud (1984) and Guy (1984) as a result of a different theoretical analysis. C) A model with two additional amphipathic domains to account for the transmembrane orientation of various sequences determined by immunochemical and other techniques. Sequence numbering is for α subunits (Noda et al., 1982). The organization of the polypeptide chain in the extramembranous domain is not specifically depicted, except to suggest that the NH_2 terminus is not exposed on the subunit surface. The carbohydrate is shown attached at α128 or 142 and α192 or 193. It is proposed that the α subunits differ in glycosylation. This may affect which disulfide occurs with α192 or α193, which may in turn account for the observation that one α is easily reduced and labeled with MBTA, while the other is difficult to sustain reduced and reactive with MBTA (Criado et al., 1985).

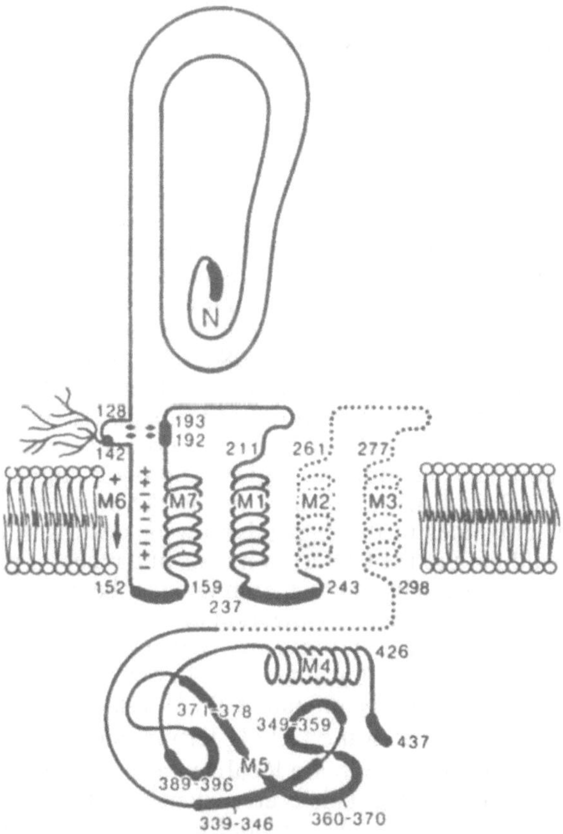

Fig. 1.9. A model of the transmembrane orientation of the polypeptide chain in the α subunit of the *Torpedo* acetylcholine receptor. This model summarizes the above results (Ratnam et al., 1986). Thick-line segments indicate sequences whose transmembrane orientation has been explicitly determined by the experiments described in the text. Thin-line segments indicate sequences whose transmembrane orientation is most likely due to constraints caused by the location of surrounding sequences. Dotted-line segments indicate sequences whose transmembrane orientation has not been determined.

In another approach, the acetylcholine receptor from chicken and rat brain has been purified using an immunoaffinity purification procedure (Whiting and Lindstrom, 1987a). This receptor is similar in size to the monomers of receptors from the *Torpedo* electric organ but contains two subunits, with an apparent M_r equal to 51,000 and 79,000. The authors conclude that the neuronal acetylcholine receptor apparently has the structure $(\alpha)_2(\beta)_2$ or $(\alpha)_3(\beta)_3$, in contrast to the structure $(\alpha)_2\beta\gamma\delta$ of the *Torpedo* acetylcholine receptor mentioned above. Further analysis suggests that the recognition center in this neuronal receptor may be located in the β rather than in the α subunit (Whiting and Lindstrom, 1987b; see Chapter 2).

The secondary structure of each subunit has been theoretically predicted from their respective amino acid sequences (see Chapter 2.3). In addition, the transmembrane organization of all five subunits has been postulated using three main approaches: 1) high-resolution electron microscopy; 2) determination of how a polypeptide chain is arranged in the membrane considering its hydrophobicity; 3) immunological methods to determine localization of individual subunit domains in the membrane (see Noda et al., 1983c; Kosower, 1983; Devillers-Thiery et al., 1983; Criado et al., 1985; Hucho, 1986). Figure 1.8A shows one of the models of the transmembrane organization of nicotinic receptor (Claudio et al., 1983; Devillers-Thiery et al., 1983; Noda et al., 1983a-c). According to this model, each of the four subunits forms four transmembrane domains (M_1–M_4), and both the NH_2 terminus and the COOH terminus of each subunit are on the extracellular surface. However, subsequent analysis of the amino acid sequences using another computerized technique and immunological approach suggested other models, which are shown in Fig. 1.8B, C. Antibodies to the synthetic peptides similar in their amino acid sequences to the respective regions of a subunit were obtained. The antibodies could bind to the COOH terminals of a receptor protein reconstituted into liposomes only after the liposome membrane had been destroyed by detergent. These results favored intracellular localization of the COOH terminus (Ratnam and Lindstrom, 1984), as illustrated by Fig. 1.8B. Other results obtained with a similar approach suggest that the NH_2 terminus may likewise be located inside the cell (Ratnam and Lindstrom, 1984; Neumann et al., 1984). Furthermore, the immunological results suggest that both the COOH terminus of the δ subunit (sequence δ501–516) and sequence β350–358 are localized at the intracellular surface of the membrane (Young et al., 1985). This is consistent with the hypothesis that there is another transmembrane domain, M_5, in addition to those shown in Fig. 1.8A, an amphipatic helix located between domains M_3 and M_4.

Finally, antibodies against the sequences α152–159 and α-235–242 which, according to the models shown in Fig.1.8A, B, are localized at the opposite sites of the membrane, exhibit competition for binding at the extracellular membrane surface (Criado et al., 1985b). This competition has been explained as being due to steric hindrance caused by the very close location of the above two sequences. Cys-192, in agreement with the results obtained by other authors, is localized at the extracellular membrane surface (Kao et al., 1984), the only reasonable explanation being that there are two additional transmembrane domains, M_6 and M_7, between the NH_2 terminus and the domain M_1, as shown in Fig. 1.8C. The domain M_7 is an amphipatic helix similar to domain M_5 (Finer-Moore and Stroud, 1984). As for the transmembrane domain M_6, it is not long enough to span the membrane in a helical conformation; therefore, its secondary structure remains unclear (Criado et al., 1985a). The model shown in Fig. 1.8C summarizes most of the theoretical and experimental results mentioned above.

However, in the most recent studies with the use of antibodies specific for the sequences α330-346, α349-364, α360-378, α379-385, and α389-408, it has been shown that the antibodies bind to receptor-rich membrane vesicles only after the vesicles become permeable, and electron microscopy indicates that the antibodies bind to intracellular surface of vesical membrane (Ratnam et al., 1986). These results suggest that all above sequences are located on the intracellular side of the membrane. The model for the transmembrane orientation of receptor subunit polypeptide chains, consistent with these results, is demonstrated by Fig. 1.9.

The functional organization of the nicotinic receptor is suggested by the following experimental results. α-Bungarotoxin protects the nicotinic receptor of electric organ from binding with specific alkylating ligands, i.e., ligands interacting probably with the site that normally binds acetylcholine (Karlin et al., 1976; Damle and Karlin, 1978). On the other hand, some toxins and local anesthetics may block the acetylcholine-induced increase in ionic permeability measured in physiological experiments, but without blocking acetylcholine binding measured directly in biochemical experiments – such as perhydrohistrionicotoxin and ceruleotoxin (Albuquerque et al., 1973; Bon and Changeux, 1977). It can be concluded that there are at least two functionally different components in the nicotinic receptor molecule – the recognition center and the ionic channel. The former binds α-bungarotoxin and specific alkylating ligands, while the latter binds perhydrohistrionicotoxin and ceruleotoxin as well as local anesthetics. Although this differentiation is somewhat oversimplified, it appears useful to systematize the numerous experimental results obtained.

Another important point is that α-toxins bind to both α subunits rather than to any other subunit (Reynolds and Karlin, 1978; Damle and Karlin, 1978), indicating that the recognition center is localized at the α subunits. It should be noted, however, that some neurotoxins can bind to more than one subunit (Ovchinnikov et al., 1983b; Tsetlin et al., 1983, 1984).

At least two molecules of acetylcholine are necessary to open the ionic channel of the nicotinic receptor, which is evidenced by a Hill number of about 2 (see Chapter 2.1). Two binding sites for acetylcholine have also been found in a nicotinic receptor reconstituted in phospholipid membrane (Lindstrom, 1980). Examination of acetylcholine binding to nicotinic receptors in equilibrium in biochemical experiments has revealed two types of binding sites, one with high affinity ($K_D = 1 \cdot 10^{-8}$ M) and one with low affinity ($K_D = 1 \cdot 10^{-4}$ M; Raftery et al., 1983).

The above results suggest that both acetylcholine and competitive antagonists interact with α subunits. In the case of noncompetitive antagonists (local anesthetics, antiarrythmic and psychogenic agents), their binding sites are not yet clear. Two distinct mechanisms of their blocking effect on the nicotinic receptor have been suggested: 1) an allosteric mechanism that accelerates the transmission from the nonactivated state of the receptor to its desensitized state, and 2) steric blockade of the ionic channel by direct binding to the channel (see Changeux et al., 1983). The latter mechanism implies binding of a blocking agent either to a closed channel (Selyanko et al., 1982) or to an open channel (see Steinbach, 1980; Adams, 1981).

The binding sites in the channel area may be of two types, possessing high and low affinity to channel-blocking agents. Examples are the sites that bind perhydrohistrionicotoxin, meproadiphen, and phencyclidine with $K_D = (2\text{–}8)\cdot10^{-7}$ M and with $K_D = 1\cdot10^{-4}$ M. In both cases each monomer binds only one ligand molecule; binding with low affinity may occur with any subunit (Changeux et al., 1983).

It is assumed that the nicotinic receptor interacts with acetylcholine or with noncompetitive antagonists as a typical allosteric protein. This interaction results in several conformational transitions whose description requires a model with at least four states. A concentration of acetylcholine as high as 10^{-4} M, close to its tentative concentration in the synaptic cleft, is required to induce fast transition from the resting state of the receptor to its desensitized state. This transition is followed by a positive homotropic interaction between the two binding sites for acetylcholine and makes the high-affinity binding sites available for noncompetitive antagonists. In addition, interaction of acetylcholine with the high-affinity binding sites ($K_D = 10^{-6}\text{–}10^{-9}$ M) is followed by at least two slow conformational transitions that result in a two-step desensitization. It is suggested that similar high-affinity transitions can be evoked in the nicotinic receptor by the interaction of noncompetitive antagonists with closed ionic channels (Changeux et al., 1983; Karlin et al., 1983; Raftery et al., 1983).

Desensitization is one of the fundamental features of nicotinic receptors. It follows a relatively long action on the receptors of an agonist at high concentration and can be observed as a gradual decrease in their integral response. In neuronal nicotinic receptors, as in the end-plate, desensitization develops with a two-exponential time course (Feltz and Trautmann, 1982; Chesnut, 1983). The corresponding two time constants, as measured in molluscan neurons, are of the order of seconds and tens of seconds (Andreev et al., 1984).

It can be concluded that interaction with acetylcholine evokes a chain of rather complex conformational changes in the receptor molecule. The increase in ionic permeability usually observed is the only result of these changes that has been studied in detail with a sufficient time resolution. For the increase in ionic permeability following application of acetylcholine, the whole complex of five subunits is necessary (Lindstrom, 1980; Mishina et al., 1984). This fact supports the above suggestion that the ionic channel is formed by all five subunits.

In addition, Ca^{2+} ions are thought to play an important role in the normal activity of the nicotinic receptor. About four to six Ca^{2+} ions are released by a receptor protein as a result of its activation with acetylcholine, while the subsequent application of α-bungarotoxin causes the reuptake of Ca^2 released. Based on these observations, it has been suggested that acetylcholine competes with Ca^{2+} ions for their binding sites in the receptor molecule (Chang and Neumann, 1976). It is commonly accepted that Ca^{2+} ions with their high coordination number are important in maintaining the conformation of the protein molecule. This makes possible their direct participation in the acetylcholine-induced conformational transitions of the receptor molecule, including the ionic channel opening.

1.4. CONCLUSIONS

Acetylcholine receptors are found in a few types of neurons, including the neurons of the autonomic ganglia, of the retina, and of the central nervous system in vertebrates, as well as the neurons in the ganglia of invertebrates. Neuronal acetylcholine receptors are involved in many important physiological processes, from autonomic regulation of visceral functions to learning. There are some functional and pharmacological differences between neuronal nicotinic acetylcholine receptors and nicotinic acetylcholine receptors in other tissues. One difference is that nicotinic receptors in neurons usually coexist with other types of receptors (muscarinic, serotoninergic, etc.). Other differences are the markedly low density of receptor distribution in the subsynaptic areas of the neuronal surface and the slow kinetics of neuronal receptor activity.

In multipolar neurons both nicotonic and muscarinic receptors are located predominantly on dendrites and, less often, on soma. The density of nicotinic receptors is much higher in synaptic than in extrasynaptic neuronal membrane (575 and 36 μm^{-2}, respectively), which correlates with the difference in sensitivity to acetylcholine.

The results obtained from nicotinic receptors of the fish electric organ suggest that, although the native nicotinic receptor is probably a dimer, each separate monomer is likewise functionally active. The monomer is an integral membrane protein molecule of cylindrical shape that penetrates the membrane perpendicular to the membrane plane and consists of five subunits of $\alpha_2\beta\gamma\delta$ (in neurons $\alpha_2\beta_2$ or $\alpha_3\beta_3$) stoichiometry, with a total molecular mass of about 268,000. The central part of the receptor molecule is thought to be an ionic channel. Two functionally different components are differentiated in each monomer: The recognition center is formed mainly by two α subunits, and the ionic channel is formed by all five subunits. The recognition center binds acetylcholine and competitive antagonists, while the ionic channel is responsible for translocation of ions through the membrane.

Chapter 2

RECOGNITION CENTER OF THE NEURONAL NICOTINIC ACETYLCHOLINE RECEPTOR

As has been mentioned in Chapter 1, two functionally different components can be distinguished in a nicotinic receptor molecule. One is a recognition component that normally binds acetylcholine, thus causing the ionic channel opening. The other component is the ionic channel. The acetylcholine-binding site, or recognition center, should be located at two α-subunits and may bind acetylcholine, other agonists, and competitive antagonists.

By contrast, noncompetitive antagonists interact with other sites in the receptor molecule, e.g., with the open or closed ionic channel; this process will be described in more detail in Chapter 5. There is evidence that binding of noncompetitive antagonists may facilitate binding of the competitive ones, and vice versa (see Changeux et al., 1983). This interaction is probably similar to the positive cooperation observed between two acetylcholine-binding sites, which will be described below in this chapter. Thus, the relations between the recognition components and the ionic channel are probably bidirectional. Moreover, acetylcholine is able to combine with the open ionic channel of a nicotinic receptor (see Lindstrom, 1980) causing, if used in high concentrations, the open-channel blockade (Colquhoun and Ogden, 1984; Sine and Steinbach, 1984).

These observations imply that the commonly accepted insight into the recognition center as a structure distinct from the ionic channel is somewhat oversimplified, and that the recognition center should probably include, at least partially, the ionic channel area. However, to make the description consecutive, we shall consider the characteristics of these two components apart from each other.

2.1. STOICHIOMETRY IN THE AGONIST–RECEPTOR INTERACTION

The approach most frequently used in studying the agonist–receptor stoichiometry is the determination of the Hill number n. This number indicates the num-

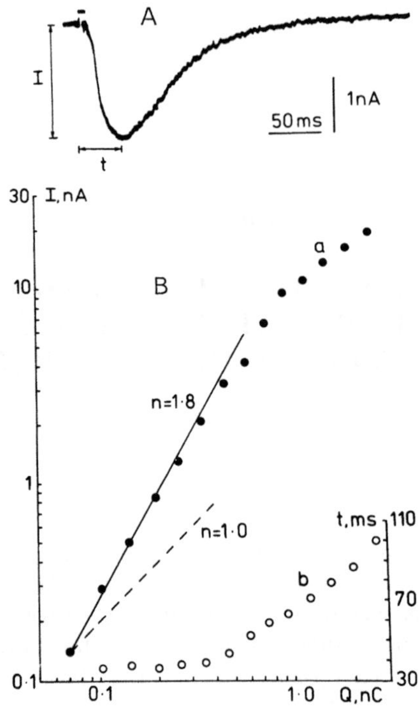

Fig. 2.1. Concentration dependence for the action of acetylcholine on rabbit superior cervical ganglion neuron. A) An example of inward membrane current elicited by short (10 ms) iontophoretic application of acetylcholine to the surface of the neuron (iontophoretic current was 35 nA). Arrows measure the amplitude (I) and rise time (t) of the response. B) Dependence of the amplitude (a) and rise time (b) of the response on the iontophoretic charge (Q) transferred through the micropipette. The continuous line indicates the maximum slope of the experimental $I(Q)$ curve shown in double logarithmic scale, equal to 1.8, whereas the interrupted line indicates the slope of the theoretical curve equal to 1. The constancy of t at $Q \leq 0.4$ nC shows that in the range of Q values indicated the surface area activated by acetylcholine and, therefore, the number of available receptors remain unchanged (see Peper et al., 1982). The data were obtained at –80 mV and 36°C.

ber of agonist molecules that must interact with one receptor molecule to trigger the ionic channel opening. More precisely, the Hill number gives the lower limit for the number of interacting agonist molecules (see Werman, 1969; Colquhoun, 1979):

$$n = \lim_{[A] \to 0} \frac{d \log I}{d \log [A]}, \tag{2.1}$$

where I is the amplitude of the cell response to the application of an agonist in concentration $[A]$. Thus, n is equal to the slope in a dose–response relationship plotted in a double logarithmic scale provided that the agonist concentration doses do not exceed a certain low limit. Figure 2.1 illustrates the relationship between the doses of acetylcholine applied iontophoretically to a neuron of rabbit superior cervical ganglion and the acetylcholine-induced current (ACh-current). The dose of acetylcholine is assumed to be proportional to the charge transferred through the iontophoretic micropipette. The n value in Fig. 2.1 is 1.8, and the mean value obtained from seven neurons of the ganglion is 1.9 ± 0.1 (see also Selyanko and Skok, 1979; see, however, Dun et al., 1976). This means that at least two molecules of acetylcholine usually interact with one receptor molecule. Other approaches also suggest that there is more than one site in the nicotinic receptor molecule that is able to bind acetylcholine or competitive antagonists. Carbachol and nicotine, but not tubocurarine, effectively inhibit the binding of α-bungarotoxin to nicotinic receptors in chick embryo ciliary ganglion cultured neurons, indicating that α-bungarotoxin and tubocurarine may bind to different sites (Messing, 1982).

A Hill number close to 2 has also been found for nicotinic receptors in other objects, e.g., in molluscan neurons (see Vulfius, 1978; Vulfius et al., 1979), frog end-plate, and fish electric organ, with acetylcholine or other full agonists used for their activation (see Colquhoun, 1979; Adams and Feltz, 1980a). A higher Hill number, 2.91, has been reported for the molluscan neurons activated with carbachol (Ziskind and Werman, 1975a, b). Thus, in this case, three ligand molecules may interact with a receptor, which is supported by the results obtained on purified nicotinic receptors (Carroll et al., 1973; Eldefrawi and Eldefrawi, 1973).

In molluscan acetylcholine receptors, the Hill number decreases from 2.91 to 1.55 following their desensitization (Ziskind and Werman, 1975a, b) and from 2.2 to 1.1 following their alkylation with a monofunctional β-haloidalkylamine (a compound containing only one alkylating group). By contrast, no decrease in the Hill number has been observed with the use of bifunctional β-haloidalkylamine (Vulfius, 1978; Vulfius et al., 1979). This difference is thought to be due to a decrease in the number of sites available for acetylcholine in each receptor in the former case, while in the latter case the number of available receptors decreases instead. Based on the distance between two quaternary ammonium groups in a bifunctional alkylating compound equal to 2.0–2.2 nm, a similar distance has been suggested to separate the points that normally bind two quaternary ammonium groups of two acetylcholine molecules (the anionic sites). This suggestion is consistent with the fact that bis-choline esters of suberic and terephthalic acids, the reversible agonists, with distances between their quaternary ammonium groups of about 2.0–2.2 nm, are more potent activators of molluscan nicotinic receptors than acetylcholine (Vulfius et al., 1969). It should be noted, however, that direct measurement of the distance between two α-bungarotoxin-binding sites in the nicotinic receptor of electric organ yields a higher value – about 5 nm (Zingsheim et al., 1982).

More evidence in favor of the concept that the nicotinic receptor possesses at least two binding sites for agonists has been obtained using the fluorescent ligands DNS-chol and quinacrine (Cohen and Changeux, 1973; Changeux et al., 1976).

DNS-chol – 1-(5-dimethylaminonaphthalene-1-sulfonamido)ethane 2-trimethylammonium iodide – is an environment-sensitive fluorophore somewhat similar in structure to cholinergic agonists and local anesthetics. Its interaction with the receptor-rich membrane fragments of electric organ is followed by increased fluorescence caused by energy transfer from the receptor protein to the ligand. Cholinergic agonists, as well as α-bungarotoxin, markedly reduce the receptor-induced fluorescence, indicating that the ligand may bind to a "secondary site" that is distinct from a "primary site," binding the ligand in the absence of agonists or α-bungarotoxin. The structural similarity between the fluorescent ligands and local anesthetics suggests that the secondary site may likewise bind local anesthetics.

The nicotinic receptor model with two binding sites for agonists has also been suggested by the results of experiments on modification of carboxyl groups in the electric organ receptor protein with trimethyloxonium salts. In these experiments carbachol exerts only a very weak protecting effect against modification of the carbachol-binding sites and actually does not protect the decamethonium-binding sites (Chao et al., 1975). The two-site model is also consistent with the finding that modification by 4-(N-maleinimido)benzyl trimethylammonium diiodide (MBTA) of only half of the α-bungarotoxin-binding sites in nicotinic receptors reconstituted into the lipid membrane completely blocks the carbachol-induced increase in sodium permeability of the lipid membrane. It is suggested that binding of an agonist to one of the above two sites is followed by the channel opening, while its binding to another site exerts only a cooperative effect. The latter site is probably able to bind local anesthetics (Lindstrom, 1980).

2.2. FUNCTIONAL ORGANIZATION AND CHEMICAL STRUCTURE OF THE RECOGNITION CENTER

It has become traditional to study the functional organization of the acetylcholine receptor by analyzing the structure–activity relationships for agonists and antagonists. Later, the effects of chemical modification of the receptor protein on its functions were analyzed, which yielded information about the particular chemical structures in the receptor protein involved in the ligand–receptor interactions. Although other, more recent approaches are now rapidly being developed (analysis of the receptor protein primary structure, the immunological method, etc.), their contribution to our knowledge of the organization of the recognition center in the neuronal acetylcholine receptor still remains small compared with that in the structure–activity method.

Much information accumulated in this field over the past half century is difficult to analyze, first because of the great heterogeneity of the objects and of the criteria for estimating the drug efficacy used in the experiments, ranging from blood pressure measurement to transmembrane current measurement. Second, the majority of works were aimed at finding new hypotensive drugs rather than elucidating how the receptor is organized.

It should be mentioned that the decrease in agonist potency observed in the experiments may be due to hydrolysis of the agonist by acetylcholinesterase rather

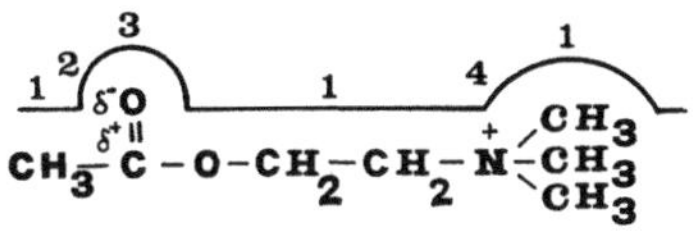

Fig. 2.2. General scheme of interactions between the acetylcholine molecule and the nicotinic acetylcholine receptor. 1) Hydrophobic interactions; 2) partial electrical charge bond; 3) dipole–dipole interactions; 4) electrostatic bond (Eldefrawi, 1974).

than to weak interaction of the agonist with a recognition center. On the other hand, the increased agonist potency may be related to its muscarinic rather than nicotinic activity, since muscarinic receptors often coexist with nicotinic ones in the same cell. As for the antagonists, they may act not on the recognition center but rather on the ionic channel. In all of these cases, information obtained from structure–activity experiments may lead to incorrect conclusions about the structure of a recognition center. This was taken into consideration when the structure–activity data were selected for further discussion.

According to quantum-chemical calculations, the acetylcholine molecule is strongly polarized. In addition to a positively charged nitrogen atom, there are at least two other functionally important partial charges in the molecule: a negative charge (–0.4 to 0.7) at the carbonyl oxygen (Pullman and Pullman, 1963; Golovanov et al., 1966) and a positive charge either at the ester oxygen (+0.1: Pullman and Pullman, 1963) or at hydrogen atoms linked to the carbon atom (+0.05: Golovanov et al., 1966). From analysis of numerous structure–activity data, the scheme of binding of acetylcholine to a nicotinic receptor shown in Fig. 2.2 has been suggested.

According to the scheme shown in Fig. 2.2, the recognition center consists of the "anionic site" (the area that interacts with the cationic head of the acetylcholine molecule) and the "esterophilic area," which interacts with the side chain of the acetylcholine molecule including the acylic group, esterophilic bond, and alkyl chain (Michelson and Zeimal, 1973; Anichkov, 1974). The positively charged nitrogen atom interacts with the anionic site by forming the ionic (electrostatic) bond. In contrast, methyl and methylene groups reveal hydrophobic interactions, while carbonyl and esterophilic oxygen atoms form hydrogen bonds with the appropriate receptor areas. It has been suggested that the ionic bond dominates among all others since its energy is higher (5–7 kcal·mole^{-1}) than that of hydrophobic interactions (0.6 kcal·mole^{-1} each) and the hydrogen bond (3 kcal·mole^{-1}) (see Eldefrawi, 1974).

The importance of the electrostatic interaction between the cationic group of acetylcholine and the anionic site of a receptor is supported by the good correlation between the experimentally determined energy for binding of aryltrimethylammonium antigen (hapten) to antibody and the calculated nitrogen charge density (see Triggle and Triggle, 1976).

2.2.1. Anionic Site

In nicotinic receptors of sympathetic ganglion neurons the anionic point is more important for acetylcholine–receptor interaction than in other types of nicotinic receptors. This is evidenced by the fact that, although their other structural components differ from those in acetylcholine, compounds possessing the cationic group of acetylcholine are selective ganglionic stimulants. Examples are tetramethylammonium (TMA, **2.1**),[*] 1,1-dimethyl-4-phenylpiperazinum (DMPP, **2.2**), and nicotine (β-pyridine-α-methylpyrrolidinium, **2.3**), all evoking excitatory effects in sympathetic ganglia at lower concentrations than in skeletal muscles.

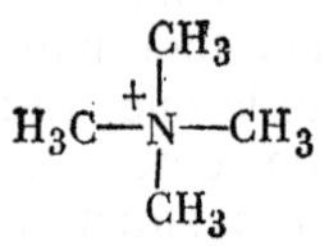

TMA, **2.1**

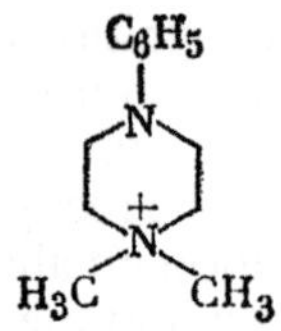

DMPP, **2.2**

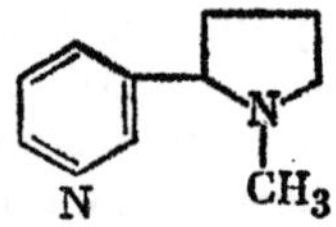

Nicotine, **2.3**

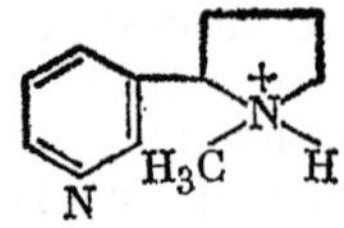

Nicotinium ion, **2.4**

Unlike TMA and DMPP, nicotine is a tertiary amine. However, at physiological pH levels, nicotine dissociates, thus forming the nicotinium ion (**2.4**) responsible for the high stimulant activity of nicotine.

Compared with acetylcholine, TMA is 4.6 times less potent in producing a depolarizing effect on sympathetic ganglion neurons (Selyanko and Skok, 1979), 4 times less potent in causing a contraction of the nictitating membrane (Lukomskaya, 1972), and 2 times more potent in causing an increase in blood pressure (Triggle and Triggle, 1976). The difference between the latter estimate and the two former ones is apparently attributed to the effect of acetylcholinesterase on acetylcholine. In skeletal muscle, TMA is 13,000 times less potent than acetylcholine, thus indicating that TMA is a selective stimulant for ganglionic nicotinic receptors.

The agonist activity strongly depends on the size of the cationic group. Substitutions for nitrogen that increase the size of the cationic group markedly reduce both nicotinic and muscarinic activities. The phosphonium, sulfonium, and arsenium derivatives are 6, 15, and 37 times less active than acetylcholine, respectively, in their stimulatory effect upon the skeletal muscle, which correlates with the increase in the size of the cationic group (see Barlow, 1955). In sympathetic ganglia, the sulfonium derivatives are completely devoid of blocking activity, in contrast to their ammonium analogs. It should be mentioned, however, that there are also very potent ganglion-blocking agents among other sulfur-containing compounds, e.g., trimetaphan (see Kharkevich, 1967; Trčka, 1980a).

Another factor that strongly reduces agonist activity in all types of nicotinic receptors is the increased size of the alkyl groups in the cationic head as compared with methyl groups. The replacement of one methyl group in acetylcholine by an ethyl group decreases activity fivefold, both in sympathetic ganglion and in skeletal

[*]Henceforth, for reasons of simplicity, the structural formulas only for cations of the charged ligands are shown; their anions (which are usually bromides, iodides, or chlorides unless specially mentioned) are usually not important in determining the ligand–receptor interaction mechanism.

muscle, while the replacement of two methyls abolishes the activity in the ganglia (Lukomskaya and Gmiro, 1982) and decreases it 300-fold in the muscle (Barlow, 1955). A difference in the same direction between the ganglia and muscle has been observed in their sensitivities to another agonist, bis-acetylcholine (**2.5**), which is inactive in the ganglia and 164 times less active in the muscle than acetylcholine:

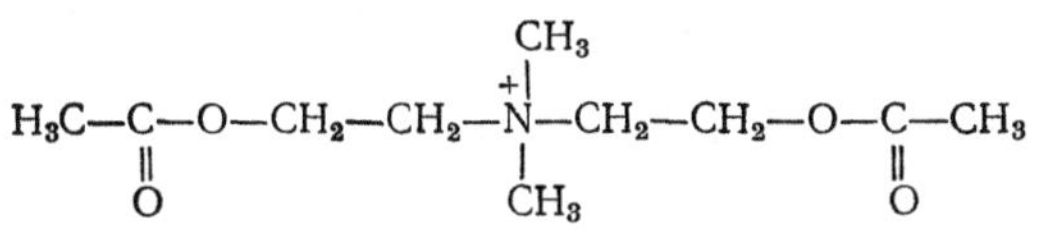

Bis-acetylcholine, **2.5**

The replacement of all methyl groups in TMA by ethyl groups (tetraethylammonium, TEA, **2.6**) turns an agonist into an antagonist. The same result can be obtained if one methyl group in TMA is replaced by a 2-hydroxycyclopentyl group (**2.7**) or by a 2-hydroxycyclohexyl group (**2.8**). The compounds thus obtained are as active as TEA:

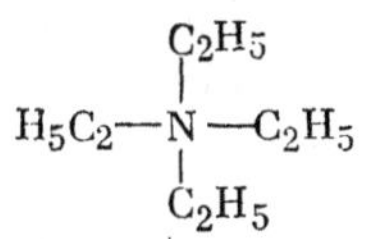

TEA, **2.6**

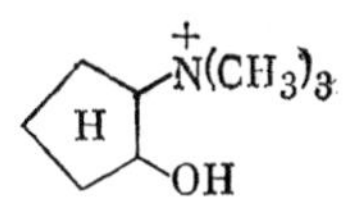

2.7

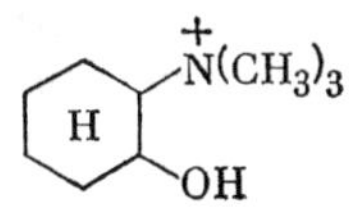

2.8

It seems possible, therefore, that the appearance of blocking activity in compounds **2.7** and **2.8** compared with agonist DMPP (**2.2**) is due to their structural component 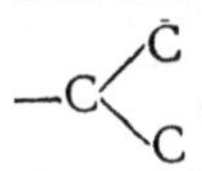, a chain branching at the alpha carbon. This structure is possessed by many other quaternary and tertiary amines that are ganglion-blocking agents (see Chapter 4).

The mechanisms responsible for the reduction of excitatory activity with the increased size of the cationic group are not yet clear. It was suggested initially (Ing, 1949) that only two methyl groups could interact with a receptor, while the third methyl group was important simply because it increased the probability that at least two methyl groups would face the receptor surface, regardless of the orientation of a ligand relative to the receptor. The receptor area interacting with the cationic head of acetylcholine was thought to be either a pocket just large enough to bring two methyl groups inside or a flat surface of the corresponding size (Barlow, 1955). According to these suggestions, the increased size of the cationic group should prevent the interacting structures from approaching each other closely enough, and thus should weaken their interaction.

However, more recent measurements performed on molecular models showed that with the replacement of three methyl groups by ethyl groups the minimal interionic distance increases by no more than 0.08 nm, which should result in a loss of binding energy of about 2 kcal·mole^{-1} and about a 50-fold decrease in affinity. This is much less than the approximately 1000-fold affinity loss observed in experiments. To account for this discrepancy, the important role of specific van der Waals binding in the interaction between the cationic group and the recognition center has been postulated, in addition to the ion-pairing interaction. Both types of interactions imply that the trimethyl group should fit the receptor surface. It is possible that the increase in the interionic distance permits the intrusion of water into the binding locus, thus increasing the dielectric constant of the microenvironment; this may be a factor accounting for the above discrepancy. An alternative approach to interpreting the stimulant activity loss caused by ethyl substitution is to admit that the stabilization of the adjacent water structure increases with increased size of the alkyl substituents. According to calculations, a transition between structure-breaking and structure-making properties occurs between the tetramethylammonium and tetraethylammonium ions: the former behaving like large alkali metal cations breaking the structure of local water, while in the latter the additional hydrocarbon residues overcome the structure-breaking effect (see Triggle and Triggle, 1976).

It should be noted that neither of the above approaches explains why the methyl-to-ethyl substitutions are followed not only by a loss in stimulant activity but also by the appearance of a blocking effect. The antagonist–receptor interaction is thought to occur at a higher affinity constant compared with the agonist–receptor interaction. It is therefore not clear why a weaker ligand–receptor interaction should correlate with the appearance of the receptor blockade.

This discrepancy, however, can be explained if one admits that the methyl-to-ethyl substitutions, besides reducing the probability of interaction of a ligand with the recognition center, also increase the probability of its interaction with some other site in the nicotinic receptor molecule, e.g., with the ionic channel, causing channel blockade. Results recently obtained with the use of membrane current recording make this interpretation highly favorable. Moreover, this approach allows one to explain another paradox: the transition from the stimulatory effect of mono-quaternary ammonium derivatives to the blocking effect of bis-quaternary ammonium derivatives (see Chapters 4–7).

The area of hydrophobic interactions that is located around the anionic site is probably of a size comparable to the size of the cationic group in acetylcholine. Assuming that the distance between the carbon atoms in the cationic head is 0.24 nm (see Michelson and Zeimal, 1973) and adding to this two covalent radii of carbon atoms (2×0.077 nm) and hydrogen atoms (2×0.003 nm), one obtains an area about 0.4 nm in diameter (see Dneprovski and Temnikova, 1979). After adding two van der Waals radii (2×0.15 nm), this yields about 0.7 nm diameter for the expected area of hydrophobic interactions.

2.2.2. Esterophilic Area

The importance of the esterophilic area in the receptor for acetylcholine binding is evidenced by the fact that acetylcholine is 4.6 times more effective than TMA in the neurons of sympathetic ganglion (see above) and 30–100 times more effective than choline in cultured cells of adrenal medulla (Holz and Senter, 1981).

The binding site for the alkyl chain of acetylcholine in sympathetic ganglia and in skeletal muscle is characterized by a rather high specificity. The introduction of a β-methyl to acetylcholine [acetyl-β-methylcholine, or methacholine (**2.9**)] abolishes nicotinic activity in both types of nicotinic receptors without affecting muscarinic activity (Takeshige and Volle, 1964b). One more example is another selective muscarinic agonist, 5-methylfurmethide, 5-MF (**2.10**).

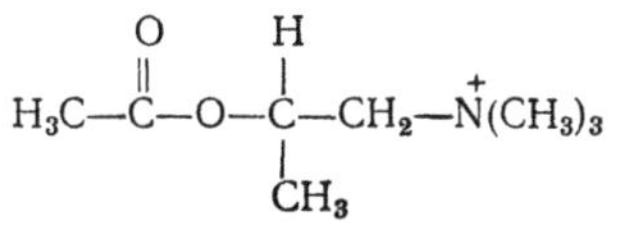

Methacholine, **2.9**

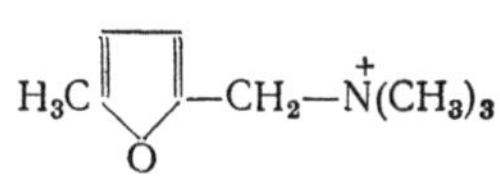

5-Methylfurmethide (5-MF), **2.10**

Binding of an ether oxygen should stabilize the ionic channel of the skeletal muscle nicotinic receptor in its open configuration (Auerbach et al., 1983). Acetylthiocholine, which contains sulfur instead of ether oxygen of acetylcholine, is 5 times more potent than acetylcholine in sympathetic ganglia (Triggle and Triggle, 1976), although its muscarinic activity is 300 times lower than that of acetylcholine (Scott and Mautner, 1967). It should be noted, however, that acetylthiocholine strongly differs from acetylcholine in the conformation of the molecule (Shefter and Mautner, 1969), which provides an alternative explanation for the difference between their nicotinic activities.

In a series of choline phenylethers (**2.11**), the most active ganglionic stimulants are compounds with R = 3,5-BrC_6H_3 and R = 5-BrC_6H_4:

$$R—O—CH_2—CH_2—\overset{+}{N}(CH_3)_3$$

2.11

These compounds are 150-200 times more potent as stimulators of nicotinic receptors than acetylcholine, with no muscarinic activity, which is accounted for by the accumulation of a positive charge in their ether oxygen caused by the electron-withdrawing capacity of the substituent (Barlow, 1964).

Binding of ether oxygen and especially of carbonyl oxygen is probably less important for the acetylcholine–receptor interaction in sympathetic ganglion than in skeletal muscle. By contrast, the high hydrophobicity of the acyl group seems to be much more important in the ganglion than in the muscle. This conclusion follows from the higher stimulant activity in the ganglion, compared with that in the muscle, of some alkyltrimethylammonium salts that are lacking both oxygen atoms. Two examples are pentyltrimethylammonium (**2.12**) and hexyltrimethylammonium (**2.13**), compounds with methyl groups in their alkyl chains that are about as distant from the positively charged nitrogen atoms as in the acylic methyl group in acetylcholine (about 0.54 nm). With longer or shorter alkyl chains, the activity drops abruptly.

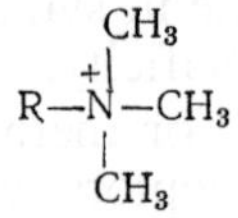

$R{=}C_5H_{11}$ $R{=}C_6H_{13}$

Pentyltrimethylammonium, **2.12** Hexyltrimethylammonium, **2.13**

Pentyltrimethylammonium is also active as an agonist in other nicotinic receptors and in muscarinic receptors. However, its activity in skeletal muscle is 25 times lower; in the smooth muscle of the gut wall it is 100 times lower than that of acetylcholine, while in sympathetic ganglion it is only 6 times lower than that of acetylcholine (see Triggle and Triggle, 1976).

Substitutions in the acetate group of acetylcholine are critical for its activity in sympathetic ganglia. Carbacholine (**2.14**), a compound with a more polar amino group than the methyl group in acetylcholine, is 7.8 times less potent than acetylcholine (Selyanko and Skok, 1979), while trimethylacetylcholine (**2.15**), possessing a less polar $C(CH_3)_3$ group, is 10 times more potent than acetylcholine in sympathetic ganglion. In contrast, the latter compound is 5 times less active, compared with acetylcholine, in skeletal muscle, and is completely devoid of muscarinic activity (see Triggle and Triggle, 1976). Therefore, carbacholine is less active in sympathetic ganglion than TMA. This differs from what has been observed in skeletal muscle, where carbacholine is about 3 times more potent compared with TMA (Magazanik, 1968). It can be concluded from these results that binding of the acylic methyl group is specific for sympathetic ganglion.

$$R{-}\overset{\overset{O}{\|}}{C}{-}O{-}CH_2{-}CH_2{-}\overset{+}{N}(CH_3)_3$$

$R{=}NH_2$ $R{=}C(CH_3)_3$

Carbacholine, **2.14** Trimethylacetylcholine, **2.15**

R= 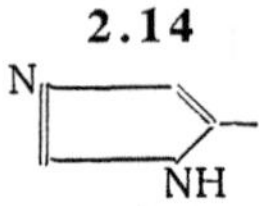 R= 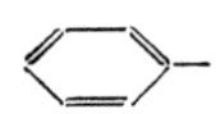

2.16 Phenylacetylcholine, **2.17**

Compounds **2.16** and **2.17** (phenylacetylcholine) are almost as active as acetylcholine in sympathetic ganglia, while in skeletal muscle their activity drops 1000-fold and 100-fold, respectively, compared with that of acetylcholine; their muscarinic activity is negligible (see Triggle and Triggle, 1976). The marked difference between the agonist potencies in the ganglia and in the muscle relative to that of acetylcholine makes it unlikely that this difference is due to the different sensitivity of the agonist to acetylcholinesterase and suggests that there is a specific hydrophobic binding area in the ganglionic nicotinic receptor. This area is located about 0.6 nm away from the anionic site, which is similar to the distance between the positively charged nitrogen and the acylic methyl group in the acetylcholine molecule. The diameter of the hydrophobic area should be about 0.4 nm, as evidenced by the activity maximum observed in trimethylacetylcholine (**2.15**).

In addition to the above results obtained with agonists, important information on the organization of the recognition center has been provided by the use of competitive antagonists. As will be shown in Chapter 5, only a few of those compounds considered in the past to be competitive antagonists are identified as such by modern experimental techniques. Among these are trimetaphan and surugatoxin. Other compounds are mixed antagonists, since they may act competitively at resting membrane potential and noncompetitively in hyperpolarized neurons. This effect has been observed for tubocurarine and pentamethonium in rabbit superior cervical ganglion and in rat submandibular ganglion (Ascher et al., 1979; Volle, 1980; Selyanko et al., 1981b). Pentamethonium is a symmetrical bis-quaternary ammonium compound (**2.18**) with five methylene groups ($n = 5$):

$$(CH_3)_3\overset{+}{N}-(CH_2)_n-\overset{+}{N}(CH_3)_3$$

Symmetrical bis-quaternary ammonium compounds,
2.18

The most probable distance between two positively charged nitrogen atoms in pentamethonium is 0.75 nm (Dyadyusha et al., 1985). It is assumed by some authors that one of these two positive charges may bind to the anionic site in the nicotinic receptor. The fact that pentamethonium is a selective ganglion-blocker suggests that there is one more negatively charged area about 0.75 nm away from the anionic site, having an adjacent area of hydrophobic interactions nearly similar in diameter to that assumed for the area of hydrophobic interactions adjacent to the anionic site, i.e., 0.4 nm, in the ganglionic nicotinic receptor. Thus, there are probably two anionic sites in the ganglionic receptor; the second site is called the "peripheral" (Triggle and Triggle, 1976) or "second" anionic site (Lukomskaya and Gmiro, 1982).

A more detailed examination of the peripheral anionic site has been performed on cat superior cervical ganglion using bis-ammonium ganglion-blocking agents of various structures (Lukomskaya and Gmiro, 1982). However, the results of recent electrophysiological experiments indicate that some of these agents may bind to the open ionic channel rather than to the recognition center (Skok et al., 1984; Gurney and Rang, 1984). For this reason, the results obtained will be discussed in more detail in Chapter 5, which characterizes blocking mechanisms.

In trimetaphan (**2.19**), the positive charge is carried by sulfur instead of nitrogen as in many other ganglion-blocking agents.

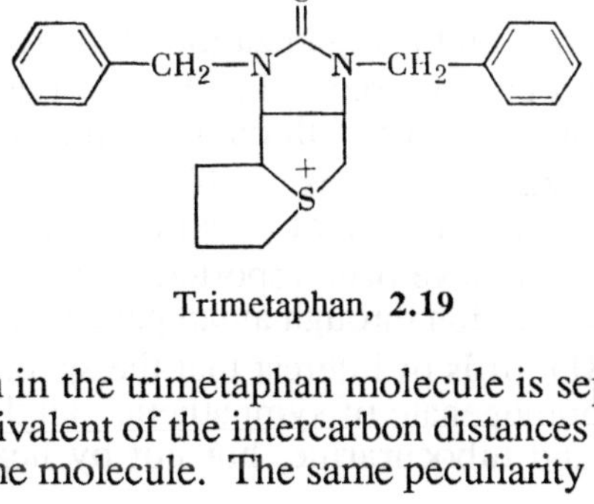

Trimetaphan, **2.19**

The oxygen atom in the trimetaphan molecule is separated from the positive charge by the same equivalent of the intercarbon distances as is the carbonyl oxygen atom in the acetylcholine molecule. The same peculiarity is possessed by two other partially competitive blockers, the curare-like substances tubocurarine (**2.20**) and dihydro-β-erythroidine; the latter is 3 times less potent than tubocurarine in sympathetic ganglia (Grob, 1967).

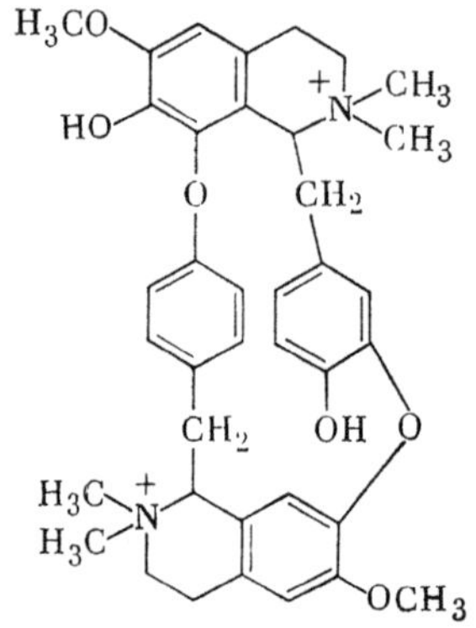

Tubocurarine, **2.20**

In addition, tubocurarine and trimetaphan have large aromatic groups, which may interact with hydrophobic areas in the nicotinic receptor.

It is interesting that under some circumstances tubocurarine can stimulate rather than block sympathetic ganglion neurons (Bülbring and Depierre, 1949). In particular, this effect of tubocurarine has been observed after treatment of the ganglia with anticholinesterase agents (Volle, 1962) or with N-ethylmaleinimide, a compound that alkylates SH groups in protein molecules (Komalahiranya and Volle, 1963). A similar transition from blocking to stimulatory effect has been observed for hexamethonium in nicotinic receptors of electric organ and skeletal muscle, after the disulfide bonds in their receptor protein have been reduced by dithiothreitol (Karlin and Winnik, 1968; Rang and Ritter, 1971). No stimulatory effect of hexamethonium has been observed in sympathetic ganglion (Brown and Kwiatkowski, 1976). According to recent results, the stimulatory effect of tubocurarine combined with its blocking effect can be seen even in native nicotinic receptors (Trautmann, 1982; Takeda and Trautmann, 1984). These features of tubocurarine support the concept that it can interact with the recognition center of a nicotinic receptor.

Additional information on the organization of the recognition center has been provided with the use of α-bungarotoxin. In electric organ and skeletal muscle, α-bungarotoxin blocks nicotinic receptors through binding to a site common with that for acetylcholine. α-Bungarotoxin does not block synaptic transmission through sympathetic ganglia, although binding of a toxin to ganglionic receptors and their partial competitive blockade have been reported (see Chapter 5). By contrast, α-bungarotoxin blocks transmission through avian parasympathetic (ciliary) ganglion (Chiappinelli et al., 1981). It is of interest that the specific binding of labeled α-bungarotoxin in tissue homogenate of sympathetic ganglia is inhibited by acetylcholine, carbacholine, and tubocurarine, but not by hexa- and decamethonium

(Fumagalli and De Renzis, 1980). A similar inhibitory effect of tubocurarine has been observed by other authors in sympathetic (Carbonetto et al., 1978) and parasympathetic ganglia (Messing, 1982), where α-bungarotoxin blocks the depolarizing action of carbacholine. The fact that tubocurarine acts similarly to agonists but not to the channel-blocking antagonists supports the above suggestion that tubocurarine in the neuronal nicotinic receptor may interact with a recognition center.

It should be emphasized that trimetaphan, tubocurarine, and dihydro-β-erythroidine, unlike acetylcholine, are large molecules of rigid structure. Thus, there should be rather extended areas of hydrophobic interactions in the receptor, adjacent to its recognition center. These areas, according to common knowledge, may increase the blocking effect by increasing ligand affinity to a receptor.

2.2.3. Functional Organization of the Recognition Center

The above results are summarized in the scheme of functional organization of the recognition center shown in Fig. 2.3, the best schematic representation at the present time. Of greatest functional importance is anionic site 1 with the adjacent area of hydrophobic interactions 2 (radius about 0.2 nm). These two structures are complementary to the quaternary ammonium nitrogen with attached methyl groups, as evidenced by selective sensitivity of a receptor to TMA and DMPP. Another interesting characteristic of this structure is that it is complementary to the methyl group in the β-position, which accounts for a reduction in nicotinic selectivity of agonists with enlarged groups in this position. There is also an area 3, which interacts with the carbonyl oxygen atom of acetylcholine and, possibly, with the oxygen atom of the competitive antagonist trimetaphan through interaction between their partial electrical charges and through the formation of hydrogen bonds. This conclusion is in agreement with the "two-point" concept of acetylcholine interaction with the nicotinic receptor: the ionic interaction of the cationic group and the formation of a hydrogen bond by carbonyl oxygen located 0.59 nm away from the cationic group (Beers and Reich, 1970; Chothia and Pauling, 1970; Baker et al., 1971). In addition, there is an area 4 having the opposite, negative partial charge about 0.3 nm away from the anionic site, interacting with the ether oxygen of acetylcholine. The existence of this area is evidenced by the binding to the receptor of acetylcholine and of the substituted phenyl ethers. In addition, there is an area 5 of hydrophobic interactions about 0.6 nm away from the anionic site, which selectively binds the nonpolar groups in agonists trimethylacetylcholine and phenylacetylcholine; this area is probably similar in size to the area surrounding the anionic site. Areas 3 and 5 are shown in Fig. 2.3 as separate ones according to the localization of the corresponding groups in the acetylcholine molecule (Chothia, 1970). The total size of the recognition center, about 1 nm, is actually similar to the sizes of active centers and central cavities in many other proteins that range from 0.6 to 2.0 nm (see Kaivyaryainen, 1980).

Finally, one cannot exclude the existence of another, hypothetical, peripheral anionic site 6 suggested by the selective binding of pentamethonium. This site should have a hydrophobic environment 7 similar in size to that in the main anionic

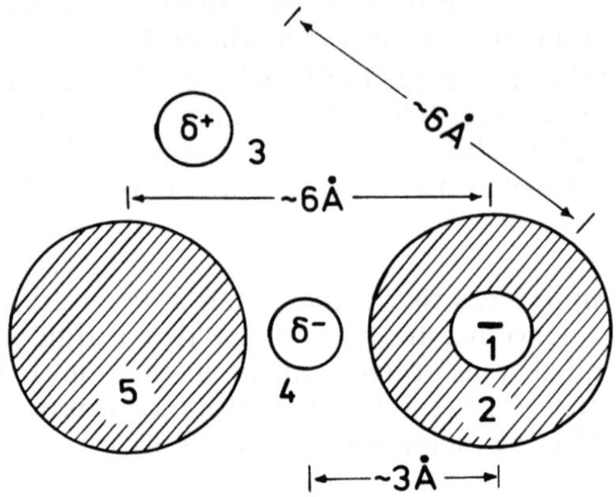

Fig. 2.3. Scheme of recognition center of nicotinic acetylcholine receptor. For explanations see text.

site. Alternatively, this anionic site can be located in a distinct part of the receptor molecule, probably in the ionic channel (see Chapter 6). Sites 6 and 7 are not shown.

Recent examination of the molecular structure of isoarecolone methiodide, a potent nicotinic agonist, when compared with that of a less potent analog, suggests that small steric and electrostatic field differences between the agonist molecules are equally important to their high agonist activity as is the $\overset{+}{N}$–C distance postulated by the above "two-point" concept (Spivak et al., 1986).

It is possible that the areas 5 and 7 are actually the same, since they are at a similar distance from the main anionic site. However, this possibility is not supported by the fact that pentamethonium or hexamethonium do not exhibit any agonist activity in sympathetic ganglion.

If one compares the scheme shown in Fig. 2.3 with the scheme of functional organization of the nicotinic receptor in skeletal muscle or in electric organ (Michelson and Zeimal, 1970; Anichkov, 1974; Triggle and Triggle, 1976), at least two differences become evident. First, the anionic site 1 and the area of hydrophobic interactions 5 are more important in the neuronal nicotinic receptor than in those of muscle and electric organ. Second, no hypothetical peripheral anionic site 6 has been suggested in muscle and electric organ receptors, at least at a distance from the main anionic site similar to that in the neuronal receptors.

The area of hydrophobic interactions 5 is of great importance for the muscarinic receptor, which binds acetylcholine by at least three points (areas 1, 3, and 5 in Fig. 2.3), in contrast to two points in the case of muscle and electric organ nicotinic receptors (areas 1 and 3 in Fig. 2.3: see Triggle and Triggle, 1976). Neuronal nicotinic receptor differs from muscarinic receptor by the lack of sensitivity to methacholine and by the increased sensitivity to the substituted phenyl ethers of choline (**2.11**) and to alkyltrimethylammonium derivatives (**2.12** and **2.13**).

It should be kept in mind that the autonomic ganglion neurons, in addition to nicotinic receptors, possess muscarinic receptors whose activation may be erroneously interpreted as an activation of nicotinic receptors, especially in experiments where no differentiation between the responses of these two receptor types (e.g., according to their latencies) is made.

2.2.4. Chemical Structure of the Recognition Center

Information on the chemical structure of the recognition center has been obtained using chemical modification of the receptor protein. Most advantageous is the affinity modification method in which the modifying group is attached to an agonist or competitive antagonist, thus ensuring that it binds to a recognition center. Another approach is to protect the receptor against the modifying agent with an agonist or competitive antagonist; this procedure ensures that it is the recognition center which is modified in the absence of a protector.

Experiments on a modification of the anionic site in neuronal acetylcholine receptors have been performed only on molluscan neurons, while those in nonneuronal acetylcholine receptors have been done on electric organ and skeletal muscle. Based on theoretical principles, one can suggest that the most probable structure for the anionic site is a carboxyl group of either of two negatively charged polar amino acids – aspartic or glutamic acid. For this reason, the modifying agents used are mostly reagents for protein carboxyl groups: carbodiimides, trimethyloxonium salts, diazo compounds, and haloidalkylamines. For example, by means of a carbodiimide, the nucleophilic reagent H–X can be attached to the protein carboxyl group Pr–COO$^-$ resulting in the formation of protein derivative Pr–COX. The reaction occurs in two steps: The first step is the formation of an intermediate product resulting from the protein–carbodiimide interaction, and the second step is a reaction between the intermediate product and the nucleophilic reagent, which finally gives Pr–COX. 1-Cyclohexyl-3,2-(N-methyl)morpholinoethylcarbodiimide reversibly blocks nicotinic acetylcholine receptors in molluscan neurons, while causing their irreversible blockade after a nucleophilic reagent, glycine methyl ether, is added. The irreversible blockade of the receptors can be prevented by using protection of a receptor by acetylcholine. The modification is strongly reduced by decreased pH ($pK_a \sim 6.7$). These and other results support the suggestion that the anionic site is a carboxyl group (Vulfius et al., 1979; Vulfius, 1978).

One of the most easily reacting groups in a protein molecule is a disulfide bond (S–S bond), which allows one to investigate the role of this bond in receptor functions. It has been found that the reduction of a disulfide bond with dithiothreitol, which results in the appearance of sulfhydryl groups (–SH), markedly decreases the response of electroplax to acetylcholine and other agonists. The initial amplitude of the response can be restored after application of SH oxidizing agents. However, the restoration is not observed after the receptors are modified with bromoacetylcholine or with maleinimide derivatives – the affinity alkylating reagents, in particular 4-(N-maleimido)benzyltrimethylammonium iodide (MBTA). Based on these results, it has been suggested that there is a functionally important disulfide bond very close to the recognition center. Experiments on protection of the receptors from the modifying agent and on their affinity modification indicate

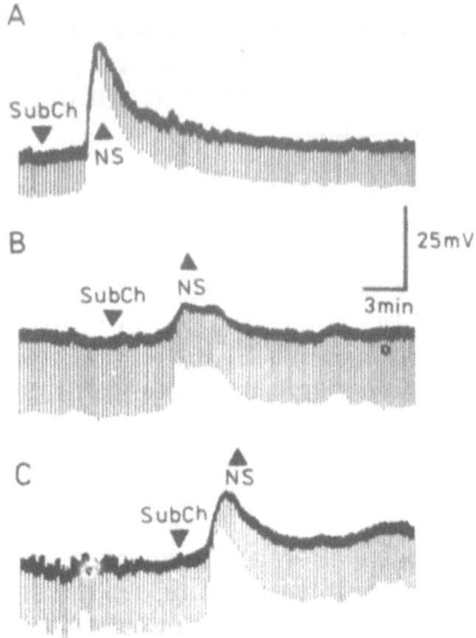

Fig. 2.4. Effects of dithiothreitol and ferricyanide on the responses evoked by suberyldicholine (SubCh, 10^{-4} M) in frog sympathetic ganglion neuron. A) Control response. B) Response elicited after 20-min washout of dithiothreitol (10^{-3} M), which was acting in the course of 10 min. C) Response elicited after 20-min washout of ferricyanide (10^{-3} M) acting in the course of 10 min. Drugs were added and removed at the moments indicated by triangles. Vertical lines are voltage deflections produced by testing current pulses (Trinus and Skok, 1979).

that the disulfide bond is localized about 0.1 nm away from the anionic site (Karlin, 1969, 1973).

Experiments on modification of disulfide bonds have also been performed on nicotinic receptors of autonomic ganglion neurons and molluscan neurons. The amplitude of depolarization evoked in rat superior cervical ganglion neurons by carbacholine has been strongly reduced by dithiothreitol. Furthermore, bromoacetylcholine, when applied after dithiothreitol, has produced irreversible depolarization, probably due to irreversible activation of acetylcholine receptors, thus indicating that the disulfide bond is functionally important in neuronal acetylcholine receptors (Brown and Kwiatkowski, 1976).

Experiments of this kind have been extended to frog sympathetic ganglion neurons (Trinus and Skok, 1979). Dithiothreitol ($1 \cdot 10^{-3}$ M) has markedly reduced the suberyldicholine-induced depolarization and increase in membrane conductance. A somewhat smaller reduction of the responses is observed if TMA is used instead of suberyldicholine. An oxidizing agent, potassium ferricyanide ($1 \cdot 10^{-3}$ M), if applied after dithiothreitol, restores the responses, although restoration has never been complete. An example of such an experiment is shown in Fig. 2.4.

If carbacholine is used instead of suberyldicholine, two-component depolarization is observed, with early and late components which are due to activation of nicotinic and muscarinic receptors, respectively. The behavior of the early component in the ganglion treated by the above modifying agents is similar to that of suberyldicholine-induced depolarization.

Acetylcholine ($5 \cdot 10^{-4}$ M) or TMA ($1 \cdot 10^{-3}$ M), if applied together with dithiothreitol, effectively protects the nicotinic receptors of sympathetic ganglion neuron against modification by dithiothreitol; the protective effect of TMA is greater than the effect of acetylcholine.

A similar effect of dithiothreitol has been observed in molluscan neurons (Vulfius et al., 1979). The effect can be removed by the oxidant dithiobischoline, and the action of dithiobischoline can be antagonized by the affinity modification of nicotinic receptors with bromoacetylcholine or with the maleinimide derivatives. Acetylcholine and some other agonists (TMA and decamethonium) protect the nicotinic receptor against modification by dithiothreitol, in contrast to tubocurarine and other antagonists. Interestingly, the protecting potencies of agonists correlate with their stimulant potencies but not with their abilities to evoke desensitization. This difference between the effects of agonists and antagonists is probably due to basic differences in the conformational transitions produced by agonists and antagonists in the receptor molecule. One should keep in mind that the irreversible modification of the receptor produced by bromoacetylcholine can be prevented by tubocurarine. Altogether, these results suggest that the functionally important disulfide bond is located very close to the recognition center.

One should keep in mind that, in general, the loss of the ability of the receptor whose disulfide bonds are modified to respond to agonist stimulatory action can be equally explained as the result of modification either of the recognition center or of the ionic channel. To differentiate between these two possibilities, one has to know whether the binding of acetylcholine and agonists by the receptor is affected by the modification procedure.

This point was recently studied using neuronal nicotinic acetylcholine receptors from the brains of chickens and rats. It has been found that after reduction with dithiothreitol the receptors can be specifically labelled with 4-(N-maleimido)benzyltri[^{3}H]methylammonium iodide, the acetylcholine binding site directed reagent ([^{3}H]MBTA), and that both [^{3}H]MBTA and another affinity alkylating reagent, bromoacetylcholine, inhibit [^{3}H] nicotine binding to these receptors (Whiting and Lindstrom, 1987a, b). These results suggest that the amino acid residues Cys 192-193, which are thought to form disulfide bonds directly adjacent to the recognition center in the *Torpedo* electric organ acetylcholine receptor (see Chapter 2), are conserved in the neuronal acetylcholine receptor. On the other hand, it is not yet clear whether the estimate of the distance between the anionic site and the functionally important disulfide bond obtained from electric organ by the chemical modification method (see above) is valid for neuronal nicotinic receptors. The reason for this is that the dependence of the modification rate on the distance between the modifying and trimethylammonium groups in maleinimide derivatives found in electric organ has not been confirmed in molluscan neurons (Vulfius et al., 1979). It should be noted that the high-affinity binding sites for nicotine and

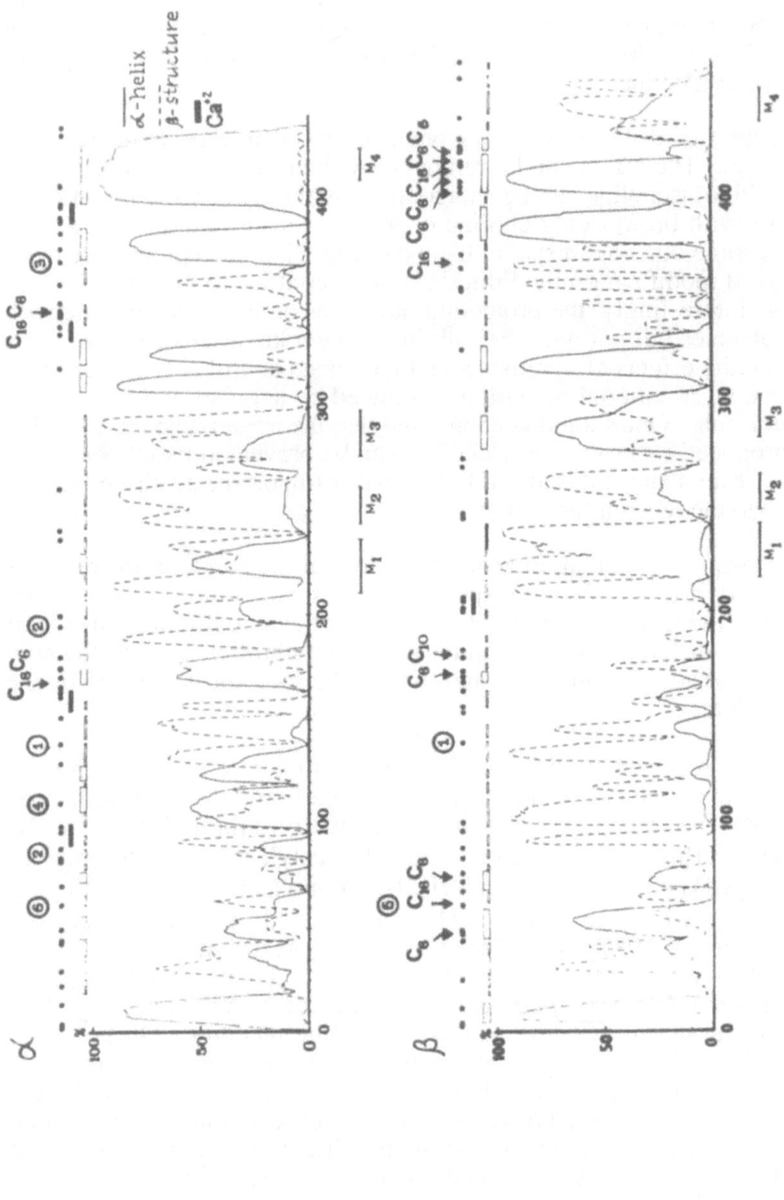

Fig. 2.5. Probabilities for secondary structures of *Torpedo californica* acetylcholine receptor α and β subunits. Large and small light rectangles indicate the most probable α helix and most probable β structure, respectively. The primary structure was taken from Noda et al. (1982, 1983b). Numbers indicate sequence numbering; the negatively charged aspartic and glutamic residues are marked with dark circles. M_1–M_4 indicate the transmembrane domains shown in Fig. 1.8A. Numbers within circles mean the assumed binding sites for acetylcholine (see text). C_6, C_{10}, and C_{16} indicate the assumed binding sites for penta- and hexamethonium, nano- and decamethonium, and hexa- to octadecamethonium, respectively, responsible for three peaks of blocking activity shown in Fig. 6.1. Dark rectangles indicate the tentative binding sites for Ca^{2+}.

acetylcholine in the vertebrate brain may not be equivalent to the ganglionic nicotinic acetylcholine receptors (see Kemp and Morley, 1986).

2.3. LOCALIZATION OF THE RECOGNITION CENTER IN THE ACETYLCHOLINE RECEPTOR MOLECULE

The results of the investigation of the recognition center mentioned above do not indicate where this center is localized in the receptor molecule or what its chemical structure is. Some information on these points can be obtained provided that the amino acid sequence and conformation of the receptor protein are known.

The amino acid sequences are known only for the acetylcholine receptors in electric organ and skeletal muscle and also for the α subunit of the neuronal acetylcholine receptor, while their conformations remain unknown. However, using theoretical predictions, the secondary structure of a protein can be predicted from its amino acid sequence, thus allowing one to estimate the possible distances between the amino residues along the polypeptide chain. By comparing the scheme of the functional organization of the recognition center shown in Fig. 2.3 with the structure of the polypeptide chain obtained by this approach, one can try to find the structure of greatest similarity – a tentative recognition center.

The probability of a certain secondary structure of nicotinic receptor protein has been predicted based on its amino acid sequence (see Ptitsyn and Finkelstein, 1983). Both remote and nearby interactions within each polypeptide chain have been considered according to the stereochemical stability theory of α and β conformations. The α or β conformation has been predicted provided that in at least 50% of the chain length in each particular area the corresponding conformation appears most probable. Otherwise the conformation of the polypeptide chain has been considered unknown (possibly irregular). Figures 2.5 and 2.6 illustrate the predicted secondary structure thus obtained for all four subunits of the nicotinic receptor in electric organ. The tentative membrane organization of each subunit is shown in Fig. 1.8C. The intrabilayer domains M_1–M_7 are indicated under the main acid ordering line. It follows from Fig. 1.8C that the polypeptide regions on the left from domain M_6 and between the domains M_7 and M_1, M_2 and M_3, M_5 and M_4 are located at an extracellular site, while the regions between domains M_6 and M_7, M_1 and M_2, M_3 and M_5, and on the right from M_4 are intracellular.

The carboxylic group that is thought to form the anionic site in a recognition center probably belongs to aspartate rather than to glutamate. The reason for this suggestion is that it is aspartic acid which is usually a carrier of a negative charge in the active center of the enzymes, probably because its carboxylic group is fixed more rigidly relative to the main chain than in glutamate (see Shulz and Schirmer, 1982). It should be noted that, in general, there are many common features between the enzyme–substrate interaction and the ligand–agonist interaction (cooperativity, competitive and noncompetitive inhibition, etc.). In this regard, it is interesting that the interaction of the enzyme with trypsin starts with the electrostatic binding of a positively charged lysine or arginine residue to the aspartate residue located at the bottom of a binding "pocket" in the enzyme molecule. This is followed by additional binding of the lysine or arginine residue, as well as the

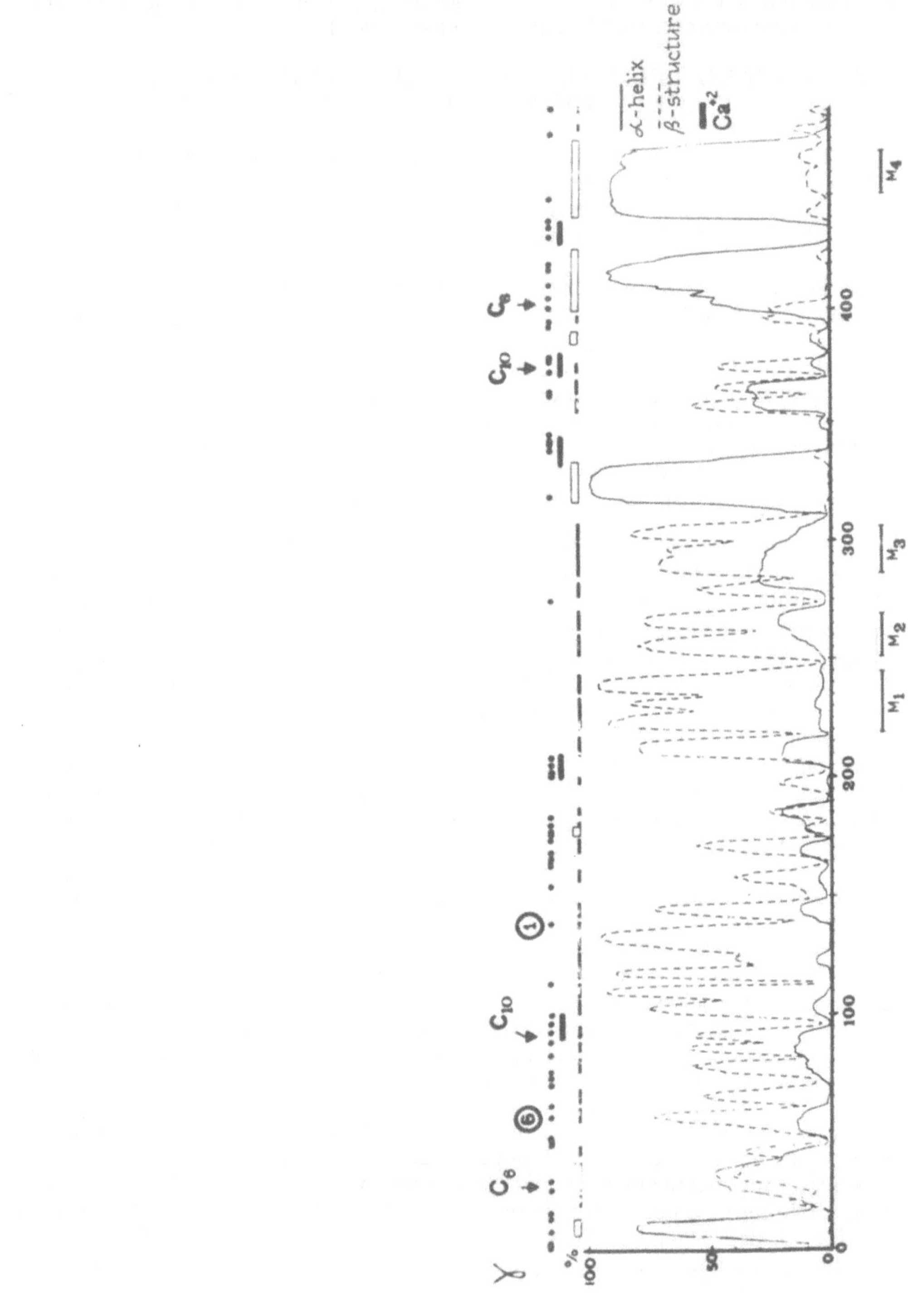
α-helix
β-structure
Ca^{+2}
C_8
C_{10}
C_{10}
C_8
M_1
M_2
M_3
M_4
γ
%
100
50
0
0
100
200
300
400

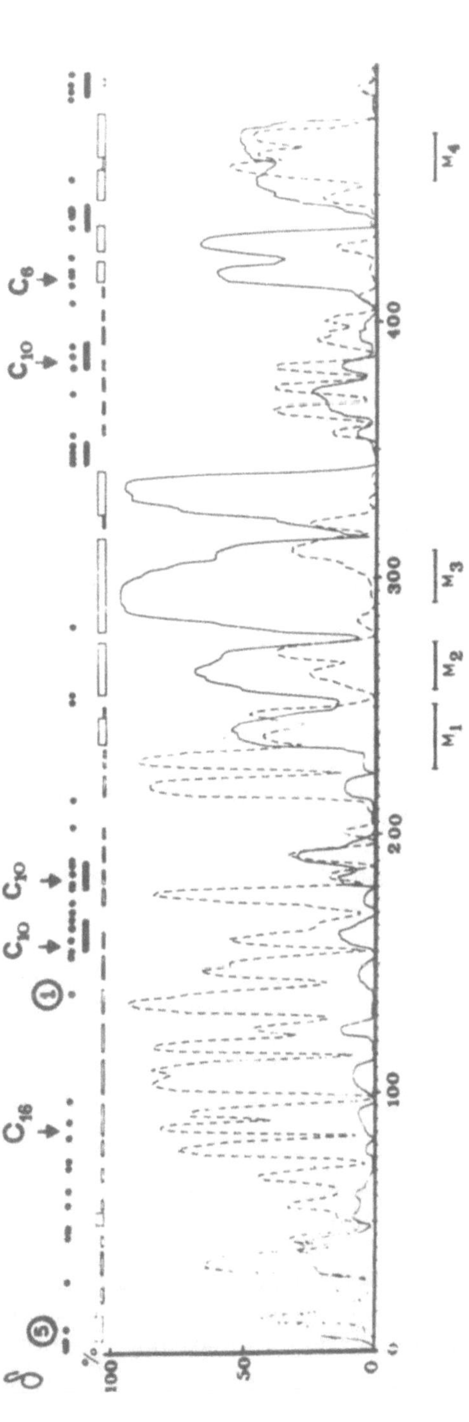

Fig. 2.6. Probabilities for secondary structures of *Torpedo californica* acetylcholine receptor γ and δ subunits. Symbols are the same as those used in Fig. 2.5.

neighboring residues in the protein molecule, to the walls of the enzyme pocket through their hydrophobic and hydrogen interactions (Stroud, 1974; Fehlhammer et al., 1977). The above-mentioned idea that there is a pocketlike structure in the recognition center of a nicotinic receptor that binds the trimethylammonium group is thus consistent with the general features of the enzyme–protein interaction.

The positive charge in the esterophilic area of a recognition center can be carried by lysine, arginine, or histidine, although one cannot exclude the possibility that the positive charge is partial and is carried, e.g., by the amine group of asparagine or glutamine.

Two opposite charges in the recognition center separated by a distance of 0.7 nm could belong either to the same short region of the polypeptide chain or to two remote regions that approach each other in a specific tertiary structure of the acetylcholine receptor protein. Insofar as the tertiary structure of the receptor protein remains unknown, only the former possibility can be discussed at present. In the β structure a distance of 0.7 nm corresponds to the distance between two neighboring amino residues localized at the same site of the β strain (0.67 nm, see Volkenstein, 1975), i.e., to the distance between each third residue in the amino acid sequence. In the α helix with averaged parameters (about 0.6 nm radius, 0.54 nm step, with 3.6 residues per step; see White et al., 1978; Shulz and Schirmer, 1982), a distance of 0.7 nm should separate the first and fourth or each first and fifth residues in their sequence (see Skok, 1985).

Several attempts to localize the recognition center in the amino acid sequence of the electric organ and skeletal muscle acetylcholine receptor protein have been made using the above approaches. All attempts have been based on a "two-point" binding scheme which assumes that one positive and one negative charge in the acetylcholine molecule 0.7 nm apart bind to the corresponding two opposite charges in the α subunit. Another assumption is that a disulfide bond should be localized within the 0.1-nm distance from the negative charge in the α subunit.

In the first attempt (Noda et al., 1982, 1983b, c), the anionic site was thought to be formed by Asp 138 while the disulfide bond was formed by Cys 142. This region of the polypeptide chain, which is indicated in Fig. 2.5 (site 1), has also attracted the attention of other investigators (Smart et al., 1984) because of one peculiar detail: it was Gln 139 in the α subunit instead of Trp 139 in three other subunits (Fig. 2.8).

According to theoretical prediction, this region is a β structure. The presence of glutamine instead of tryptophan is thought to be responsible for an orientation of the amino residues in which residues of Asp 138 and Cys 142 appear at the same side of the β strand, thus facilitating the binding of acetylcholine (Fig. 2.7). In the case of Trp 139 instead of Gln 139, the above residues are directed to different sides of the β strand, which makes acetylcholine binding impossible. Another advantage of this region is that two cysteines, Cys 128 and Cys 142, may form the functionally important disulfide bond.

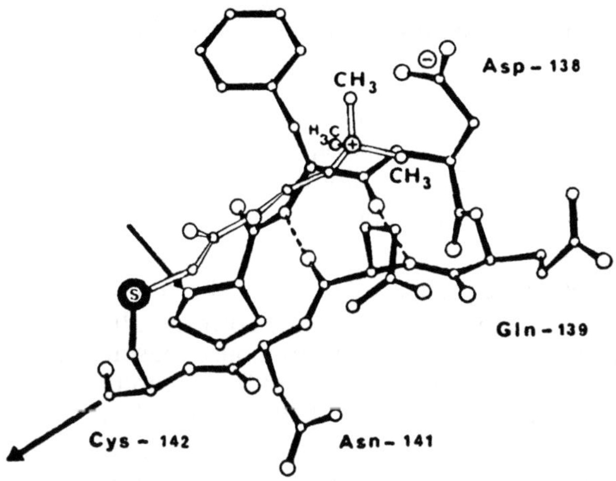

Fig. 2.7. Assumed binding site for acetylcholine in the α subunit of the eel electric organ acetylcholine receptor. The acetylcholine molecule is indicated by light lines. For further details see text (Smart et al., 1984). Arrows indicate joints to peptide chain.

N^+ ↓ (anionic site, at D 139); numbering: 130 at I, 140 at Q

Sequence	Subunit	Source
S C K I D V T Y F P F D Y Q N C T	α	rat neurones
Y C S I I V T H F P F D Q Q N C T	α	fish electric organ
S C T I K V M Y F P F D W Q N C T	β	fish electric organ
T C P I A V T Y F P F D W Q N C S	γ	fish electric organ
S C P I N V L Y F P F D W Q N C S	δ	fish electric organ
Y C E I I V T H F P F D E Q N C S	α	calf muscle
S C S I Q V T Y F P F D W Q N C T	β	calf muscle
S C P V S V T P F P F D W Q N C S	γ	calf muscle
Y C E I I V T H F P F D E Q N C S	α	mouse muscle
S C S I S V T Y F P F D W Q N C S	γ	mouse muscle

Fig. 2.8. Amino acid sequences in the region of the α subunit which presumably corresponds to a recognition center in nicotinic acetylcholine receptors of rat neurons (Boulter et al., 1986), of the eel electric organ (Noda et al., 1982, 1983a) and of calf (Noda et al., 1983b; Tanabe et al., 1984; Shibahara et al., 1985) and mouse muscle (Clarke et al., 1985; see Boulter et al., 1986). Other subunits are also shown. Sequence numbering is that for rat neuronal α subunit. Anionic site is marked by arrow.

Recently, region Lys 125 to Gly 147 was synthesized. The synthetic peptide exhibited an increased ability to bind ^{3}H-acetylcholine and ^{125}I-α-bungarotoxin, although this ability was somewhat lower than in the purified acetylcholine receptor protein (McCormick and Atassi, 1984). However, other authors have not observed binding of α-bungarotoxin to a 126-143 sequence in the α subunit (Neumann et al.,

1985). The fact that α-bungarotoxin abolishes the ability of the acetylcholine receptor to bind agonists does not yet prove that the toxin binds to a recognition center. The agonist-binding can be blocked simply because of the steric hindrance produced by the large toxin molecule attached to a site (or to many sites) that is contiguous to a recognition center in a tertiary or quaternary structure of the receptor molecule but is remote from a recognition center in the amino acid sequence or even belongs to another subunit. This possibility is consistent with the multisite binding of α-bungarotoxin discussed in Chapter 5.

Figure 2.8 shows sequence 127-143 in the α subunit of the rat neuronal acetylcholine receptor. For comparison, the corresponding regions of α and other subunits of the acetylcholine receptor from mouse muscle, fish electric organ, and calf muscle are also shown (Noda et al., 1982). Theoretical prediction performed using Ptitsyn's and Finkelstein's method (Ptitsyn and Finkelstein, 1983) has confirmed the above statement that sequence 138-142 in the α subunit of electric organ acetylcholine receptor is a β structure (the conformation for this region in neuronal acetylcholine receptors is discussed below). As one can see in Fig. 2.8, 11 of 15 amino residues between Cys 128 and Cys 142 are homologous in all four proteins compared.

Another interesting detail is that there is much more homology in α subunits between the muscle and electric organ than between both these objects, on the one hand, and nervous tissue on the other hand. This difference correlates with the fact that in their pharmacology nicotinic receptors in skeletal muscle are more similar to those in electric organ than to those in autonomic ganglia. Assuming that the anionic site is formed by Asp 138, the Gln 140, which is located about 0.6–0.7 nm away (this follows from the prediction that the region in question is a β structure), could form by its amide group and by partial charge of its side chain the points 3 and 7, respectively (see the above).

In the second attempt (Kosower, 1983), three sites that may bind acetylcholine were identified in α subunits of the electric organ acetylcholine receptor. Each site is assumed to be formed by two parallel β strands from two α subunits. In each of two binding sites (A and B), negative charges are carried by Asp 83 of one α subunit and by Asp 195 of another α subunit (sites 2 in Fig. 2.5), while positive charges are carried by Lys 76 and Lys 125 of two α subunits. The third binding site (C) is formed by negative charges of Asp 371 and by positive charges of Lys 354 from both α subunits. Cys 193 forms a functionally significant disulfide bond. Site C, according to the author's suggestion, may also bind noncompetitive blocking agents.

Thus, the second attempt suggests that the acetylcholine molecule can simultaneously interact with regions that are localized far away from each other in the amino acid sequence but approach each other in the hypothetical tertiary and quaternary structures. The third site (C), in contrast to the other two sites, is localized on the cytoplasmic surface of the membrane (site 3 in Fig. 2.5).

Later results obtained with the use of site-directed mutagenesis (Mishina et al., 1985) showed that cysteine residues 128 and 142 in all four subunits of the *Torpedo* acetylcholine receptor are important for α-bungarotoxin and acetylcholine binding activities of the receptor, which is consistent with the above results. Moreover, in this study an additional site important to acetylcholine binding (but not to α-bungarotoxin binding), which are the α192 and α193 cysteine residues, has been found. The authors speculate that cysteine residues 192 and 193 of the α subunit may play a specific role in agonist binding and also in signal transduction, whereas the cysteine residues 128 and 142 of all the four subunits, probably forming a disulfide bridge in each subunit, serve to maintain the proper conformation of the extracellular region of acetylcholine receptor molecule (Mishina et al., 1985).

Most recently, the importance of the region α185-196, which includes the above cysteins 192 and 193, for binding of acetylcholine was confirmed by analyzing the ligand binding to the synthetic dodecapeptide α185-196 of the *Torpedo* acetylcholine receptor (Neumann et al., 1986). In addition to acetylcholine, α-bungarotoxin has also been found to bind selectively to this peptide. This is consistent with other results indicating that α-bungarotoxin binds to the sequence α173-204 in the *Torpedo* acetylcholine receptor (Wilson et al., 1985).

There are also other sites in the acetylcholine receptor molecule that correspond to the requirements mentioned above, in particular to the existence of negative and positive charges about 0.7 nm away from each other and, therefore, can be considered as the tentative recognition centers. These sites can be identified assuming that a distance of 0.7 nm corresponds to the interval between each first and fourth or fifth amino residues in the α helix, and between each first and third amino residues in the β structure. Analysis of the theoretically predicted α helical regions (see Skok, 1985) from this point of view reveals in the extracellular part of the α subunit site Lys 109–Lys 115 (Fig. 2.5, site 4), which may bind acetylcholine either by the pair Asp 111–Lys 107 or by the pair Asp 111–Lys 115, and which has no analogous sites in other subunits. The δ subunit possesses the pair Arg 6–Asp 10 (Fig. 2.6, site 5). Another interesting site is Asp 62–Arg 64, which is a predicted β structure. This site is present in α, β, and γ subunits. In the α subunit, Asp 62 and Arg 64 are separated by nonpolar valine, and Asp 62 is adjacent to nonpolar isoleucine. Both nonpolar residues have branching at the C_β atom, which should make the main chain rigid. On the other hand, the tryptophane residue that is adjacent to the aspartic residue located at the same side of the β strand is most hydrophobic, thus providing a possible structure for the hydrophobic interaction in close vicinity to the anionic site. Finally, the nonpolar isoleucine, valine, and leucine residues are commonly thought to have preferable orientation toward the inside of the protein molecule, which means that aspartate and arginine residues, on the contrary, should be located at the surface of the molecule. By contrast, in all other subunits, the aspartic and arginine residues in the corresponding region have only polar residues adjacent to them. Therefore, the region Asp 62–Arg 64 in the α subunit corresponds to at least some requirements for a tentative recognition center, although there is no cysteine adjacent to this region (Fig. 2.5, site 6).

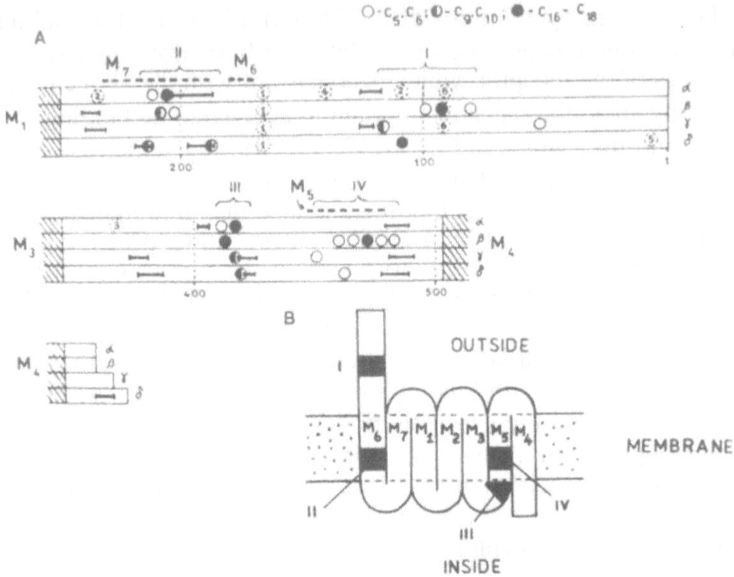

Fig. 2.9. Probable binding sites for acetylcholine and for some blocking agents in *Torpedo californica* acetylcholine receptor. A) The amino acid sequences of four subunits (α, β, γ, and δ) are aligned to attain maximum homology (Noda et al., 1983b). The binding sites for acetylcholine (numbers within circles), for blocking agents (see inset), and for calcium ions (horizontal bars) are similar to those shown in Figs. 2.5 and 2.6. One can see four binding areas (I–IV). Numbers at the bottom indicate the amino acid sequence numbering. B) Hypothetical model of the transmembrane arrangement of the acetylcholine receptor molecule according to Fig. 1.8C.

The examples just considered indicate that there are a few regions in the α subunit that may provide a possible structure for a recognition center. Their total number is five (excluding site 3 in Fig. 2.5, which is probably located on the cytoplasmic surface of the membrane).

An interesting feature of all the sites described above is that none of them contains a second negative charge, or a "peripheral" anionic site shown in Fig. 2.3, about 0.7 nm away from the main one. This implies that neither penta- nor hexamethonium may interact with a recognition center in the electric organ and skeletal muscle acetylcholine receptor, which is consistent with experimental results.

Furthermore, all five tentative binding sites for acetylcholine are localized in the region of the α subunit which, according to the model illustrated by Fig. 1.8C, is extracellular (see also Fig. 2.5). Figures 2.5 and 2.6 also show the tentative binding sites for bis-ammonium blocking agents, which are described in detail in Chapter 6.

In Fig. 2.9A the amino residues of all four subunits are aligned to obtain their maximum homology (Noda et al., 1983b). The hypothetical arrangement of the subunits in the membrane is illustrated by the scheme in Fig. 2.9B, which is made

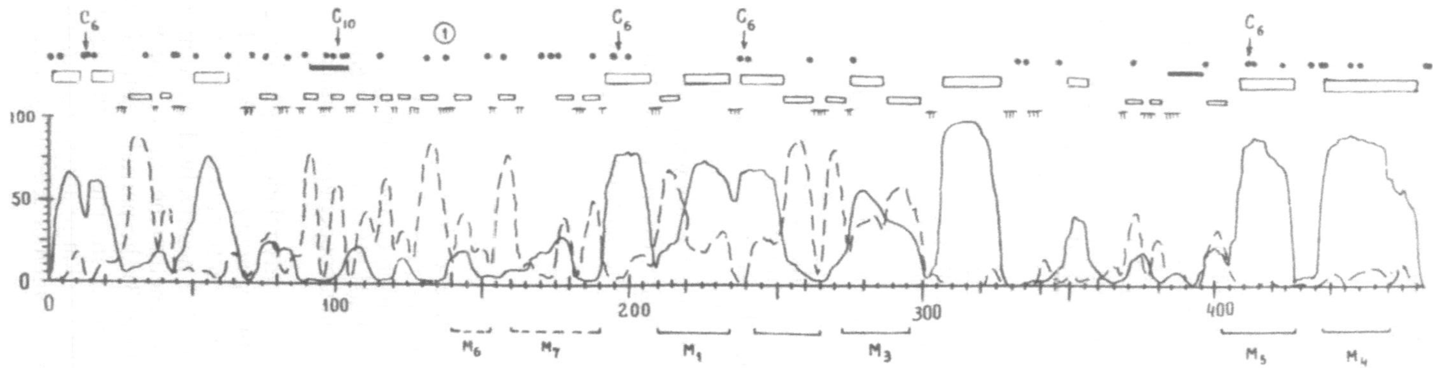

Fig. 2.10. Probabilities for secondary structures of rat neuronal acetylcholine receptor α subunit (see Boulter et al., 1986) calculated as in Figs. 2.5 and 2.6. Symbols are the same as those used in Figs. 2.5 and 2.6 (T means the turn in polypeptide chain).

according to the model shown in Fig. 1.8C; as in Figs. 2.5 and 2.6, the tentative binding sites for bis-ammonium blocking compounds are indicated. One can see that these sites form four areas: area I is at the extracellular, while area III is at the intracellular surface of the membrane; two other areas (II and IV) are in the transmembrane domains. Only area I overlaps with the tentative binding sites for acetylcholine.

Figure 2.10 illustrates the results of theoretical prediction of the secondary structure for the only subunit of neuronal nicotinic acetylcholine receptor whose primary structure is known at the present time – the α subunit. The approach used is the same as that demonstrated in Figs. 2.5 and 2.6 (in addition, the turns in the polypeptide chain are indicated in Fig. 2.10). One can see from a comparison of Figs. 2.10 and 2.5 that the α subunits of electric organ acetylcholine receptor and neuronal acetylcholine receptor are actually very similar in their secondary structures. It should be emphasized, however, that the region Asp 138–Cys 142, which is most probably the binding site for acetylcholine, is localized in the turn area in the neuronal α subunit, while in the electric organ α subunit it is probably a regular β structure. The tentative binding sites for other ligands shown in Fig. 2.10 will be discussed in Chapter 6.

Recent findings suggest that, in general, localization of the recognition center in the neuronal acetylcholine receptor may differ from that in the nicotinic receptors of the electric organ and muscle. Evidence for this has been obtained in recent experiments on neuronal acetylcholine receptors from chicken and rat brain (Whiting and Lindstrom, 1987a). These receptors have been found to consist of only α and β subunits, and their labelling with [^{3}H]MBTA suggests that the acetylcholine binding site, as well as the amino acids which may be homologous to Cys 192-193 of the α subunits of the acetylcholine receptors from the electric organ and muscle, are located on the β subunit of the neuronal acetylcholine receptor.

2.4. CONCLUSIONS

The results obtained by two independent approaches – estimation of the Hill number in normal and modified nicotinic acetylcholine receptors and examination of the effect produced by receptor activation on the fluorescence of the receptor-bound ligands – suggest that there are at least two (or three) agonist-binding sites in the receptor molecule that may interact in a cooperative manner.

Examination of the structure–activity relationship indicates that the recognition center (the site that normally binds acetylcholine) of the neuronal nicotinic acetylcholine receptor contains five functional components: 1) the anionic site responsible for ionic interaction with the positively charged nitrogen atom of acetylcholine; 2) the hydrophobic area attached to the anionic site, which binds methyls of the cationic group of acetylcholine; 3) the partial electrical charge about 0.6 nm away from the anionic site, which interacts with the carbonyl oxygen of acetylcholine; 4) the partial negative charge about 0.3 nm away from the anionic site, which interacts with the ether oxygen of acetylcholine; 5) the hydrophobic area about 0.6 nm away from the anionic site, which binds the methyl of the acylic group of acetylcholine. In addition, some authors suggest another (peripheral) anionic site localized at a distance of nearly 0.7 nm from the above (main) anionic site, which may selectively bind penta- and hexamethonium; however, the relation of the second anionic site to the recognition center for acetylcholine is debatable (see Chapter 5).

The most specific features of the recognition center in neuronal nicotinic acetylcholine receptors, compared with those in nicotinic acetylcholine receptors of other tissues, are the increased functional importance of its two components: the main anionic site with its attached hydrophobic area, which interacts with the cationic group of acetylcholine, and the hydrophobic area interacting with the methyl of the acylic group of acetylcholine.

The recognition center is apparently localized in the α subunit. An examination of the primary structure and the theoretically predicted secondary structure of the α subunit reveals a few tentative acetylcholine-binding sites which form two areas, e.g., a region Asp 138–Cys 142, with the carboxylic group of Asp 138 forming the anionic site and Cys 142 forming the functionally important disulfide bond.

Chapter 3

IONIC CHANNEL OF THE NICOTINIC ACETYLCHOLINE RECEPTOR

3.1. CHANNEL IONIC SELECTIVITY AND ENERGY PROFILE

Activation of nicotinic acetylcholine receptors in all the cells studied results in an increase in membrane conductance. Depending on whether the reversal potential is less negative or more negative than the resting potential, nicotinic response appears as membrane depolarization (excitation) or membrane hyperpolarization (inhibition), respectively. In invertebrate neurons, both excitatory and inhibitory nicotinic responses have been observed, whereas in vertebrate neurons they are solely excitatory.

The ionic mechanisms underlying excitatory and inhibitory nicotinic responses are different. These responses result from the increase in membrane conductance for cations and anions, respectively. They can be differentiated using pharmacological tests: the increase in cationic conductance can be selectively inhibited by hexamethonium and Tris ions, whereby the increase in anionic conductance can be blocked by anthracene-9-carbonic acid or by furosemide; anionic conductance can be selectively activated by suberylcholine (Ger et al., 1979a, 1980; Zeimal and Katchman, 1983; Witte et al., 1985a). Thus, the increase in acetylcholine-induced conductance for cations and anions seems to be completely independent of each other.

A detailed quantitative study of acetylcholine-induced conductance is possible only when one type of channel, either cationic or anionic, is present in a neuron tested. The anionic conductance has been studied in detail in molluscs of different species (Adams et al., 1976; Kurchikov and Kazachenko, 1979; Bregestovsky and Iljin, 1980; Gardner and Stevens, 1980; Simonneau et al., 1980; Chemeris et al., 1982; Katchman and Zeimal, 1982; Finkel, 1983). Channel selectivity for chloride

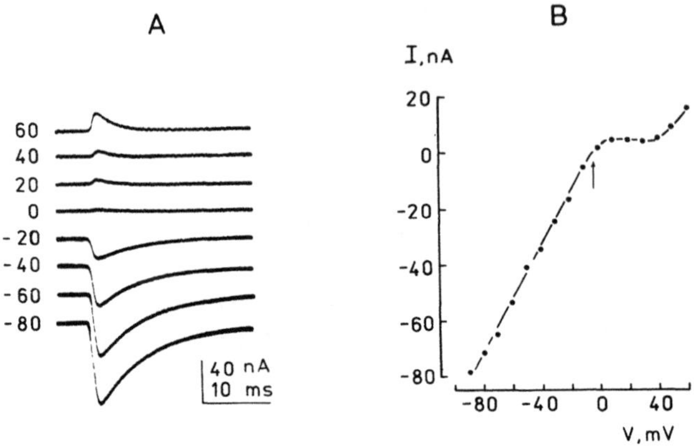

Fig. 3.1. Effect of membrane potential on fast excitatory postsynaptic current (e.p.s.c.) recorded from rabbit superior cervical ganglion neuron. A) An example of e.p.s.c.'s recorded at different levels of membrane potential (indicated in mV near each record). B) Dependence of e.p.s.c. amplitude on membrane potential. The level of e.p.s.c. reversal potential (–3 mV) is indicated by the arrow. The data were obtained at 36°C (Derkach et al., 1983).

ions is very high since 1) the reversal potential for anionic nicotinic response corresponds exactly to the chloride equilibrium potential E_{Cl} (in molluscan neurons E_{Cl} is several millivolts more negative than the resting membrane potential); 2) the reversal potential for anionic response depends only on the chloride concentration on both sides of the membrane and does not depend on the presence of cations.

In contrast to anionic conductance, the cationic conductance increases for several species of cations simultaneously. This is indicated by the observation that the reversal potential for cationic response is not similar to the equilibrium potential for any cation. In autonomic ganglia, the reversal potential for cationic nicotinic response is intermediate between E_{Na} and E_K and ranges from –10 mV to 0 mV (Nishi and Koketsu, 1960; Dennis et al., 1971; Selyanko and Skok, 1979a; Selyanko et al., 1979; Kuba and Nishi, 1979; Rang, 1981). An example of the reversal potential measurement for excitatory postsynaptic current (e.p.s.c.) is shown in Fig. 3.1. The reversal potential for nicotinic response is very stable and is not affected, for instance, by culturing of autonomic neurons (Obata, 1974) or by their denervation (Dun et al., 1976).

In molluscan neurons, early estimates of the reversal potential for cationic nicotinic response, obtained by the extrapolation method, gave much more positive values than those mentioned above (80 mV: Marty et al., 1976, or 20 mV: Ascher et al.,, 1978a). Based on these results, it was suggested that nicotinic acetylcholine receptor channels in molluscan neurons were selective for sodium ions. However, this suggestion has not been confirmed by later investigations. On the one hand, the reversal potential estimates obtained in snail neurons by more accurate interpolation methods have yielded values close to those found in other neurons (–4

TABLE 3.1. Permeability Ratios for Monovalent Metal Ions of Acetylcholine Receptor Channels in Nerve and Muscle Cells

Preparation	Permeability ratio	Method	References
	Neurons		
Mammalian sympathetic ganglion	$Na^+ < K^+$ (1:1.16)	Reversal potential shifts	Skok and Selyanko,1979
Molluscan ganglion	$Na^+ < K^+$ (1.0:1.6)	Same	Witte et al., 1985b
	Muscle cells		
Mammalian	$Na^+ > K^+$ (1:0.6)	Same	Ritchie and Fambrough, 1975
	$Li^+ > Na^+ > K^+$ (1.1:1:0.8)	Same	Linder and Quastel, 1978
Amphibian	$K^+ > Na^+$ (1.1:1)	Same	Lewis, 1979
	$Cs^+ > K^+ > Na^+ > Li^+$ (1.25:1.1:1.0:0.9)	Same	Gage and Van Helden, 1979
	$Cs^+ > Rb^+ > K^+ > Na^+ > Li^+$ (1.4:1.3:1.1:1.0:0.9)	Same	Adams et al., 1980
Avian	$Cs^+ > Rb^+ \cong K^+ > Na^+$ (1.9:1.5:1.5:1)	Movement of labeled ions	Huang et al., 1978

to −3 mV: Witte et al., 1985b). On the other hand, in experiments with varied extracellular medium content (see below), an increase in conductance for cations other than sodium ions has been observed.

In autonomic ganglia, the reversal potential for cationic nicotinic response depends on the external concentration of sodium, potassium, and calcium ions and is independent of the presence of chloride ions (Koketsu, 1969; Selyanko and Skok, 1979). Under normal conditions, when extracellular calcium concentration is relatively low, the current through nicotinic channels is carried mostly by sodium and potassium ions. The permeability ratio for sodium and potassium ions (P_{Na}:P_K) can be estimated from the Goldman–Hodgkin–Katz equation (Goldman, 1943; Hodgkin and Katz, 1949), based on estimates of the reversal potential for the response and the equilibrium potentials for sodium and potassium ions. In rabbit superior cervical ganglion, where the mean value of the reversal potential for e.p.s.c. is −10 mV (Selyanko et al., 1979), mean E_{Na} and E_K values are 37 mV and

–92 mV, respectively, as calculated from $[Na^+]_{in}$ = 37.4 mM and $[K^+]_{in}$ = 163 mM (Woodward et al., 1969); P_{Na}:P_K is 0.86. Thus, in the ganglion neurons nicotinic acetylcholine channels are more permeable for potassium ions than for sodium ions. However, because the electromotive force for sodium ions is higher than for potassium ions, the current flowing through the channels at resting potential level is carried mostly by sodium ions. The conclusions on the selectivity of nicotinic cationic channels drawn from electrophysiological experiments are consistent with the results obtained by other methods. It has been found using labeled ions (Brown and Scholfield, 1974) or ion-sensitive microelectrodes (Ballanyi et al., 1984) that activation of nicotinic receptors in sympathetic ganglion neurons is followed by an increase in intracellular sodium concentration and a decrease in intracellular potassium concentration.

Nicotinic receptor channels in autonomic ganglia are also permeable for calcium ions. This becomes apparent when sodium ions in extracellular medium are partially or completely replaced with Ca^{2+} ions. Under such conditions, acetylcholine still increases membrane conductance, although the increase in conductance is smaller in calcium solution than in sodium solution (Pappano and Volle, 1966; Koketsu et al., 1968). However, even in normal solution, when extracellular calcium concentration is rather low, there is a substantial flow of calcium ions inside the cell, indicated by the appearance of calcium-activated potassium conductance following acetylcholine response (Ascher et al., 1978a; Tokimasa and North, 1984).

In molluscan neurons, as in autonomic ganglia neurons, cationic nicotinic channels are completely impermeable for anions (Ascher et al., 1978a; Witte et al., 1985b). Channel selectivity in these neurons has been studied by two methods. Some authors (Ascher et al., 1978a; Marchais and Marty, 1979; Marty, 1980) have measured single-channel characteristics (current amplitude and open time) in the presence of different permeable ions. Others (Witte et al., 1985b) have measured the effect of a change in the extracellular medium on the reversal potential for acetylcholine response. In the latter case, it has been found that the permeability ratio for potassium, sodium, and calcium ions is 1.6:1:0.1. Table 3.1 shows that in molluscan neurons the permeability ratio for sodium and potassium ions ($Na^+ < K^+$) is similar to that in amphibian and avian muscle cells, whereas in mammalian muscle cells the permeability ratio is different ($K^+ < Na^+$). In mammalian sympathetic ganglion neurons, as in molluscan neurons, this ratio is in favor of potassium ions.

As can be seen from Table 3.2, relative single-channel conductance and open time depend on the species of permeable monovalent cation. Despite the fact that only single-channel current, but not conductance, can be measured in molluscan neurons (because the reversal potential in these studies was uncertain), there is good agreement, both qualitative and quantitative, between the current ratios for different ions in neuronal and muscular membranes. At the same time, the ratio of channel open times, obtained in the presence of different permeable ions, differs in molluscan neurons from that in amphibian muscle cells. This ratio for lithium and sodium ions also differs between molluscan and sympathetic neurons (see Table 3.2).

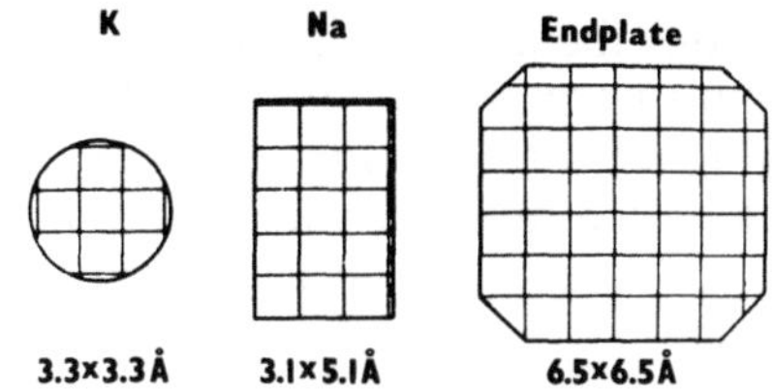

Fig. 3.2. Hypothetical cross section of frog end-plate channel. For comparison, cross sections of voltage-dependent sodium and potassium channels are shown. Grids with lines separated by 1-Å intervals restrict the areas for ion diffusion. The minimum size of each channel is estimated from the sizes of permeant and impermeant ions (Dwyer et al., 1980).

TABLE 3.2. Conductance and Open Time Ratios for Single Acetylcholine Receptor Channels of Nerve and Muscle Cells in the Presence of Various Monovalent Metal Ions

Preparation	Conductance ratio	Open time ratio	References
Amphibian sympathetic ganglion neurons		$Li^+ < Na^+$	Parsons et al., 1984
Molluscan neurons	$Li^+ < Na^+ < Cs^+$ (0.5:1.0:1.7)	$Cs^+ > Li^+ > Na^+$ (1.9:1.3:1.0)	Ascher et al., 1978a; Marchais and Marty, 1979
Amphibian muscle fibers	$Li^+ < Na^+ < Cs^+ < NH_4^+$ (0.5:1.0:1.2:1.3)	$Li^+ > Na^+ > Cs^+$ (1.3:1.0:0.9)	Van Helden et al., 1977; Gage and Van Helden, 1979;
			Takeda et al., 1980
Amphibian muscle fibers	$Li^+ < Na^+ < Cs^+ < NH^+$ (0.5:1.0:1.1:1.6)	$Li^+ > Na^+ > Cs^+$ (1.1:1.0:0.6)	Adams et al., 1981

The similarity in relative conductance for monovalent cations between neurons and muscle cells indicates that the abilities of channels to pass ions in these two types of cells are essentially similar. However, the difference in relative open times indicates that permeable ions interact with these channels in different ways.

The flow of ions through nicotinic cationic channels has been studied in more detail in skeletal muscle fibers than in neurons. It has been found that in muscle cells the channels are permeable for at least 75 different cations, including organic and mono- and divalent metal cations (Dwyer et al., 1980; Adams et al., 1980). The channel permeabilities for different organic cations, estimated from the reversal potential shifts, are inversely proportional to ion size (Dwyer et al., 1980). This indicates that for organic ions the channel is an aqueous pore with which they do not interact. The minimum pore size as estimated from the size of the largest permeable organic cation is 6.5 × 6.5 Å (Fig. 3.2; see Dwyer et al., 1980). Since the conductance of acetylcholine receptor channels in neurons is similar to that in muscle cells (see Chapter 3.2), one may suggest that the pore sizes in these two types of channels are similar.

The permeability of the end-plate channels for monovalent metal cations is proportional to the ion crystal radius (Gage and Van Helden, 1979; Adams et al., 1980; see Table 3.1). Because a similar relationship exists between the ion radius and its aqueous mobility, it was suggested that for monovalent metal cations the channel is an electroneutral pore.

In contrast to monovalent metal cations, divalent metal cations show the opposite relationship between ion permeability and ion radius (concentration): the larger the ion radius (concentration), the lower the permeability (Lewis, 1979; Adams et al., 1980). This observation indicates that divalent cations pass the channel in a different way compared to that in monovalent cations. One possible explanation of this difference is to assume that a negative surface charge exists at the vicinity of the pore (Lewis, 1979). This charge can interact with divalent cations, thus changing their local concentration at the entrance to the channel. This effect should be stronger with divalent than with monovalent cations.

Further knowledge of the mechanisms of permeability of end-plate channels has been obtained by studying the effect of ion species on single-channel characteristics – its conductance and open time (Van Helden et al., 1977; Gage and Van Helden, 1979). Both channel characteristics appear to be affected by ions. Surprisingly, for almost all metal and organic ions the directly measured values of channel conductance appeared markedly lower than those calculated from the corresponding values of their permeabilities. Channel conductance for organic cations depends inversely on their hydrophobicity. This indicates that there is a saturating hydrophobic binding site within a channel that interacts with organic cations. For monovalent metal cations such a saturation has not been observed (Adams et al., 1981).

The above-mentioned data suggest that different species of permeable ions (organic, mono-, and divalent metal cations) see the end-plate channel in different ways. Thus, a comprehensive view on ion flow through the end-plate channel does not yet exist.

However, such a comprehensive model explaining the flow of all kinds of ions permeable through nicotinic acetylcholine channels has been suggested for cationic nicotinic channels in molluscan neurons (Ascher et al., 1978a; Marchais and Marty, 1979). These authors have noticed that the effect of permeable ions on channel open time is very similar to that produced by channel-blocking agents. It is known that channel-blocking agents interact with binding sites within a channel (see Chapter 5). By analogy, two proposals have been incorporated in the model: first, before entering the cell all permeable ions must bind within a channel and, second, the channel cannot close until the ion remains bound to it.

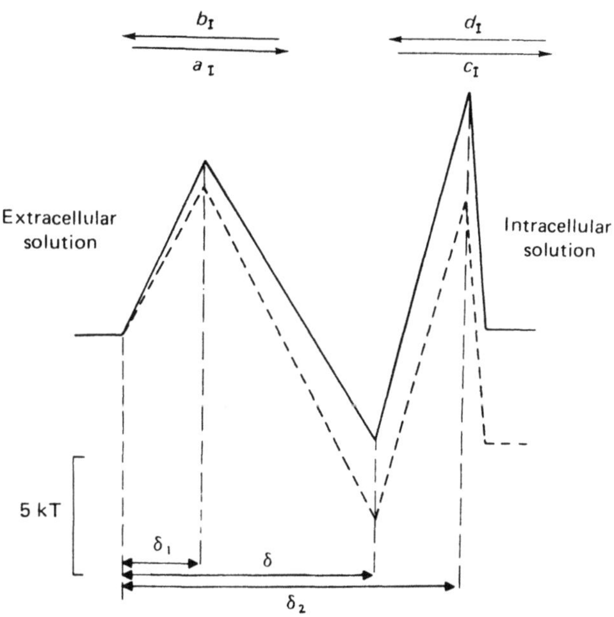

Fig. 3.3. Energy profile for Na^+ ion in the nicotinic receptor channel of *Aplysia* ganglion neuron (Marchais and Marty, 1979).

Based on these ideas, the channel energy profile for sodium ions has been proposed (Marchais and Marty, 1979), as illustrated by Fig. 3.3. The energy profile contains only one energy minimum, corresponding to the binding site for the permeable ion. This minimum is separated from external and internal solutions by two energy barriers. In this respect the energy profile in the acetylcholine receptor channel differs from those in voltage-sensitive channels, where more than one, usually two or three, binding sites are postulated (see Chizmadjev et al., 1973; Kostyuk and Krishtal, 1981; Hille, 1984).

It has also been suggested that the barrier height (but not the positions of the barriers in the membrane) depends on the permeable ion species and on the membrane potential. Figure 3.3 shows the energy profiles calculated for the solution containing only monovalent ions at different membrane potential levels: at the nicotinic response reversal level (20 mV, continuous line) and at the hyperpolarized level (–100 mV, dashed line). The ion movement through the membrane is characterized by $a_I I_e$, the rate constant for entering the channel by ions from external solution (I_e is the external concentration of ion I), by b_I and c_I, the rate constants for leaving the channel by ions for external and internal solutions, respectively, and by $d_I I_i$, the rate constant for entering the channel by ions from internal solution (I_i is the internal ion concentration). The effect of an electric field on the passage of ions through energy barriers is proportional to the voltage drop between well and barrier (barrier and well positions are indicated in the figure by δ_1, δ_2, and δ).

The ion–channel interaction can be described by the following reaction scheme, similar to that describing the interaction of a channel with ganglion-blocking drugs (see Adams, 1976; Ascher et al., 1978b):

$$\mathrm{R} \underset{q_0}{\overset{p}{\rightleftharpoons}} \mathrm{R}^* \underset{b_I + c_I}{\overset{a_I I_e + d_I I_i}{\rightleftharpoons}} \mathrm{R}^* I\,, \tag{3.1}$$

where R is the inactive (closed) state of the receptor–channel complex; R* and R*I are the open states of this complex without or with ion bound, respectively; p and q_o are rate constants for channel opening or closing, respectively, in the absence of permeant ions. The second part of this reaction should be much faster than the first part. Indeed, as follows from the value of the single-channel current at very negative membrane potential levels, the rate constant c_I should be about $10^7\ \mathrm{s}^{-1}$. The rate constant of ion binding within a channel $a_I I_e$ is likewise thought to be high. Assuming that a_I is similar to the rate constant that characterizes the effect of channel-blocking agents (about $10^6\ \mathrm{M}^{-1}\cdot\mathrm{s}^{-1}$, see Ascher et al., 1978b), it has been found that for sodium ions present in the external solution, $a_I I_e$ is $5\cdot 10^{-5}\ \mathrm{s}^{-1}$. The approximate estimates for c_I and $a_I I_e$ are several orders higher than an averaged channel closing rate q_o ($50\ \mathrm{s}^{-1}$, see Ascher et al., 1978a).

In general terms Eq. (3.1) can be written as

$$\mathrm{R} \underset{q}{\overset{p}{\rightleftharpoons}} (\mathrm{R}^* + \mathrm{R}^* I). \tag{3.2}$$

Here the rate constant of channel closing q is proportional to the probability (E) of the channel being in an open state without ion bound

$$q = q_o E,$$

where

$$E = \frac{\mathrm{R}^*}{\mathrm{R}^* + \mathrm{R}^* I}\,.$$

Then

$$q = q_o\,\frac{b_I + c_I}{a_I I_e + d_I I_i + b_I + c_I}\,. \tag{3.3}$$

At low concentrations of acetylcholine, when the probability of channel activation is very small ($p \ll q$), q and q_o correspond to the rate of channel closure in the presence or absence of permeant ions, respectively. Then the mean channel open time in the presence of permeant ions ($\tau = 1/q$) is

$$\tau = \tau_o \left(1 + \frac{a_I I_e + d_I I_i}{b_I + c_I} \right), \tag{3.4}$$

where τ_o is the mean channel open time in the absence of ions ($\tau_o = 1/q_o$).

The model suggests that the electric field within the membrane does not affect the first stage of the reaction (3.1), which was confirmed experimentally.

At the same time, the rate constants of ion binding with the channel are voltage-dependent and can be described by the following equations:

$$\begin{aligned} a_I &= a_I(0)\exp\left(\frac{\delta_1 z_I e_o V}{kT}\right), \\ b_I &= b_I(0)\exp\left(\frac{(\delta - \delta_1) z_I e_o V}{kT}\right), \\ c_I &= c_I(0)\exp\left(\frac{(\delta_2 - \delta) z_I e_o V}{kT}\right), \\ d_I &= d_I(0)\exp\left(\frac{(1 - \delta_2) z_I e_o V}{kT}\right). \end{aligned} \tag{3.5}$$

Here k and T are the conventional thermodynamic quantities; e_o is the electronic charge; z is the valence of the ion; V is the membrane potential; $a_I(0)$, $b_I(0)$, $c_I(0)$, and $d_I(0)$ are the corresponding rate constants at $V = 0$.

The effect of the membrane potential on τ is described by Eqs. (3.5) and (3.4).

The single-channel current i_{el} is equal to the number of ions entering the cell through the channel per unit time, multiplied by the unit charge and by the ion valence. The number of ions entering the cell is equal to the difference between the number of ions passing through the channel inside and outside the cell. Then the single-channel current is given by

$$i_{el} = e_o z_I \frac{c_I a_I I_e - b_I d_I I_i}{b_I + c_I + a_I I_e + d_I I_i}. \tag{3.6}$$

In the limiting cases, when $V = V_r$ and $V = -100$ mV, Eqs. (3.4) and (3.6) can be simplified. At $V = V_r$, when $i_{el} = 0$, the relationship between d_I and other kinetic parameters is given by

$$c_I(V_r)a_I(V_r)I_e = b_I(V_r)d_I(V_r)I_i.$$

At $V = -100$ mV, one can assume that $a_I I_e \gg d_I I_i, b_I, c_I$. Then Eqs. (3.4) and (3.6) will be simplified to

$$\tau = \tau_o \frac{a_I I_e}{b_I + c_I}, \tag{3.7}$$

$$i_{el} = e_o z_I c_I \,. \tag{3.8}$$

It follows from Eq. (3.8) that the current flowing through the single channel does not depend on the concentration of the permeable ion and exponentially increases with membrane hyperpolarization [c_I is voltage dependent; see Eq. (3.7)].

When the membrane is strongly hyperpolarized, the assumption $b_I \ll c_I$ appears to be true. Then Eq. (3.7) becomes even more simplified:

$$\tau = \tau_o \frac{a_I I_e}{c_I}, \tag{3.9}$$

which gives an estimate of $a_I\tau_o$, since c_I is known from Eq. (3.8). The experimentally measured τ values are voltage independent when the membrane is strongly hyperpolarized. Therefore, the voltage dependences of a_I and c_I should be similar, which in turn means that $\delta_1 \cong \delta_2 - \delta$ [see Eq. (3.5)]. Thus, one can accept that $\delta_1 \cong \delta_2 - \delta$.

In a strongly depolarized membrane, $c_I \ll b_I$, and Eqs. (3.7) can be transformed to

$$\tau = \tau_o \frac{a_I I_e}{b_I}. \tag{3.10}$$

The values of b_I and δ can be derived from a comparison of the voltage dependence of τ, obtained experimentally at low membrane potentials, with that predicted by Eq. (3.10).

Now only two parameters from Eqs. (3.5) and (3.7) remain to be found. Since $\delta_1 \cong \delta_2 - \delta$ (see above), the value of δ_1 can be deduced from the voltage dependence of c_I [see Eq. (3.5)]. The latter can be calculated from the experimentally measured voltage dependence of i_{el} [Eq. (3.8)].

The value of the product $\tau_o a_I$ can be obtained from Eq. (3.9), from which the parameters τ and c_I have already been found. Although τ_o and a_I cannot be estimated independently, both have a very narrow range of possible values.

Taken altogether, Eqs. (3.4)–(3.6) represent the complete mathematical model of monovalent ion flow through the neuronal nicotinic receptor channel. Using all these equations, together with their modifications, Marchais and Marty (1979) have obtained good agreement with experimental relationships $\tau(V)$ and $i_{el}(V)$ found in the presence of either monovalent cations or a mixture of mono- and divalent cations. As the authors themselves point out, they have obtained a numerical solution for the energy profile shown in Fig. 3.3, although there might be several such solutions.

The values of the model parameters used to fit experimental data at –100 mV are as follows: $\delta_1 = 0.22$, $\delta = 0.7$, $\delta_2 = 0.9$; $\tau_o = 1$ ms. The values of a, b, and c are $5.1 \cdot 10^8$ $M^{-1} \cdot s^{-1}$, $5.0 \cdot 10^6$ s^{-1}, $1.0 \cdot 10^7$ s^{-1}, and $9.6 \cdot 10^8$ $M^{-1} \cdot s^{-1}$, $7.0 \cdot 10^5$ s^{-1}, $1.9 \cdot 10^6$ s^{-1} for mono- and divalent cations, respectively.

Thus, the effect of various permeable ions on characteristics of the neuronal acetylcholine receptor channel, such as its open time or conductance, can be explained based on the effect of ion binding within a channel. However, the model described above seems to be even more comprehensive. It is possible to explain with this model the channel-blocking effect of various drugs, as will be shown in Chapter 5, and also to describe the behavior of other channels. For instance, this model describes voltage dependences of the end-plate channel open time and current (Marchais and Marty, 1979). Interestingly, in the latter case it is necessary to assume that the distance δ_1 is two times longer than the distance $\delta_2 - \delta$ (for the neuronal acetylcholine receptor channel they are similar). However, some properties of end-plate channels appear to be inconsistent with the ion-binding model: For example, the end-plate channel open time and its voltage dependence are almost unchanged with increased external calcium concentration (Bregestovski et al., 1979). On the other hand, no correlation has been found between end-plate channel conductance and open time (Adams et al., 1981). This indicates that there are fundamental differences in some mechanisms underlying the flow of ions between neuronal and muscular acetylcholine receptor channels.

3.2. CHANNEL CONDUCTANCE

Single-channel conductance can be estimated using two different approaches. One approach is to analyze current fluctuations induced by continuous action of the agonist and recorded from the whole cell at a fixed level of the membrane potential (Anderson and Stevens, 1973). Because the current results from the activity of a large number of channels, analysis of the fluctuations allows one to obtain their averaged characteristics. An alternative approach is to record directly the activity of a single channel from a small membrane area (the patch-clamp method, see Neher and Sakhmann, 1976; Hamill et al., 1981).

TABLE 3.3. Characteristics of Single Neuronal Acetylcholine Receptor Channels

Channel selectivity	Preparation	γ (pS)	τ (ms)	H (mV)	Temp. (°C)	E (mV)	Method	References
Cationic	Rabbit superior cervical ganglion		4.5	−140	34–37	−80	e.p.s.c.	Derkach et al., 1983
			25		20–22			
			7.9, 44.7		22–25	−80	relaxations	Derkach, 1986
		36	1.1, 5.0		34–37	−80	fluctuations	Derkach et al., 1983
			6.0, 29.6		20–22			
Cationic	Rat superior cervical ganglion	20, 50	0.2; 2.6, 8.5		24–25	−110	patch-clamp	Derkach et al., 1987
	Rat lumbar sympathetic ganglion		3.6–4.0		37	−55 to −65	e.p.s.c.	Hirst and McLachlan, 1984
	Frog sympathetic ganglion B-neurons		4.3	−524	20–25	−60	e.p.s.c.	Kuba and Nishi, 1979
			5.3	−256	21–23	−50	e.p.s.c.	MacDermott et al., 1980
			5.5		21–23	−50	e.p.s.c.	Marshall, 1985
		17.3	5.2				fluctuations	
	Frog sympathetic ganglion C-neurons		10.2				e.p.s.c.	
		16.9	9.8				fluctuations	
	Rat submandibular ganglion		19, 45	−100	20	−80	e.p.s.c.	Rang, 1981
			5–9, 50		20	−80	relaxations	
		31	8, 51		20	−80	fluctuations	

	Chick ciliary ganglion neurons in culture	28	1.36–1.92		30	–60	fluctuations	Ogden et al., 1984
		38–42	1.08	≥0	30	–60	patch-clamp	
	Aplysia ganglia	8	27		12	–80	fluctuations, relaxations	Ascher et al., 1978a
		14	5.4		21			
Anionic	*Aplysia* ganglia		12	–156	20	–40	i.p.s.c.	Adams et al., 1976
	Limnaea stagnalis ganglia		32	–185	20	–50	relaxations	Kurchikov and Kaza-chenko, 1984b
		3.5	17.2	–250	20–22		fluctuations	Bregestovsky, 1980
	Aplysia ganglia	2	10		20	–80	fluctuations	Simonneau et al., 1980
			19		20	–150 to –130	i.p.s.c.	Gardner and Stevens, 1980
		2.8–16	3.4, 20.4		20	–150 to –130	fluctuations	
	Helix aspersa ganglia		4.5, 34	+191	12	–120	relaxations	Finkel,1983
		0.58 at E_{Cl} = –26 mV; 2.7 at E_{Cl} = –62 mV	3.9, 45		12	–100	fluctuations	
		18	100			–50 to –100	patch-clamp	Ascher and Erulkar, 1983
			5, 140			–50 to –100	fluctuations	

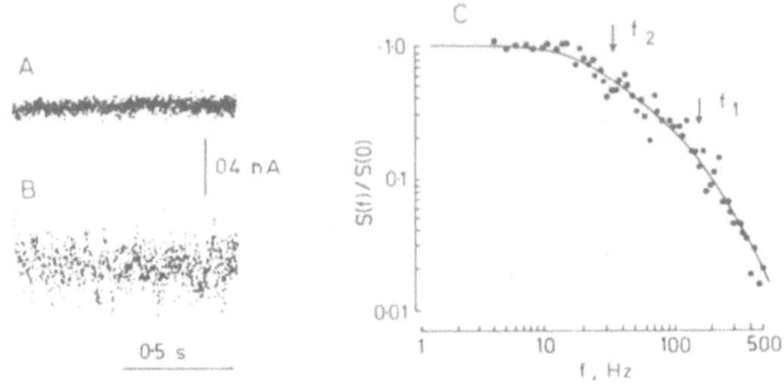

Fig. 3.4. Power spectrum of acetylcholine (ACh)-induced current fluctuations recorded from rabbit superior cervical ganglion neuron. Examples of current records obtained before (A) and during (B) application of ACh. Steady-state ACh-induced current was –3.8 nA. Power spectrum of ACh-induced current fluctuations (C) was obtained by subtraction of an averaged power spectrum for four current records similar to those shown in A, from an averaged power spectrum for four records similar to those shown in B. The spectrum was fitted by the sum of two Lorentzian curves [see Eq. (3.14)] at $S_1(0)/S_2(0) = 0.3$ and at cut-off frequencies $f_1 = 145$ Hz and $f_2 = 32$ Hz, corresponding to $\tau_{f1} = 1.1$ ms and $\tau_{f2} = 4.9$ ms. The data were obtained at 36°C (Selyanko et al., 1981).

Assuming that the channels are homogeneous, that they open randomly, and that only a small number of channels are open at any given time, the channel conductance γ can be estimated from an analysis of current fluctuations (Katz and Miledi, 1972; Anderson and Stevens, 1973). The value of γ can be obtained from the variance (σ^2) and steady-state current value (I) as

$$\gamma = \frac{\sigma^2}{I(V - V_r)}, \tag{3.11}$$

where V is the holding membrane potential and V_r is the reversal potential for I.

It was suggested in earlier studies that channel open time varies exponentially and that the spectrum of current fluctuations is Lorentzian as given by

$$S(f) = \frac{S(0)}{1 + (f/f_1)^2}, \tag{3.12}$$

where $S(f)$ is the spectral density, f is frequency, f_1 is the cut-off frequency, or the value of f corresponding to $S(f) = {}^1/_2 S(0)$. The estimate for γ can also be obtained from spectral analysis of current fluctuations, using the equation

$$\gamma = \frac{\pi S(0) f_1}{2I(V - V_r)}. \tag{3.13}$$

Equations (3.11) and (3.13) give similar estimates for γ provided that the spectrum of current fluctuations is not distorted by the limited frequency bandwidth of the recording system. However, the precise estimate of γ can be obtained from Eqs. (3.11) and (3.13) when only one population of channels having similar values of γ and channel open time contributes to the response. Indeed, for end-plate channels the estimates of γ are similar when obtained either from analysis of fluctuations (Anderson and Stevens, 1973) or from single-channel activity (Neher and Sakmann, 1976). However, in many types of neurons the spectra of current fluctuations can be satisfactorily fitted only with the sum of two Lorentzian functions

$$S(f) = \left\{S_1(0)\Big/\left(1 + (f/f_1)^2\right)\right\} + \left\{S_2(0)\Big/\left(1 + (f/f_2)^2\right)\right\}. \tag{3.14}$$

This suggests that at least two populations of acetylcholine receptor channels, having different open times, may exist in these neurons (see below). If two populations of channels have different mean values of γ, then Eq. (3.11) should give a γ estimate intermediate between those for each population; independent estimation of γ values for each population of channels is very difficult in this case. However, if the conductances are similar in both populations of channels, then the value of γ can be calculated according to Eq. (3.13) modified as follows:

$$\gamma = \frac{\pi}{2I(V - V_r)}\{S_1(0)f_1 + S_2(0)f_2\}. \tag{3.15}$$

Spectral analysis of acetylcholine-induced current fluctuations has been carried out in the neurons of rabbit superior cervical ganglion (Derkach et al., 1983). Figure 3.4 shows an example of the results obtained. It can be seen from Fig. 3.4 that acetylcholine produces an increase in the level of background noise. The spectrum of these extra current fluctuations is satisfactorily fitted by Eq. (3.14). Channel conductances estimated according to Eqs. (3.11) and (3.15) are essentially similar. Assuming that $V_r = -10$ mV (the value of the reversal potential for the synaptic current in these neurons, see Derkach et al., 1983), the mean value of γ is 36 pS at –80 mV and 34-37°C (see Table 3.3).

Two-component spectra, similar to that shown in Fig. 3.4, have been obtained from the majority of sympathetic neurons, the remaining neurons showing one-component spectra. In these two groups of neurons the γ values are similar, confirming the above hypothesis.

In Table 3.3, the values of the conductance of acetylcholine receptor channels in different types of neurons are shown. Mean values of γ close to those obtained in sympathetic neurons have been found in parasympathetic neurons of rat (Rang, 1981) and chick (Ogden et al., 1984). A somewhat lower value of γ (14 pS) has been found in molluscan neurons (Ascher et al., 1978a). The conductance of the end-plate channels seems to be similar to that in neurons (see Colquhoun, 1979).

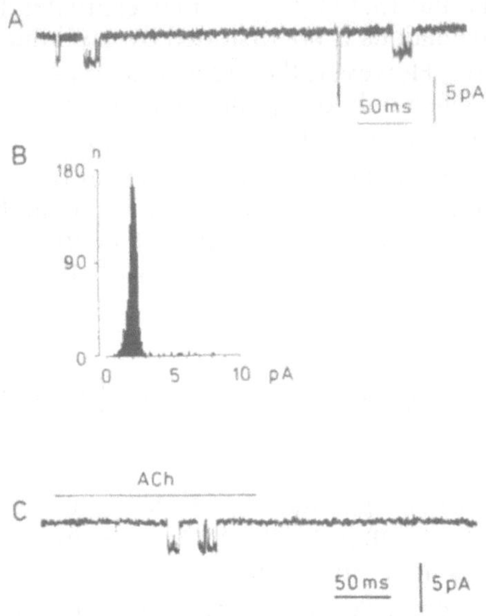

Fig. 3.5. Activity of single channel induced by acetylcholine (30 μM) and recorded, using the patch-clamp method, from rat superior cervical ganglion neuron. A) Example of currents recorded from "cell-attached" membrane patch at the assumed membrane potential level of –110 mV (resting membrane potential level was taken as –50 mV). B) Amplitude distribution of currents, some of which are shown in A. Mean amplitudes are 2.2 and 5.6 pA for low- and high-amplitude currents, respectively. C) Activity of single channel recorded from "outside-out" membrane patch and elicited by threshold dose of acetylcholine (ACh) at the membrane potential of –140 mV. Temperature was 23°C (Derkach et al., 1987).

Estimates for γ made in anionic channels using analysis of current fluctuations are similar in the neurons of different molluscan species although they are about one order lower than in cationic channels (see Table 3.3).

In a few studies the values of γ have been estimated with single-channel recording (Ascher and Erulkar, 1983; Ogden et al., 1984; Derkach et al., 1985).

Figure 3.5 shows an example of single-channel activity induced by acetylcholine in rat superior cervical ganglion neuron. As is evident from the record of activity and from the current amplitude distribution, two kinds of currents with different amplitudes are observed. Both kinds of currents are reversed at the same membrane potential (about 0 mV), and their channel conductances are about 20 and 50 pS, respectively. Further analysis of the activity has shown that high-amplitude currents are not the low-amplitude ones superimposed in a random fashion and that the two kinds of currents probably represent the activity of two different types of channels. This is indicated by the following observations:

1) high-amplitude currents are not multiples of low-amplitude ones;

2) the contribution of high-amplitude currents to the overall single-channel activity is much higher than one would expect from a random superposition of low-amplitude currents.

Thus, the value of γ obtained earlier from an analysis of current fluctuations (Derkach et al., 1983) appears to be intermediate between two γ estimates given by single-channel recording. It should be pointed out that the contribution of high-amplitude currents to the overall single-channel activity was rather small, varying from 5 to 30% (Derkach et al., 1985, 1987). This is consistent with the fact that the value of γ obtained from an analysis of current fluctuations is markedly closer to the value of γ for low-conducting channels than for high-conducting channels estimated from single-channel measurements.

In cultured chick parasympathetic neurons (Ogden et al., 1984), as well as in molluscan neurons (Ascher and Erulkar, 1983), differences have also been revealed between the values of γ obtained by different methods (see Table 3.3). In both cases the values of γ are higher when obtained from single-channel recordings than when obtained from an analysis of current fluctuations.

Ogden et al. (1984) suggested that the value of γ obtained from an analysis of current fluctuations is underestimated due to insufficient voltage control at high frequencies on the remote (dendritic) membrane areas of a neuron. According to their calculations, based on cell geometry, this could lead to underestimation of the cut-off frequency in the spectrum of acetylcholine-induced current fluctuations by 40–50%, which is about the value of underestimation of γ.

In molluscan neurons, the reason for the low values of γ obtained from an analysis of current fluctuations may be different. On the one hand, the value of γ may vary in different neurons of the same ganglion (Gardner and Stevens, 1980; see Table 3.3). On the other hand, the conductance of anionic channels seems to be very sensitive to the intracellular concentration of chloride ions, more sensitive than is predicted by the Goldman–Hodgkin–Katz equation. For instance, with increasing intracellular concentration from 9 to 36 mM, one could expect only a 2.5-fold increase in γ, while the increase obtained in γ was fivefold (Finkel, 1983). This implies that even small variations in intracellular chloride concentration may essentially affect the estimation of γ for chloride channels.

Figure 3.6 shows the voltage dependences of elementary currents (i_{el}) flowing through cationic acetylcholine receptor channels in rabbit (A) and rat (B) sympathetic ganglion neurons. In the former case (A), the values of γ are estimated from an analysis of current fluctuations, while in the latter case (B) they are estimated from single-channel recordings. It is seen that both relationships are linear, indicating that single-channel conductance is voltage-insensitive.

The lack of rectification in acetylcholine receptor channels has also been shown in other types of neurons (Ascher et al., 1978a; Kurchikov and Kazachenko, 1984a; Ogden et al., 1984), although some degree of nonlinearity in current–voltage relationships is predicted by either the Goldman–Hodgkin–Katz equation or the ion-binding model (Marchais and Marty, 1979). In the latter case the nonlinearity may result from the voltage dependence of c_I, the rate constant of the entry of the ion into the cell from the channel [Eq. (3.8)], which in turn exponentially increases with membrane hyperpolarization [see Eq. (3.5)].

One could suggest that in the cases mentioned above the channel rectification actually exists, although it is too weak to be resolved experimentally. Indeed, in at least one study an essential rectification has been found in anionic acetylcholine receptor channels (Finkel, 1983).

By comparing single-channel currents and currents produced by a single quantum of acetylcholine, one may, dividing the latter value by the former one, estimate the number of channels activated by a single quantum. Using this approach it has been found that a single quantum of acetylcholine opens about 100 channels either in parasympathetic or sympathetic neurons (Rang, 1981; Derkach et al., 1983) and about 500 channels in molluscan neurons (Simonneau et al., 1980). Both estimates are much smaller than that obtained at the end-plate (2000: Large and Rang, 1978).

The number of acetylcholine molecules per quantum is 8000–12,000 and seems to be similar in sympathetic ganglion and at the end-plate (Nishi et al., 1967). Thus, only a small number of acetylcholine molecules (about 1% of those contained in a quantum) is able to open channels in sympathetic ganglion. This number is about 20 times smaller in autonomic ganglia than at the end-plate. Interestingly, a similar difference has been found between the densities of α-bungarotoxin binding sites in synaptic areas of sympathetic neurons (575 per 1 μm^2: Smolen, 1983) and of muscle fibers (18,000–30,000 per 1 μm^2: Fertuck and Salpeter, 1976; Matthews-Bellinger and Salpeter, 1983). Therefore, it is likely that the difference in efficiencies of a single acetylcholine quantum in autonomic ganglia and muscle may result from the difference in the densities of acetylcholine receptors in neurons and in muscle fibers.

3.3. CHANNEL KINETICS

3.3.1. Sequential Model

According to the Del Castillo–Katz sequential model (Del Castillo and Katz, 1957; Katz and Miledi, 1972), a receptor (R) can be activated by acetylcholine (A) by forming an inactive acetylcholine–receptor complex

$$nA + R \underset{k_{-1}}{\overset{k_{+1}}{\rightleftharpoons}} A_nR \underset{\alpha}{\overset{\beta}{\rightleftharpoons}} A_nR^*, \tag{3.16}$$

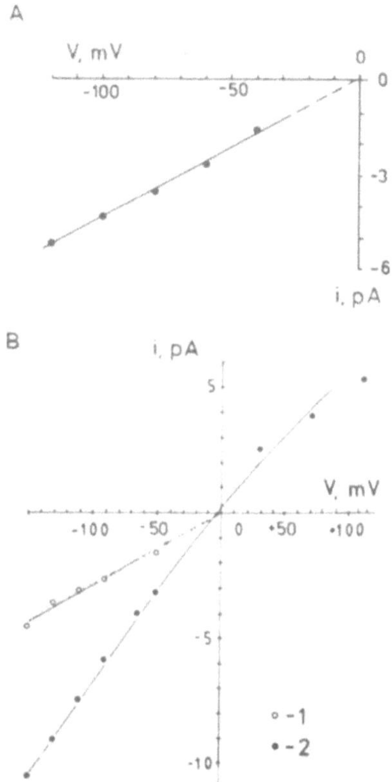

Fig. 3.6. Voltage dependence of current (i) flowing through single nicotinic receptor channel in rabbit (A) and rat (B) superior cervical ganglion. The values of i were obtained from spectral analysis of acetylcholine-induced current fluctuations (A: Derkach et al., 1983) and from recording of acetylcholine-induced single-channel activity (B: Derkach et al., 1987). In B1, 2 the values of i for low- and high-amplitude currents, respectively, are plotted against the membrane potential level.

where A_nR and A_nR^* are inactive (nonconducting) and active (conducting) acetylcholine–receptor complexes, respectively; k_{+1}, k_{-1}, β, and α are the corresponding rate constants; n is the number of acetylcholine molecules activating a single channel. Although Eq. (3.16) was initially suggested in order to describe the concentration dependence of acetylcholine action, it is now widely used for the kinetic analysis of acetylcholine–receptor interactions.

The key question in the kinetic analysis is which transitions in the given scheme are rate-limiting. According to theoretical considerations (Colquhoun and Hawkes, 1977) with three states of a receptor, no more than two kinetic components can be revealed in the activity of receptors recorded from the whole cell. However, usually only one, the slowest component, can be revealed under these conditions. This is because faster components are far from the recording frequency bandwidth and their contributions to the activity are too small. By contrast, single-channel recording provides much higher frequency and amplitude resolution allowing one to resolve very fast and low-contributing components.

According to the initial suggestion made by Stevens and co-workers (Magleby and Stevens, 1972b; Anderson and Stevens, 1973; Dionne and Stevens, 1975), among the states of a receptor with agonist bound, an active one (A_nR^*) is much longer than an inactive one (A_nR). This suggestion has been made based on the analogy between agonist–receptor and enzyme–substrate interactions. The analogy implies that the first stage of ligand–protein interaction (binding) is much faster than the second stage (isomerization). Although later experiments showed that this suggestion is not justified for acetylcholine–receptor interactions, it is useful for describing the kinetic measurements obtained from the whole cell.

Another suggestion (which was confirmed later) made by Stevens and co-workers is that the forward rate constant (β) of channel transition between inactive and active receptor states with agonist bound is much faster than the backward rate constant (α). However, it should be pointed out that as only the slowest rate constants are identified, the transitions between the first and last states of a receptor represented in Eq. (3.16) can be characterized only by the apparent rate constants.

According to the previous suggestion that the lifetime of the state A_nR^* is much longer than that of the state A_nR, two inactive states, R and A_nR, can be conditionally replaced by one state, transforming to or from the active state with the apparent rate constant β_a (or α_a)

$$nA + R \underset{k_{-1}}{\overset{k_{+1}}{\rightleftharpoons}} A_nR \underset{\alpha_a}{\overset{\beta_a}{\rightleftharpoons}} A_nR^*. \tag{3.17}$$

The slowest kinetic component revealed in channel activity recorded from the whole cell should therefore correspond to the apparent rate of channel closing (α_a). This is true at low agonist concentrations [A], when $\beta_a[A] \ll \alpha_a$. At higher agonist concentrations the slow kinetic component is equal to ($\alpha_a + \beta_a[A]$). Since channel open time varies exponentially (see below), the mean apparent open time is

$$\tau_a = \frac{1}{\alpha_a + \beta_a[A]}, \tag{3.18}$$

or

$$\tau_a = \frac{1}{\alpha_a} \tag{3.18a}$$

at $[A] \to 0$.

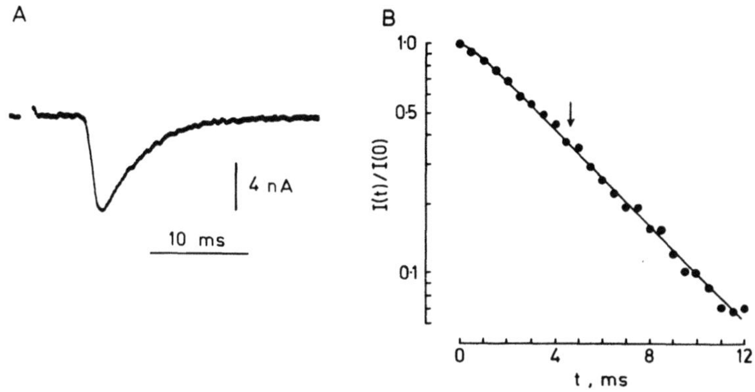

Fig. 3.7. Fast e.p.s.c. with single-exponential decay recorded from rabbit superior cervical ganglion neuron. A) Fast e.p.s.c. recorded at membrane potential level of –80 mV and temperature of 36°C. B) Semilogarithmic plot of the decay of fast e.p.s.c. shown in A; the arrow indicates time constant 4.6 ms (Skok et al., 1982).

Channel kinetics is usually investigated using four different approaches: 1) analysis of synaptic current decay ("concentration-jump" method); 2) analysis of acetylcholine-induced current fluctuations, or 3) relaxations produced by membrane voltage steps; and 4) single-channel recording. The first three approaches provide an estimate of apparent channel kinetics, whereas the latter approach enables one to estimate true channel kinetics. The results obtained from neurons using all these methods are described below.

3.3.2. Synaptic Current Decay

It has been found that the action of acetylcholine on the subsynaptic membrane after its release from nerve endings is very short (about 0.1 ms at the endplate, see Magleby and Stevens, 1972b). Since acetylcholine peak concentration is very high (about 10^{-4} M), this results in the almost instantaneous and synchronous opening of a large number of acetylcholine receptor channels. The synaptic current decay is therefore the curve fitting the channel open time distribution. Insofar as this distribution is exponential and is obtained at $[A] \rightarrow 0$ (see above), then the mean channel open time, or "lifetime," should be equal to the time constant of the synaptic current decay τ_{syn}. However, in earlier studies it was not clear whether the channel was open through the entire time equivalent to τ_{syn} or whether this long opening was interspersed with brief closings unresolved due to the limited frequency bandwidth. Since the latter possibility was later shown to be true, τ_{syn} obviously corresponds to the apparent mean channel open time α_a^{-1}.

Two types of postsynaptic currents can be recorded from neurons, either excitatory (e.p.s.c.) or inhibitory (i.p.s.c.). The latter type of currents can be observed only in invertebrate neurons.

Figure 3.7 shows an example of e.p.s.c. recorded from rabbit superior cervical ganglion neuron. E.p.s.c. is characterized by the fast rising phase and slower

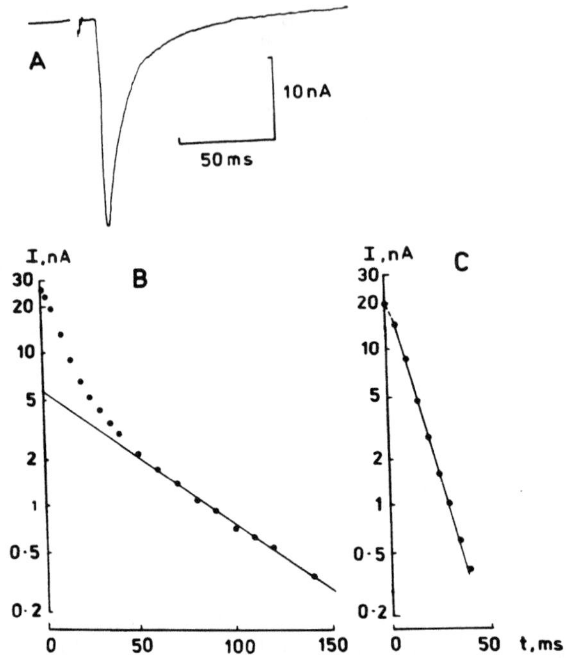

Fig. 3.8. E.p.s.c. with biexponential decay recorded from rat submandibular ganglion neuron. A) E.p.s.c. recorded at the membrane potential level of –80 mV and temperature of 20°C. B) Semilogarithmic plot of the decay of e.p.s.c. shown in A. The straight line is the slow component of e.p.s.c. decay obtained after the "tail" of e.p.s.c. was fitted by a single-exponential curve. C) The fast component of e.p.s.c. decay obtained by subtraction of the e.p.s.c. "tail" from the overall current curve (Rang, 1981).

single-exponential decay. The latter is indicated by the linear plot of e.p.s.c. decay in semilogarithmic scale. The mean values of τ_{syn} in these neurons is 4.5 ms at –80 mV and 36°C (see Table 3.3). Similarly, the decays of e.p.s.c. in amphibian sympathetic neurons (Kuba and Nishi, 1979; MacDermott et al., 1980; Marshall, 1985) and of i.p.s.c. in molluscan neurons (Adams et al., 1976; Gardner and Stevens, 1980) are single-exponential. Surprisingly, the decay of e.p.s.c. in rat parasympathetic neurons is obviously biexponential (Rang, 1981; see Fig. 3.8). This is the only example of postsynaptic current with biexponential decay at either cholinergic or noncholinergic synapses.

Channel kinetics can be deduced from postsynaptic current decay if the decay is not slowed down by acetylcholine release, hydrolysis, or diffusion. In neuromuscular junction the synaptic current decay does not seem to be affected by these factors (Takeuchi and Takeuchi, 1959; Magleby and Stevens, 1972b; Anderson and Stevens, 1973). Since the synaptic current decay is much slower in neurons than in neuromuscular junction, one could assume that this is due to the longer presence of acetylcholine in the synaptic cleft.

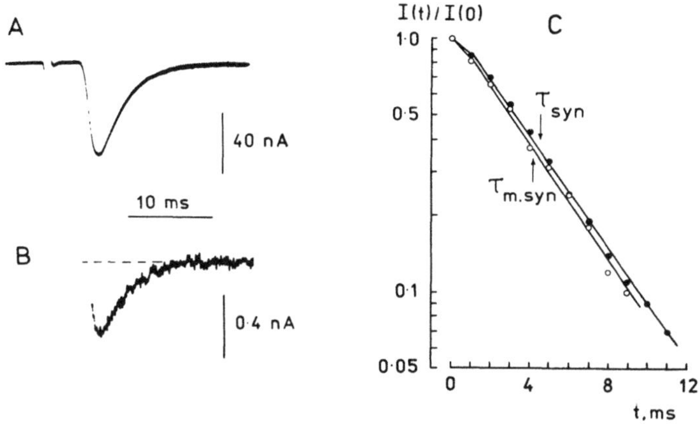

Fig. 3.9. Comparison of kinetics of the decay in fast e.p.s.c. and in spontaneous miniature e.p.s.c. in rabbit superior cervical ganglion. Fast e.p.s.c. elicited by single orthodromic stimulus (A) and spontaneous miniature e.p.s.c. (B) were recorded from the same neuron at the membrane potential level of –80 mV and temperature of 36°C. C) Semilogarithmic plot of the decays of fast e.p.s.c. and of spontaneous miniature e.p.s.c.; their time constants, τ_{syn} and $\tau_{m.syn}$, respectively, are indicated by arrows (Derkach et al., 1983).

The rate of acetylcholine release from nerve endings can be prolonged due to either nonsynchronous quanta release or nonsynchronous activation of acetylcholine receptors by individual quanta. In order to verify the first possibility, one should compare the decays of multiquantal postsynaptic currents elicited by a single preganglionic stimuli with that of miniature postsynaptic currents produced by a single spontaneously released quantum of acetylcholine. Figure 3.9 shows the e.p.s.c. and miniature e.p.s.c. recorded in the same neuron of rabbit superior cervical ganglion. It is seen that both currents decay with similar time courses. As follows from the ratio of the amplitudes of e.p.s.c. and miniature e.p.s.c. shown, e.p.s.c. is produced by about 100 quanta of acetylcholine. Despite such a high quantal content of e.p.s.c. its decay is almost identical to that of miniature e.p.s.c., which indicates that the release of acetylcholine quanta from nerve endings is very synchronous in sympathetic ganglion.

Another approach for studying the e.p.s.c. time course has been used in amphibian sympathetic ganglion neurons, where the quantal content of e.p.s.c. has been lowered several times by repetitive orthodromic stimulation. Consistent with the above data, it has been found that e.p.s.c. decay is not affected by this procedure (Kuba and Nishi, 1979; MacDermott et al., 1980).

In mammalian parasympathetic ganglion neurons the decay in miniature e.p.s.c. is single-exponential corresponding to the fast component in e.p.s.c. decay in these neurons (Rang, 1981). In some parasympathetic ganglion neurons miniature e.p.s.c.'s with biexponential decays have been recorded (Rang and Gurney,

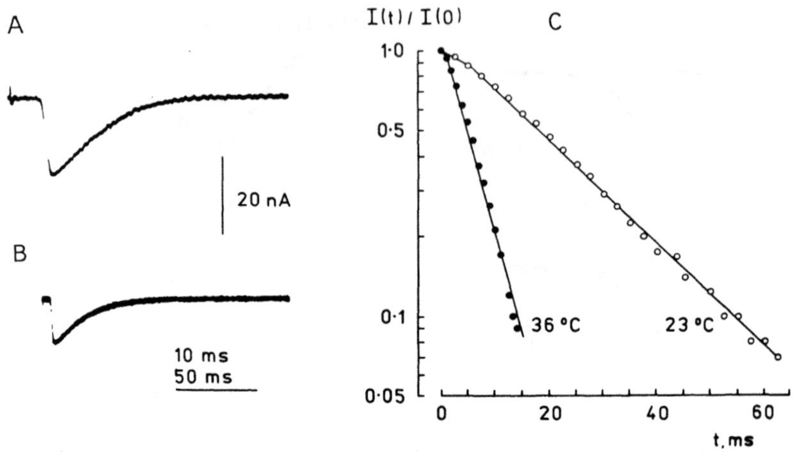

Fig. 3.10. Effect of temperature on fast e.p.s.c. in rabbit superior cervical ganglion neuron. Fast e.p.s.c.'s are shown as recorded at the membrane potential level of –80 mV and temperatures of 36°C (A) and 23°C (B). Time scale is 10 ms for A and 50 ms for B. C) Semilogarithmic plot of decays of fast e.p.s.c.'s shown in A and B (Derkach et al., 1983).

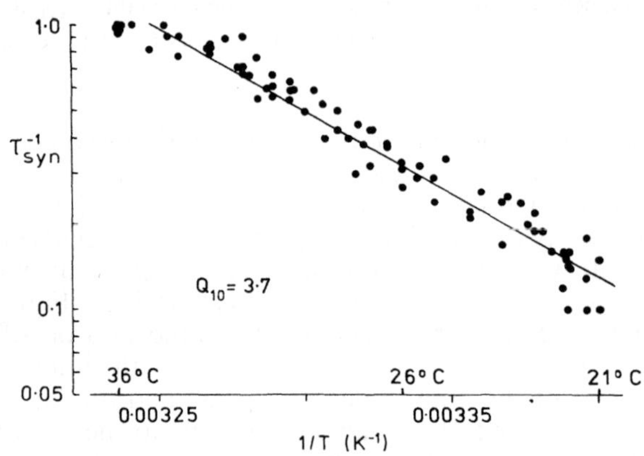

Fig. 3.11. Arrhenius plot for the rate constant (τ_{syn}^{-1}) of fast e.p.s.c. decay in rabbit superior cervical ganglion neuron. The data were obtained from five neurons. The values of τ_{syn}^{-1}, obtained from each neuron at various temperatures, were normalized to those at 36°C (Derkach et al., 1983).

1982). Further analysis of e.p.s.c. decay in parasympathetic neurons has been carried out at low $[Ca^{2+}]_o$ to $[Mg^2]_o$ ratio, when e.p.s.c. quantal content is reduced to 1/10 of the control value. In order to verify whether factors determining fast and slow components of e.p.s.c. decay depend on one another, Rang and Gurney (1982) have estimated the fluctuations in contributions of the two components into e.p.s.c. decay. They have found a significant, although weak, correlation between the two components and have concluded that miniature e.p.s.c.'s should contain both components with a variable ratio between them. Indeed, an analysis of individual miniature e.p.s.c.'s shows that some of them contain predominantly one component, either fast or slow.

The results obtained from sympathetic and parasympathetic ganglia show that e.p.s.c.'s with either single- or biexponential decays are not slowed down by the dispersion in quanta release from nerve terminals.

The possible dependence of postsynaptic current decay on the rate of acetylcholine diffusion from the synaptic cleft has been tested by varying the temperature. It was expected that if the rate of postsynaptic current decay is determined by the rate of acetylcholine diffusion, the temperature coefficient Q_{10} for both events should be similar, i.e., close to 1. However, the temperature dependence of postsynaptic current decay appears to be much higher than that of diffusion.

Figure 3.10 shows the effect of temperature on the synaptic current decay in rabbit superior cervical ganglion neuron, and Fig. 3.11 shows the temperature dependence of the e.p.s.c. decay time constant τ_{syn}. The values of Q_{10} obtained from different types of neurons are similar: 3.7 in rabbit superior cervical ganglion (Derkach et al., 1983), 3–6 in frog sympathetic ganglion (Kuba and Nishi, 1979; MacDermott et al., 1980), and 5.2 for the decay of i.p.s.c. in molluscan ganglia (Gardner and Stevens, 1980). The high temperature dependence of the postsynaptic current decay in neurons indicates that the diffusion of acetylcholine is not involved in determining its time course. A similar approach applied to the end-plate current gave identical results (Magleby and Stevens, 1972).

The rate of acetylcholine hydrolysis at the synapse cannot be estimated directly although its role in determining postsynaptic current decay can be deduced from the changes in current produced by acetylcholinesterase inhibition. It has been suggested that if acetylcholine hydrolysis is weak or slow, then acetylcholinesterase inhibition will not essentially affect the postsynaptic current time course. However, in amphibian sympathetic ganglion neurons acetylcholinesterase inhibitors, such as physostigmine (eserine) or neostigmine, at low concentrations (less than 10^{-5} M), cause a lengthening of the e.p.s.c. decay which remains single-exponential (Kuba and Nishi, 1979; MacDermott et al., 1980). At higher concentrations ($>10^{-5}$ M) anticholinesterase drugs produce further lengthening of the e.p.s.c. decay, which becomes biexponential (MacDermott et al., 1980). Neostigmine (10^{-5} M) also produces a lengthening of e.p.s.c. decay in rat parasympathetic ganglion neurons, which is 3.5-fold and 2.5-fold for fast and slow components, respectively (Rang, 1981). In molluscan neurons, physostigmine (10^{-5} M) causes a 1.7-fold lengthening of the e.p.s.c. decay, which remains single-exponential (Gardner and Stevens, 1980).

Thus, an essential lengthening of the postsynaptic current decay produced by acetylcholinesterase inhibitors implies that the time acetylcholine spends at the synapse is essentially restricted due to its hydrolysis. However, it should be pointed out that in some cases the changes produced by acetylcholinesterase inhibitors in postsynaptic current decay cannot be entirely explained by their action on acetylcholinesterase. For instance, the change in the decay of e.p.s.c. in amphibian

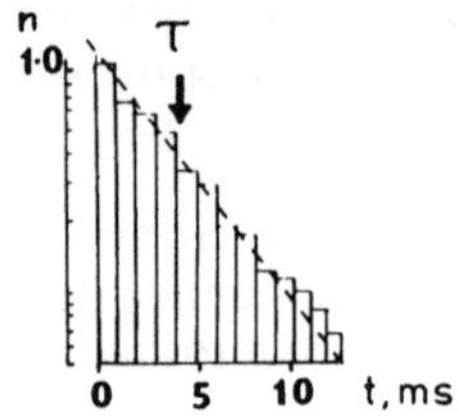

Fig. 3.12. Schematic representation of fast e.p.s.c. decay as a histogram of durations of single-channel currents evoked by acetylcholine; *n* is normalized number of open channels, and τ is the time constant of the e.p.s.c. decay.

sympathetic ganglion neurons produced by high concentrations of physostigmine, i.e., its transformation from single- to biexponential (MacDermott et al., 1980), is very similar to the changes in end-plate current decay produced by local anesthetics (Beam, 1976a, b; Ruff, 1977). Since the action of local anesthetics on the end-plate current is associated with their effects on end-plate channel kinetics (Neher and Steinbach, 1978), one can assume that high concentrations of acetylcholinesterase inhibitors may also change the kinetics of acetylcholine receptor channels in neurons. This suggestion is consistent with the fact that the temperature dependence of i.p.s.c. decay is not essentially affected by physostigmine (Gardner and Stevens, 1980). The idea of this experiment is that if acetylcholinesterase is inhibited, then postsynaptic current decay should be predominantly determined by the diffusion of acetylcholine from the synaptic cleft, which has low temperature sensitivity. However, the retaining of the high temperature dependence of postsynaptic current decay indicates that its slowing down by physostigmine may result, besides from acetylcholinesterase inhibition, from other effects, e.g., from slowing down of channel kinetics. Indeed, more detailed studies carried out on *Aplysia* neurons have shown that physostigmine may cause lengthening of the mean open time of acetylcholine receptor channels. This is indicated by lengthening of the slow kinetic component in the spectrum of acetylcholine-induced current fluctuations, which is analogous to the component contained in i.p.s.c. decay (White and Gardner, 1983).

Summing up all the data on the nature of the postsynaptic current decay in neurons, one may conclude that events such as transmitter release, diffusion, or hydrolysis are not involved in determining its time course. Therefore, an alternative explanation seems more likely, i.e., that postsynaptic current decay in neurons, as at the end-plate (Magleby and Stevens, 1972b; Anderson and Stevens, 1973), is determined by the rate of channel closure. At the same time, the faster phase of postsynaptic current, its growth, may be essentially affected by acetylcholine release or removal at the synapse.

Early studies carried out at the end-plate (Neher and Stevens, 1977) showed that the exponential decay of postsynaptic current is associated with the exponential distribution of the end-plate channel open time. Figure 3.12 schematically shows the relationship between the postsynaptic current decay and the channel open time distribution. It is seen that postsynaptic current decay is virtually a fitting curve for open time distribution. Single-exponential postsynaptic current decay can be described by the following equation:

$$I(t) = I(0)\exp(-t/\tau_{syn}) \tag{3.19}$$

where $I(0)$ and $I(t)$ are current amplitudes at times 0 and t, respectively, from the beginning of the postsynaptic current decay; τ_{syn} is the time constant of the decay. As was mentioned above, τ_{syn} is equal to the mean channel apparent open time. Since acetylcholine concentration at the synapse during the decay of postsynaptic current approaches 0, τ_{syn}, according to Eq. (3.18a), is close to α_a^{-1}. Table 3.3 shows the values of τ_{syn} obtained from different types of neurons. It is seen that at room temperature τ_{syn} varies from 5 to 45 ms. These values are several times higher than those obtained at the end-plate where, also at room temperature, τ_{syn} is about 1 ms both in amphibian and mammalian muscle fibers (see Colquhoun, 1979). The reason for this discrepancy becomes clear following single-channel analysis of these two types of acetylcholine receptor channels. Interestingly, the values of τ_{syn} for e.p.s.c. and i.p.s.c. obtained from neurons are essentially similar. This indicates that there is no obvious link between the channel kinetics and its selectivity.

3.3.3. Acetylcholine-Induced Current Fluctuations

Fluctuations of current induced by the continuous application of acetylcholine were initially observed at the end-plate (Katz and Miledi, 1972; Anderson and Stevens, 1973). Current fluctuations result from random fluctuations in the number of channels open at any given time. The kinetics of fluctuations are determined by the kinetics of single channels and, since channel open time varies exponentially, the spectrum of current fluctuations is Lorentzian [see Eq. (3.12)]. Then the mean channel open time τ_f, obtained by fitting the fluctuations spectrum with the Lorentzian function, is

$$\tau_f = \frac{1}{2\pi f_1}, \tag{3.20}$$

where f_1 is the cut-off frequency [see Eq. (3.12)]. It has been found that in some types of neurons (see Table 3.3) the spectrum of current fluctuations can indeed be described by a single Lorentzian function [see Eq. (3.12)], whereas in other types of neurons the spectrum can be described only by the sum of two Lorentzian functions [see Eq. (3.14)]. With two kinetic components, fast and slow, present in channel activity, their time constants (τ_{f_1} and τ_{f_2}, respectively) can be calculated from the following equations:

$$\tau_{f1} = \frac{1}{2\pi f_1} \text{ and } \tau_{f2} = \frac{1}{2\pi f_2}, \tag{3.21}$$

where f_1 and f_2 are cut-off frequencies corresponding to the fast and slow components, respectively [see Eq. (3.14)]. Figure 3.4 shows an example of a two-component spectrum obtained from rabbit superior cervical ganglion neuron. The mean values of τ_{f1} and τ_{f2} are 1.1 and 5.0 ms, respectively (at –80 mV and 34–37°C; see Table 3.3). One-component spectra have been obtained from some ganglion neurons with τ_f close to τ_{f1} in the two-component spectra.

Interestingly, two-component spectra have been obtained from neurons where the postsynaptic current decay is single-exponential (see Table 3.3), with τ_{syn} close to τ_{f2}. In parasympathetic ganglion neurons, the spectra of acetylcholine-induced current fluctuations are also two-component, with time constants corresponding to those in two-component e.p.s.c. decay (Rang, 1981). In other types of neurons studied, in cultured chick ciliary ganglion neurons (Ogden et al., 1984) and in molluscan neurons of different species (Ascher et al., 1978a; Bregestovski, 1980; Simonneau et al., 1980), the spectra are solely one-component. Since the spectra of current fluctuations, either in cationic or anionic channels, can be either one- or two-component, the number of components in the spectra is apparently not associated with acetylcholine receptor channel selectivity.

In amphibian sympathetic ganglion neurons both spectra of current fluctuations and e.p.s.c. decay possess only one component, with τ_f and τ_{syn} close to each other (Marshall, 1985). At the same time, the values of τ_f and τ_{syn} appear to be about two times longer in C-neurons (~10 ms) than in B-neurons (~5 ms; see Table 3.3). Since B- and C-neurons are innervated by preganglionic fibers of different types (B and C, respectively), it has been concluded that the kinetics of acetylcholine receptor channels may be determined by the properties of innervating fibers. This suggestion is supported by the finding that when B-cells are denervated and then reinnervated by C-fibers only, they express acetylcholine receptor channels with properties normally seen in C-cells (Marshall, 1985).

The coincidence of τ_{syn} and τ_{f2} found in different types of neurons is consistent with the conclusion that the synaptic current decay is determined by channel kinetics. It is also consistent with similar temperature dependences of τ_{syn} and τ_{f2} (Derkach et al., 1983).

The presence of two kinetic components in the spectra of acetylcholine-induced current fluctuations and only one component in the synaptic current decay may have two different explanations. According to one explanation, exogenous acetylcholine, unlike internally released acetylcholine, acting on both synaptic and extrasynaptic membrane areas, may activate two types of acetylcholine receptors, having different mean channel open times. Then τ_{f1} and τ_{f2} should correspond to the mean open times of the extrasynaptic and synaptic acetylcholine receptor channels, respectively. This explanation appears to be valid for two-component spectra obtained from denervated muscle fibers (Dreyer et al., 1976). The spectra obtained from normally innervated muscle fibers are one-component when acetylcholine is applied either to synaptic or extrasynaptic membrane areas, with τ_f longer for the extrasynaptic than for the synaptic receptors. At the same time the spectra obtained from denervated muscle fibers are two-component, with time constants

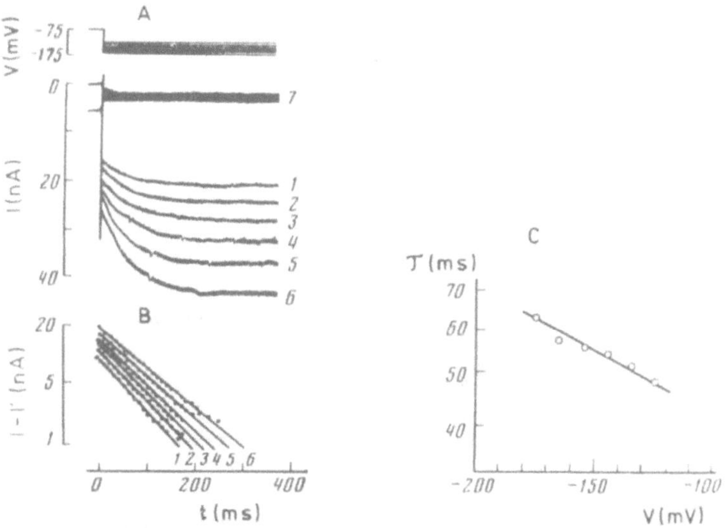

Fig. 3.13. Single-exponential relaxation of acetylcholine-induced current recorded from molluscan neuron. A) Relaxations of current (I) in the absence (7) and in the presence of 0.1 μM acetylcholine (1–6) produced by hyperpolarizing voltage (V) steps by 50–100 mV. B) Semilogarithmic plot of current relaxations shown in A. I' is the instantaneous current jump. C) Semilogarithmic plot of the time constant (τ_r) of current relaxation versus membrane potential (Kurchikov and Kazachenko, 1984b).

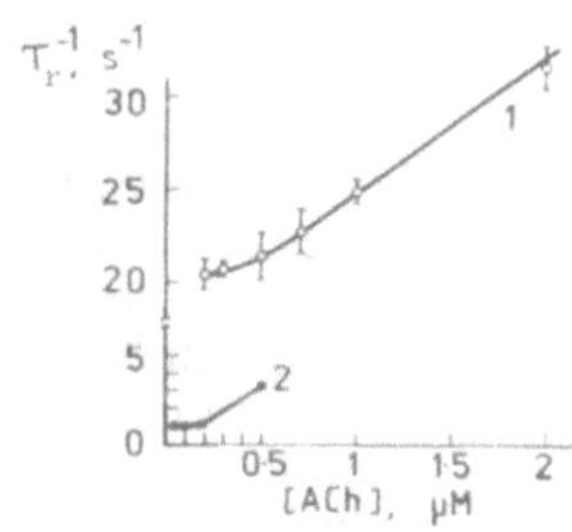

Fig. 3.14. Dependence of the time constant (τ_r) of the current relaxation on acetylcholine concentration in molluscan neuron. Current relaxations were induced by voltage steps from –60 to –160 mV at 22°C (1) or 3°C in another cell (2) (Kurchikov and Kazachenko, 1984b).

corresponding to the mean open times for extrasynaptic and synaptic channels in normally innervated muscle fibers. Thus it has been concluded that the two-component spectra obtained from denervated muscle fibers result from activation of two different types of channels, synaptic and extrasynaptic, with different mean open times.

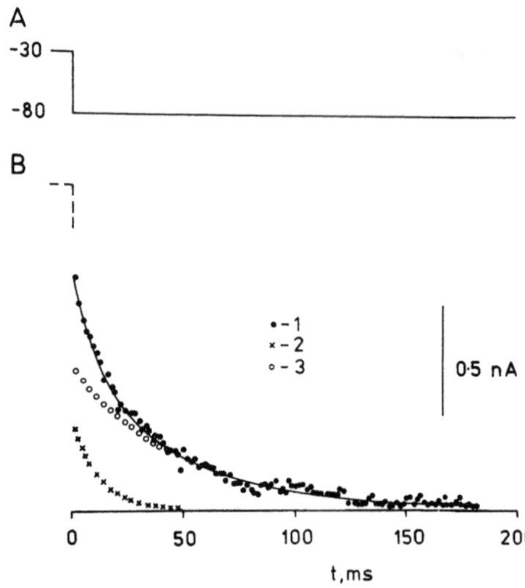

Fig. 3.15. Biexponential relaxation of acetylcholine-induced current in rabbit superior cervical ganglion neuron. A) Voltage step from –30 to –80 mV. B) Current change following voltage step as observed in the presence of acetylcholine. Interrupted line indicates the conditional level of current at –30 mV and the beginning of current ohmic jump caused by membrane potential stepping to –80 mV. 1 is the current relaxation at –80 mV; continuous line is its fitting curve which is the sum of two exponential curves, 2 and 3, with time constants of 11.7 and 47.8 ms. The temperature was 23°C (Derkach, 1986).

Another possible explanation of the two-component spectra in neurons could be that the components are derived from two states of similar receptors, e.g., short inactive (A_nR) or long active (A_nR^*) states. This explanation appears to be valid for the two-component spectra obtained from slow snake muscle fibers (Dionne, 1981). As in many types of neurons, the spectra in snake muscle fibers are two-component, whereas the synaptic current decay is one-component. These fibers are almost devoid of extrasynaptic receptors (Kuffer and Yoshikami, 1975; Dionne, 1981). Therefore, the presence of the fast component is unlikely to be related to the activity of extrasynaptic receptors opening for a shorter time than synaptic receptors.

The suggestion that the inactive state of the acetylcholine–receptor complex may contribute to the spectrum of current fluctuations is not inconsistent with the absence of a fast component in the synaptic current decay. Since the state A_nR is "silent," preceding the active state A_nR^*, it may contribute to the rising phase of the synaptic current rather than to its decay.

Crucial evidence in favor of the latter interpretation of two-component spectra has been obtained in snake muscle fibers from an analysis of the current relaxation produced by voltage steps in the presence of acetylcholine. In these experiments the delay has been revealed in the onset of relaxation apparently resulting from a "silent" state A_nR (Dionne, 1981). Using Eq. (3.16) and assuming that the contribution of the state A_nR to the channel activity is essential, Dionne has obtained good agreement between experimental and calculated kinetic data.

In neurons, two-component spectra could also be derived from the activity of two different populations of channels. This is consistent with: i) an essential density of extrasynaptic receptors, as revealed by α-bungarotoxin binding in autonomic ganglia (Smolen, 1983; Loring et al., 1985), and ii) the two-component spectra and postsynaptic current decay in parasympathetic neurons (Rang, 1981). Thus, the explanation of a fast component in spectra obtained from neurons by the "silent" state of a receptor A_nR seems unlikely. However, in order to make a final decision on whether this explanation is true, relaxation measurements have been carried out.

3.3.4. Relaxation of Acetylcholine-Induced Current Caused by Membrane Voltage Step

The relaxation technique was first applied to the study of acetylcholine receptors in frog end-plate (Adams, 1975; Neher and Sakmann, 1975). This technique is based on the voltage dependence of the channel open time. If the channel open time was not dependent on the membrane potential, a voltage step in the presence of acetylcholine would only result in the instantaneous (ohmic) shift in membrane current caused by a change in the e.m.f. However, what is actually observed in addition to the ohmic shift is a slow change of membrane current, or relaxation, which is due to a change in the total number of channels open per unit time, attributed to the voltage dependence of α_a.

The rate of current relaxation is proportional to α_a, while the direction of relaxation depends on the sign of the voltage dependence of α_a and on the direction of the membrane potential change. Relaxation kinetics, in general, does not depend on the level from which the membrane potential is stepped to a new level. Thus, the channels have no memory. Since channel open time exhibits an exponential distribution, the relaxation time course is likewise exponential and can be described by

$$I(t) = I(\infty) + [I(0) - I(\infty)] \exp(-t/\tau_r), \qquad (3.22)$$

where $I(0)$ and $I(t)$ stand for the current values at times 0 and t after the voltage step, respectively; $I(\infty)$ is the steady-state current, and τ_r is the time constant of relaxation. The value of τ_r depends on the concentration of the agonist, as follows from Eq. (3.18). When the agonist concentration is low, τ_r approaches α_a^{-1}.

Figure 3.13A demonstrates the relaxations of the acetylcholine-induced chloride current that are evoked by membrane hyperpolarizations in a neuron of the mollusc *Limnaea stagnalis*. The relaxations exhibit a linear time course in a semilogarithmic plot, i.e., are single-exponential (Fig. 3.13B). No significant change in τ_r has been observed in neurons of this particular type as acetylcholine concentration is increased from $5 \cdot 10^{-8}$ to $2 \cdot 10^{-7}$ M, although τ_r decreases at acetylcholine concentrations higher than $2 \cdot 10^{-7}$ M at 22°C and somewhat lower than this concentration at 3°C (Fig. 3.14). The mean value of α_a^{-1} estimated as τ_r at low concentrations of acetylcholine (18–22°C and –50 mV) is 32 ms (Kurchikov and Kazachenko, 1984b). Values of τ_r close to that have been found in other neurons of invertebrates, as shown in Table 3.3. It should be noted that the dependence of τ_r on acetylcholine concentration predicted by Eq. (3.18) has not been observed in many other types of neurons (Ascher et al., 1978a; Rang, 1981; Derkach, 1986). One explanation for this discrepancy is that the probability of channels being activated by acetylcholine is very low in these neurons.

The relaxations of the cationic current obtained from *Aplysia* neurons in the presence of acetylcholine are single-exponential, as are those of the chloride current (Ascher et al., 1978a). In other neurons the relaxations of the cationic current are fitted by a sum of two exponential curves (see Table 3.3). Figure 3.15 shows an example of a biexponential relaxation obtained from a neuron of rabbit superior cervical ganglion. The time course of such a relaxation can be described by the following equation:

$$I(t) = I_1(\infty) + [I_1(0) - I_1(\infty)] \exp(-t/\tau_{r1}) + I_2(\infty) + [I_2(0) - I_2(\infty)] \exp(-t/\tau_{r2}), \quad (3.23)$$

where the meanings of the symbols are the same as in Eq. (3.22) and correspond to two relaxation components – fast (1) and slow (2). As has been found from a comparison of the kinetic components present in relaxation, e.p.s.c. decay, and the acetylcholine noise spectrum, τ_{r1} corresponds to τ_{f1}, while τ_{r2} corresponds to τ_{syn} and τ_{f2}. Biexponential relaxations have also been observed in neurons of rat parasympathetic ganglion (Rang, 1981) and *Aplysia* ganglia (Finkel, 1983).

In Eq. (3.23) the two relaxation components, fast and slow, are added to each other. A different situation has been found in snake slow muscle fibers (Dionne, 1981), where the fast relaxation component is subtracted from the slow one:

$$I(t) = I(\infty) + [(I(0) - I(\infty))/(\tau_{r2} - \tau_{r1})][\tau_{r2} \exp(-t/\tau_{r2}) - \tau_{r2} \exp(-t/\tau_{r1})]. \quad (3.24)$$

This result has been explained by the comparatively long time the "silent" state A_nR exists [Eq. (3.16)]. The fact that in the neurons the fast relaxation component is added to the slow one suggests that the fast component is not related to the A_nR state. One possible explanation is that this component is due to the activity of extrasynaptic acetylcholine receptors, which open for a shorter period of time than do the synaptic receptors.

Supposing that fast and slow channels have similar conductances, one can evaluate their contributions to the total channel activity recorded from the whole cell. The contributions can be found from an analysis of either current fluctuations or current relaxations induced by acetylcholine provided that the two components are due to the activity of two separate populations of channels. The contribution of the fast channels to the acetylcholine-induced steady-state current is equal to their contribution σ_1^2 to the total variance of current fluctuations:

$$\sigma_1^2 = \frac{S(0)}{4\tau_{f1}}. \tag{3.25}$$

It has been found using this equation that in rabbit superior cervical ganglion neurons the contribution of fast channels to the total variance of current fluctuations varies from 32% to 91%, with a mean of 66% (Derkach et al., 1983). The value obtained from relaxation analysis is 90% (Derkach, 1986). In rat submandibular ganglion neurons the fast and slow channels make equal contributions both in e.p.s.c. and in acetylcholine-induced current (Rang, 1981).

3.3.5. Kinetics of Single Acetylcholine-Activated Channels

The acetylcholine-induced activity of single channels was first recorded at the frog end-plate by the patch-clamp method (Neher and Sakmann, 1976). This activity was represented by rectangular current pulses of similar amplitudes and variable durations exhibiting an exponential distribution. The idea that single-channel activity is attributed to only two channel states, either open or closed, and that the open times are distributed exponentially had been expressed earlier, during an analysis of acetylcholine-induced current fluctuations (Anderson and Stevens, 1973), and then supported by the results of single-channel recording. Further investigations have shown that these principles of single-channel activity are common for a great variety of both chemically and electrically gated channels.

The high temporal resolution of the patch-clamp method allows one to estimate the mean times that the acetylcholine receptor channel exists in each of three states shown in Eq. (3.16). Since current pulse durations and the durations of interpulse intervals are distributed exponentially, their mean values can be easily found as the time constants of exponential curves, yielding the mean open time and mean closed time, respectively. These, in turn, can be used to estimate the rate constants of the transitions shown in Eq. (3.16), based on the following simple rule (Colquhoun and Sakmann, 1983):

$$\left.\begin{array}{l}\text{Mean time acetylcholine}\\ \text{receptor is in one particu-}\\ \text{lar state}\end{array}\right\} = \left\{\begin{array}{l}\text{Reciprocal of the sum of the}\\ \text{rate constants that diverge}\\ \text{from this particular state.}\end{array}\right. \tag{3.26}$$

In this chapter, the kinetic characteristics of single acetylcholine receptor channel activity recorded from neurons are described, and this activity is analyzed according to the rule (3.26). Three types of neuronal acetylcholine receptor channels have been studied with this approach: the chloride channels in molluscan neurons (Ascher and Erulkar, 1983), and the cationic channels in cultured chick ciliary ganglion neurons (Ogden et al., 1984) and in rat superior cervical ganglion neurons (Derkach et al., 1985, 1987). A detailed kinetic analysis has been made only in the latter two works.

The recordings from rat superior cervical ganglion neurons have been made in either outside-out or cell-attached patch configurations (see Hamill et al., 1981). In the outside-out configuration, the extracellular side of the membrane patch is facing the solution outside the pipette, which allows transient application of acetylcholine to the receptors and thus allows one to check whether the observed activity is reliably attributed to acetylcholine action. In the cell-attached configuration, the extracellular side of the patch is turned toward the internal pipette solution. Only continuous action on the receptors of acetylcholine contained in the pipette can be studied in this configuration, although the signal-to-noise ratio is higher and the recording is more stable than in the outside-out configuration, because of tighter contact between the pipette and the membrane. For this reason, the results described below have been mostly obtained from cell-attached membrane patches. The following two observations indicate that the activity observed is due to the action of acetylcholine: 1) no activity is observed in cell-attached membrane patches if no acetylcholine is present inside the pipette, and 2) the response to acetylcholine iontophoretically applied to outside-out membrane patches in a few experiments is similar to the activity recorded from cell-attached membrane patches (Fig. 3.5A, C).

As has already been mentioned in Chapter 3.2, two types of single-channel currents, with low and high amplitudes, are observed as a result of acetylcholine application. Only low-amplitude currents are used for further kinetic analysis because they are much more numerous than the high-amplitude currents. The currents are recorded at 24°C and –110 mV.

Figure 3.5 illustrates that single-channel currents appear nonregularly. The distribution of the time intervals between channel openings (closed time τ_c) has at least four components, each of which can be fitted by an exponential curve, with mean time constants $\tau_{c1} = 0.15 \pm 0.02$ ms, $\tau_{c2} = 1.78 \pm 0.38$ ms, $\tau_{c3} = 82.2 \pm 27.3$ ms, and $\tau_{c4} = 1082 \pm 291$ ms. The shortest interpulse time interval τ_{c1} is the mean closed time between single openings that appear in bursts, while τ_{c2} is the mean interburst time interval, and τ_{c3} and τ_{c4} are short and long time intervals between the groups of bursts, respectively (Fig. 3.16A).

It has been assumed that the shortest closed time τ_{c1} corresponds to the inactive complex A_nR in Eq. (3.16). Based on this assumption, τ_{c2}, τ_{c3}, and τ_{c4} can be attributed to different substates of R, with different probabilities of being activated by agonist. The long components τ_{c3} and τ_{c4} approximately correspond to the time constants that characterize desensitization of nicotinic acetylcholine receptors (Magleby and Pallotta, 1981; Feltz and Trautmann, 1982; Andreev et al., 1984). Thus, two long closed times are apparently attributed to desensitized acetylcholine receptor states, with low probability of being activated by acetylcholine. In this case, τ_{c2} most probably corresponds to the state R.

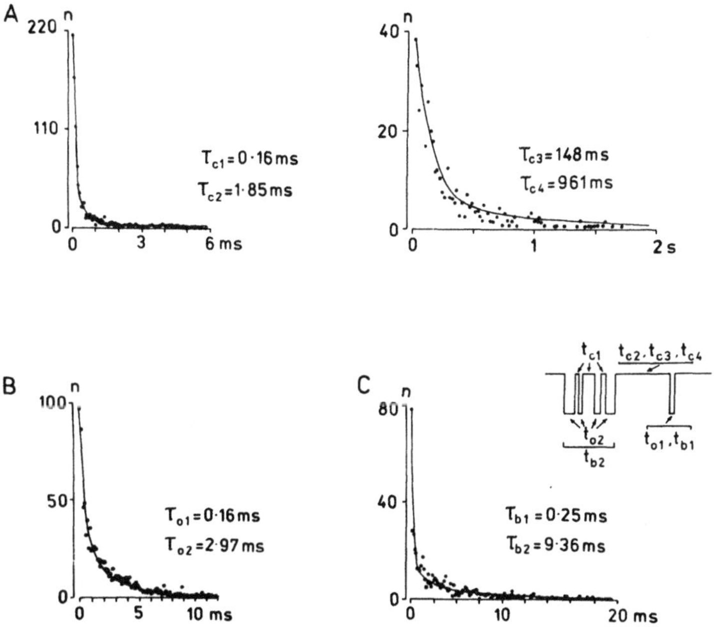

Fig. 3.16. Kinetic analysis of the acetylcholine-induced activity recorded with cell-attached patch-clamp method from single nicotinic receptor channel in rat superior cervical ganglion neuron. A) Distributions of interpulse time intervals shown in short (0–6 ms, left) and expanded (6–2000 ms, right) time scales. B) Distribution of current pulse durations. C) Distribution of current burst durations. Each distribution was fitted by the sum of two exponential curves. The value of the time constant for each exponential curve is indicated near each plot. Inset: schematic representation of single-channel activity formed by a sequence of current pulses. For further explanations see text (Derkach et al., 1987).

The distribution of current pulse durations, i.e., of channel open times, is biexponential, with time constants $\tau_{o1} = 0.21 \pm 0.04$ ms and $\tau_{o2} = 2.61 \pm 0.11$ ms (Fig. 3.16B). This means that there should be at least two open states of a channel, which is not predicted by Eq. (3.16). A possible explanation for this result will be given below.

Assuming that each current pulse within a burst corresponds to the open state A_nR^* and each interpulse time interval within a burst corresponds to the closed state A_nR the burst duration should be equal to the total time the agonist is bound to a receptor. In order to estimate the mean duration of the burst, channel closures longer than $4\tau_{c1}$ were ignored (this should include 98% of the exponentially distributed events having mean τ_{c1}). The distribution of bursts obtained following this procedure has been fitted by the sum of two exponential curves (Fig. 3.16C), with time constants $\tau_{b1} = 0.32 \pm 0.06$ ms and $\tau_{b2} = 8.54 \pm 1.22$ ms. The difference between the values of τ_{b1} and τ_{o1} appears to be insignificant, which means that short open times and short bursts are attributed to the same events, namely, short channel openings separated from each other and from long openings by long closed times (longer than $4\tau_{c1}$). A component with a time constant τ_{b2} corresponds to the bursts containing one or more long openings. A scheme showing the total sequence of pulses comprising single-channel activity is illustrated in the inset of Fig. 3.16C.

It is possible that the short openings of mean duration τ_{o1} or τ_{b1} may result from the binding to the receptor of only one molecule of acetylcholine, while the binding of two acetylcholine molecules results in long openings of mean duration τ_{o2} (Colquhoun and Sakmann, 1983). Based on this assumption, one would expect the contribution of short openings to the total channel activity to decrease at increased concentrations of acetylcholine, as the probability of the binding of two molecules instead of one increases. Indeed, the contribution of the fast component to the total distribution of single pulse and burst durations decreases when the acetylcholine concentration is raised from 30 to 250 μM. Similar observations have been made at the end-plate (Colquhoun and Sakmann, 1983, 1985), although there are data inconsistent with this hypothesis (Sine and Steinbach, 1984, 1986).

It should be pointed out that the value of τ_{o2} estimated from open time distributions is likely to be too large due to the fact that the fastest interpulse intervals (shorter than 150 μs) remain unresolved. The factor of overestimation for τ_{o2} is about two. The corrected value of τ_{o2} can be estimated using the approach described by Colquhoun and Sakmann (see Colquhoun and Sakmann, 1985, p. 505). The mean value of τ_{o2} thus obtained was 1.3 ± 0.2 ms, and it is this value which was used in further considerations.

It follows from the above analysis that the sequence of only long current pulses of mean duration τ_{o2} can be described by Eq. (3.16). Using the rule (3.26), the rate constants of Eq. (3.16) can be estimated from the following set of equations (see Colquhoun and Hawkes, 1983):

$$k_{+1} = \frac{1}{\tau_{c2}[A]}, \tag{3.27}$$

$$k_{-1} = \frac{1}{\tau_{c1}(n+1)}, \tag{3.28}$$

$$\beta = \frac{1}{\tau_{c1}} - k_{-1} = \frac{n}{\tau_{c1}(n+1)}, \tag{3.29}$$

$$\alpha = \frac{1}{\tau_{o2}}, \tag{3.30}$$

where n is the mean number of interpulse intervals per burst which can be expressed as

$$n = \frac{\tau_{b2} - \tau_{o2}}{\tau_{c1} + \tau_{o2}}\,. \tag{3.31}$$

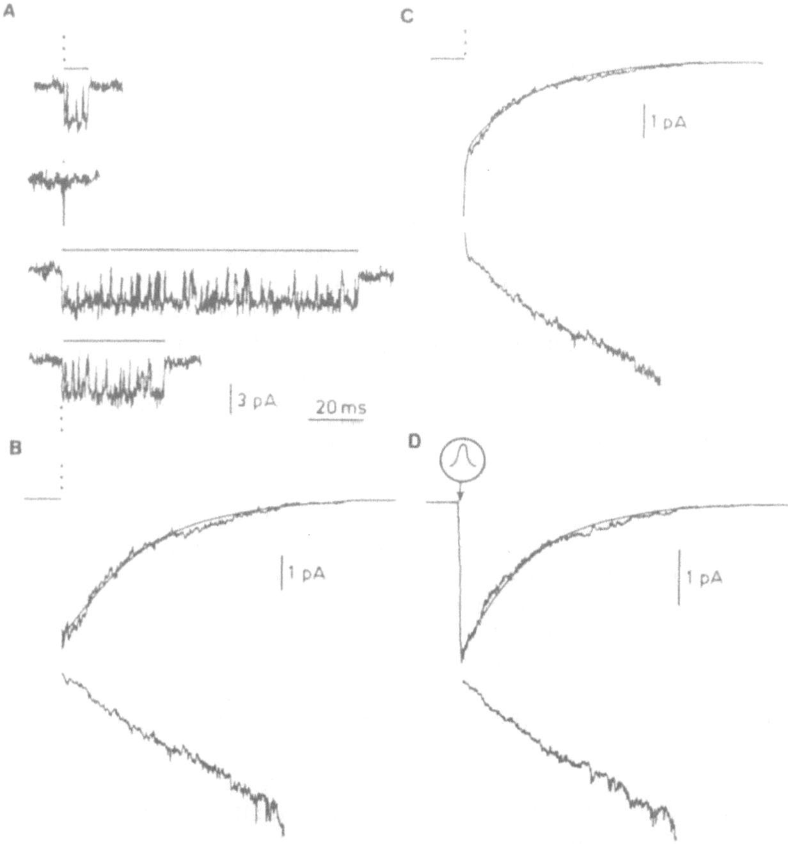

Fig. 3.17. Reconstructed fast e.p.s.c.'s obtained by averaging of currents that flow through a single nicotinic receptor channel in rat superior cervical ganglion neuron at –110 mV (calculated membrane potential) and 23–25°C. A) Examples of single-channel currents, identified as bursts, including single-channel openings. Each burst duration is indicated by a bar above the record. B, C, D) The result of synchronous summation of only "true" bursts, consisting of two or more current pulses (B), as well as of synchronous (C) and nonsynchronous (D) summation of all bursts (including single current pulses). In B, C, and D each upper record is the reconstructed e.p.s.c. and the lower record is its logarithm. The continuous line indicates the fitting curve for reconstructed fast e.p.s.c. decay with one (B, D) or with the sum of two (C) exponential components. The time constants for fitting curves are 29.2 (B), 0.5 and 28.2 (C), and 25.6 ms (D). For further details see text.

The mean number of channel openings during a burst (n + 1) would thus be on the average 6.2. The total time the receptor remains closed (state A_nR) per burst includes, in addition to n measured τ_{c1} periods, two "invisible" τ_{c1} periods, one starting and another finishing the period of occupancy of R by A.

The mean rate constants calculated using Eqs. (3.27)–(3.30) from four separate measurements (24°C, –110 mV) are

$$k_{+1} = (2.33 \pm 0.46)\cdot 10^7\ M^{-1}\cdot s^{-1};\ k_{-1} = 1235 \pm 261\ s^{-1};$$
$$\beta = 6293 \pm 259\ s^{-1};\ \alpha = 894 \pm 185\ s^{-1}.$$

The concentration of acetylcholine in the synaptic cleft following its normal release from the nerve endings is thought to be about 300 μM (Kuffler and Yoshikami, 1975), which predicts τ_{c2} to be about 0.14 ms at k_{+1} equal to the above value of $2.3\cdot 10^7\ M^{-1}\cdot s^{-1}$. This value of τ_{c2} is within the limits of the experimentally found rise time of the miniature postsynaptic current (0.3 ms: Gage, 1976) and within the limits of the shortest synaptic delay in mammalian sympathetic ganglion (1.6 ms: Skok and Heeroog, 1975). On the other hand, the values of k_{+1}, β, and α obtained do not support the above suggestion that binding of acetylcholine to a receptor is much faster than the isomerization of the acetylcholine–receptor complex, although they do support the point that the channel closing is the slowest stage in the acetylcholine–receptor interaction.

It is important to know which of the open states found in the activity of a single acetylcholine receptor channel corresponds to the apparent channel open time estimated, e.g., from the e.p.s.c. decay. The time constant of e.p.s.c. decay measured in rat superior cervical ganglion at a similar temperature (23°C) and membrane potential level (–110 mV) is 13.9 ± 0.6 ms (n = 4) (Derkach, Selyanko, and Skok, unpublished). In rat lumbar sympathetic ganglia, this time constant is 4.2 ms at –60 mV and 37°C (Hirst and McLachlan, 1984). Measurements made under similar conditions in other ganglia and in other animal species yield values ranging from 5 to 45 ms (Kuba and Nishi, 1979; MacDermott et al., 1980; Rang, 1981; Derkach et al., 1983). It is obvious that the value of burst duration τ_{b2} obtained from single-channel activity more closely resembles these estimates for the apparent channel open time than do the estimates obtained from single-channel openings (τ_{o1} or τ_{o2}).

Thus, at –110 mV and 25°C the complex of acetylcholine with a receptor would last, on the average, about 9 ms ($\tau_{b2} + 2\tau_{c1}$) and during this time the channel would open and reclose on the average 6.2 times. This means that acetylcholine normally released by preganglionic fibers as a result of a single preganglionic impulse causes not a single opening but a burst of openings in a single postsynaptic acetylcholine receptor channel.

It is known that the burstlike single-channel activity produced by acetylcholine in the presence of local anesthetics may result in the biexponential decay of the end-plate current (Neher and Steinbach, 1978). Thus, one would expect that the decay of the e.p.s.c. recorded from sympathetic ganglion neurons would likewise be split up into two exponential components, which in fact has not been observed in the experiments described above. It is of interest, therefore, to determine whether the sum of many bursts of variable durations taken from the same records of channel activity to simulate the e.p.s.c. curve would show a single- or biexponential decay.

Figure 3.17 illustrates the results obtained by such an approach. When the onsets of 106 bursts consisting of two or more single openings each are synchronized (Fig. 3.17A), the resultant decay curve is single-exponential (Fig. 3.17B). The decay curve becomes biexponential when 66 separate single openings, in addition to 106 bursts, are included in the summed pool (Fig. 3.17C). However, only one (slower) component remains after a small desynchronization (made by a computer programmed by Gaussian law with $\sigma = 0.5$ ms) is introduced in the onset time of all events before they were summed (Fig. 3.17D). Such desynchronization is consistent with the S-shaped rising phase of the miniature synaptic currents observed in experiments (see Gage, 1976). Thus, the results of simulation of e.p.s.c. decay do not contradict the above conclusion that normally released acetylcholine induces a burst of openings rather than a single opening in the acetylcholine receptor channel.

Three channel closed times approximately equal to the values of τ_{c1}, τ_{c2}, and τ_{c3}–τ_{c4} mentioned above have been found in the activity of single nicotinic receptor channels in cultured chick ciliary ganglion neurons (Ogden et al., 1984). The distribution of the channel open time in chick ciliary ganglion neurons, unlike that in rat sympathetic ganglion neurons, is single-exponential, with a time constant close to τ_{o2} in sympathetic neurons.

The activity of single chloride channels induced by suberylcholine was recently studied in the neurons of *Helix aspersa* (Ascher and Erulkar, 1983). The activity contains long-lasting bursts of channel openings (about 100 ms long at room temperature). Two kinetic components, slow and fast, have been found from the spectral analysis of either single-channel activity or current fluctuations recorded from the whole cell. The slow component corresponds to the burst duration, whereas the nature of the fast one (10–15 times shorter than the slow one) remains unclear.

The burstlike kinetics were not revealed in early studies of single end-plate channels (Neher and Sakmann, 1976) because of the limited time resolution in the initial modification of the patch-clamp method. In subsequent studies, the burstlike channel kinetics were first observed when induced by high (desensitizing) concentrations of acetylcholine (Sakmann et al., 1980). Later this kind of kinetics was found in the end-plate channels even at low concentrations of agonist (Colquhoun and Sakmann, 1983). The mean number of channel openings within a burst is about four at the end-plate, i.e., almost the same as in sympathetic ganglion neuron, although the durations of single openings and single closings of end-plate channels within a burst are much shorter, which results in higher mean values of the rate constants: k_{+1}, k_{-1}, β, and α equal to 10^8 $M^{-1}\cdot s^{-1}$, 10,000 s^{-1}, 30,000 s^{-1}, and 1300 s^{-1}, respectively. Since experiments on the end-plate and on sympathetic ganglion neurons have been performed at different temperatures – 10°C and 25C, respectively – the difference in rate constants is expected to be even more pronounced at the same temperature.

In spite of the difference in rate constants, the affinity K of acetylcholine to a receptor estimated as $K = k_{+1}/k_{-1}$ is similar in the end-plate and in sympathetic ganglion neurons (about 100 μM^{-1}), as well as the effectiveness of acetylcholine in the channel opening, estimated as β/α.

The molecular mechanism responsible for the burstlike kinetics of a single acetylcholine receptor channel remains unknown. However, it becomes obvious that this type of activity is related to conformational transitions in the whole receptor molecule. This suggestion follows from our recent results obtained from the nicotinic acetylcholine receptors of the rat superior cervical ganglion neurons with their disulfide bonds reduced by dithiothreitol (1 mM). In contrast to normal acetylcholine receptors, the single-channel activity of these receptors, as evoked either by reversibly bound acetylcholine (30 μM) or by irreversibly bound bromoacetylcholine (100 μM), exhibits single-channel openings without bursts.

Recently, another kind of channel kinetics was found, associated with the channel open state. As many investigators pointed out, a number of channels, including the acetylcholine receptor channel, reveal "extra" current fluctuations during their single openings. It was initially thought that these additional current fluctuations result either from fast channel closings, unresolved due to the limited frequency bandwidth, or from the shot noise, associated with the passage of discrete ions through the membrane (Stevens, 1972). Spectral analysis of the fluctuations of currents flowing through single acetylcholine receptor channels in muscle cells reveals a component which is relatively slow and much larger than expected for shot noise (Sigworth, 1985). The time constant of this component depends on both temperature (with $Q_{10} \approx 3$) and membrane potential (with a high positive value of H). Based on these results, it has been concluded that the fluctuations of current flowing through a single acetylcholine receptor channel may result from conformational fluctuations in the channel protein (Sigworth, 1985), with the frequency restricted by interactions between domains forming the ionic channel (Laüger, 1985).

As has been mentioned above, a single quantum of acetylcholine released from the presynaptic nerve fiber opens a number of channels which is about one order less in sympathetic ganglion neuron than at the end-plate. On the other hand, both a single opening of the channel and a burst of openings are a few times longer in sympathetic ganglion neuron than at the end-plate. This means that the electric charge transferred through the membrane as a result of the action of a single acetylcholine quantum should be similar in both objects. Since it is the electric charge transferred through the postsynaptic membrane that determines the amplitude of the excitatory postsynaptic potential (see Chapter 7), the effectiveness of a single acetylcholine quantum in both objects should likewise be similar.

3.3.6. Voltage Dependence of Channel Kinetics

In the past it was thought that chemically driven membrane channels are insensitive to membrane potential, in contrast to electrically driven membrane channels, which are responsible for the electrical excitability of the membrane. This view remains valid in the sense that chemically driven channels, in general, need an agonist to be opened and cannot be opened by just a shift in the membrane potential (although, under certain conditions chemically driven channels may open spontaneously even in the absence of an agonist: Hille et al., 1983). On the other hand, the kinetics of chemically driven channels may strongly depend on the membrane potential.

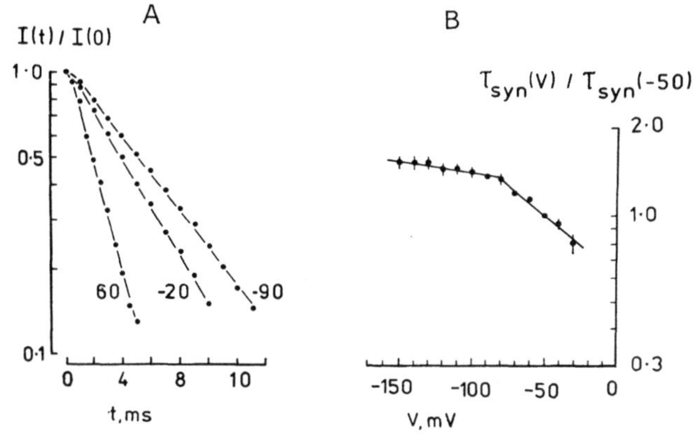

Fig. 3.18. Effect of membrane potential on kinetics of fast e.p.s.c. recorded from rabbit superior cervical ganglion neuron at 36°C. A) Semilogarithmic plot of fast e.p.s.c. decays as recorded at various membrane potentials (60, –20, and –90 mV). B) Dependence of the time constant of fast e.p.s.c. decay on membrane potential. Averaged data (means and SE) obtained from nine neurons are shown. The values of the time constant at the membrane potential –50 mV are taken as 1 (Derkach et al., 1983).

The effect of the membrane potential on channel kinetics was first observed in frog end-plate as an increase in duration of the end-plate current decay produced by membrane hyperpolarization (Gage and Armstrong, 1968; Kordas, 1969, 1972a, b; Magleby and Stevens, 1972a, b). Since the time constant of end-plate current decay is equal to α_a^{-1}, this means that α_a is voltage dependent. The voltage dependence of α_a is described by the equation

$$\alpha_a^{-1} = \alpha_a^{-1}(0)\exp\left(\frac{V}{H}\right), \tag{3.32}$$

where V is the membrane potential and H is the shift in V corresponding to an e-fold change of α_a^{-1}. In the end-plate $H \cong -100$ mV (the negative value of H means that α_a^{-1} increases as the membrane potential decreases, i.e., as the membrane becomes hyperpolarized).

The voltage dependence of acetylcholine receptor channel kinetics has been studied in a variety of objects. In these studies α_a^{-1} is estimated mainly from the e.p.s.c. decay or from relaxation analysis. The values of H obtained are shown in Table 3.3. It is evident from Table 3.3 that for cationic nicotinic channels, in both vertebrates and invertebrates, H is negative, whereas it may be positive in some anionic nicotinic channels.

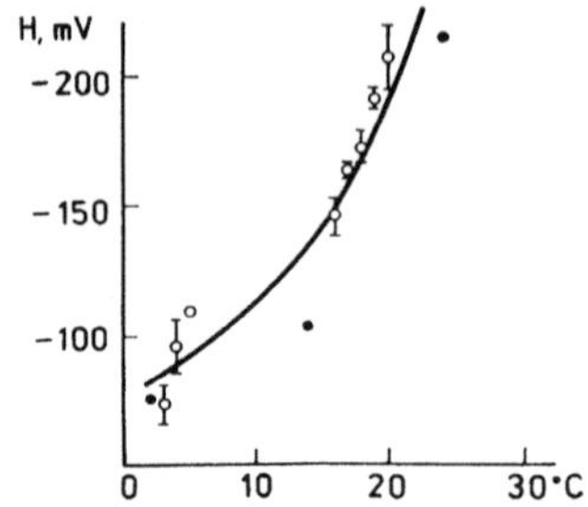

Fig. 3.19. Effect of temperature on voltage dependence of the time constant of current relaxation recorded in the presence of acetylcholine from molluscan neuron. Voltage dependence of current relaxation is expressed by H [see Eq. (3.32)]. Open circles indicate the averaged values of H obtained from 43 neurons; closed circles indicate the values obtained from the same cell (Kurchikov and Kazachenko, 1984b).

An example of the voltage dependence of α_a^{-1} measured as e.p.s.c. decay time constant τ_{syn} is shown in Fig. 3.18. E.p.s.c.'s were recorded from rabbit superior cervical ganglion neuron. The plot of normalized τ_{syn} versus the membrane potential clearly exhibits two parts: the voltage dependence of τ_{syn} is stronger at membrane potentials less negative than –80 mV, compared to that at more negative membrane potentials. A similar "saturation" in the $\alpha_a^{-1}(V)$ relationship has been observed in rat parasympathetic neurons (Rang, 1981) and molluscan neurons (Ascher et al., 1978a) and has been interpreted as the result of binding of permeant ions within a channel as described above.

If H is measured in the range of membrane potentials corresponding to the maximal voltage dependence of α_a^{-1}, the values of H in rabbit superior cervical ganglion neurons (Selyanko et al., 1979) and rat submandibular ganglion neurons (Rang, 1981) are –140 mV and –100 mV, respectively, which is close to what was found in frog end-plate (Magleby and Stevens, 1972a, b; Anderson and Stevens, 1973). In contrast to this, the absolute values of H are very high in frog sympathetic neurons: –524 mV (Kuba and Nishi, 1979) or –256 mV (MacDermott et al., 1980). The reason for such a difference between mammalian and amphibian autonomic neurons is not yet clear. In some neurons of frog sympathetic ganglia, acetylcholine receptor channels have what is called "memory": the value of α_a^{-1} measured as an e.p.s.c. decay time constant after the membrane potential has been changed from one (conditional) level to another (testing) level appears to be dependent on the conditional level of the membrane potential (MacDermott et al., 1980).

The voltage dependence of α_a^{-1} in chloride acetylcholine receptor channels in molluscan neurons varies greatly (see Table 3.3). In some neurons H is negative, while in others it is positive or infinite. Therefore, the sign and value of H

do not depend on permeant ion or animal species (in the same mollusc H may be positive for anionic channels and negative for cationic channels: Finkel, 1983).

The voltage dependence of single acetylcholine receptor channel kinetics has been investigated only in cultured chick ciliary ganglion neurons (Ogden et al., 1984). Surprisingly, the mean channel open time τ_o appears to decrease somewhat with membrane hyperpolarization, in contrast to what has been observed in all previous studies of cationic acetylcholine receptor channel kinetics.

The voltage dependence of α_a^{-1} has been shown to increase at low temperature. This was observed in molluscan neurons (Ascher et al., 1978a; Kurchikov and Kazachenko, 1984b) and frog sympathetic ganglion neurons (Kuba and Nishi, 1979), as well as in frog end-plate (Anderson and Stevens, 1973). Some authors, however, have not confirmed this result (MacDermott et al., 1980).

Figure 3.19 illustrates the effect of temperature on the voltage dependence of α_a^{-1}, estimated as the time constant of acetylcholine-induced chloride current relaxation in molluscan neuron. It is evident from Fig. 3.19 that the voltage dependence of α_a^{-1} markedly increases at low temperature.

Based on the above suggestion that a shift in membrane potential changes the probability of channel closing rather than the probability of channel opening, one would expect that the voltage dependence of α_a^{-1} would correlate with that of steady-state acetylcholine-induced current. Indeed, such a correlation has been observed and, in accordance with what should be expected, appears to be independent of temperature.

It seems reasonable to suggest that the voltage dependence of acetylcholine receptor channel kinetics is attributed to the voltage dependence of a rate-limiting step in acetylcholine–receptor interaction. According to an early hypothesis, the rate-limiting step is a conformational transition of a channel from an open to a closed state [a transition from the A_nR^* state to the A_nR state; see Eq. (3.16)], which is followed by a change in the dipole moment of the acetylcholine receptor molecule (Magleby and Stevens, 1972a, b; Anderson and Stevens, 1973). In this case, the voltage dependence of the channel-closing rate may result from a change in the dipole position within the membrane. An alternative hypothesis considers the dissociation of acetylcholine from a receptor, i.e., a receptor transition from the A_nR state to the R state, as a rate-limiting step (Kordas, 1969; Adams and Sakmann, 1978; Kurchikov et al., 1985). From this point of view, the voltage dependence of channel kinetics may result from the fact that the acetylcholine binding site is located deep in the membrane and, therefore, its interaction with acetylcholine may be affected by the membrane electric field.

Either one of these two hypotheses by itself fails to provide a complete interpretation of all the experimental results obtained. For example, the dependence of α_a^{-1} on the type of agonist can easily be explained by the latter hypothesis but not by the former one. On the other hand, it is difficult to explain why only α_a^{-1}, but not its voltage sensitivity, is dependent on the type of agonist (Kurchikov et al.,

1985), and why the sign of H may differ in different acetylcholine receptors activated by the same agonist.

One possible solution to this problem is that both of the above hypotheses may be valid, i.e., that the voltage dependence of acetylcholine receptor channel kinetics results from the voltage dependence of both binding and isomerization steps. This conclusion is consistent with the result obtained from single channels that the rates of these steps in acetylcholine receptor are similar.

3.4. LOCALIZATION OF IONIC CHANNEL IN ACETYLCHOLINE RECEPTOR MOLECULE

In Chapter 2 the region in the amino acid sequence of the α subunit of the acetylcholine receptor molecule which, most probably, is a recognition center is identified. The question arises whether it is possible, using a similar approach, to identify the regions that form the ionic channel of the acetylcholine receptor.

According to the results of experiments performed on reconstituted acetylcholine receptors (see Chapter 3.1), the ionic channel is able to function only if the whole complex of subunits of the stoichiometry $\alpha_2\beta\gamma\delta$ is present. This implies that all subunits participate in channel formation, in contrast to the recognition center, which is located mostly (or exclusively) at the α subunit (Lindstrom, 1980; Kistler et al., 1982).* It is commonly accepted that the ionic channel is in the center of a circle formed by all five subunits (Fig. 1.7). However, it is not yet clear which portion of the amino acid sequence of each subunit participates in channel formation. In recent studies, models are preferred where the channel is formed by specific amphipathic helical portions of each subunit. The advantage of an amphipathic helix is that it can face the lipid membrane environment by its apolar residues while facing the water inside the channel by its polar residues, thus providing the most energetically stable structure. An example of such a structure is a model proposed by Guy (1984) having the internal surface of the channel formed by the homologous portion M_5 (also called portion L) of all five subunits in the acetylcholine receptor of the electric organ. According to theoretical predictions, this portion is an α-helix both in electric organ (Figs. 2.9) and in the neuronal (Fig. 2.11) acetylcholine receptor molecules.

Recently a new approach was developed to study the functions of the acetylcholine receptor subunits in hybrids composed of subunits from two different animal species and expressed in the *Xenopus* oocytes through the use of the cloned complementary DNAs (Mishina et al., 1985; Sakmann et al., 1985). It has been found that this is a δ subunit that determines the open time of the channel, i.e., the channel-closing step. Furthermore, it has been found by altering different sites in the δ subunit with site-directed mutagenesis that the sites responsible for channel

*Most recent observations suggest that the neuronal nicotinic acetylcholine receptors from chicken and rat brain consist of only two types of subunits, α and β, and that the recognition center for binding of acetylcholine is located in the β subunit rather than in the α subunit (Whiting and Lindstrom, 1987b; see Chapter 2).

activity are located in the M_2 segment of the δ subunit. It is suggested, therefore, that the portions which correspond to the M_2 segments might line the ionic channel (B. Sakmann, personal communication). This is consistent with most recent models of acetylcholine receptor arrangement in the membrane. According to these models, the M_2 segment spans the membrane while the transmembrane arrangement of the portions with greater sequence numbers is questionable (Ratnam et al., 1986). According to other authors the compounds that block the acetylcholine receptor channel may bind to serine 269 which is in the M_2 segment (Hucho, 1986). This is consistent with the above results.

3.5. CONCLUSIONS

Activation of neuronal nicotinic acetylcholine receptors is followed by opening of their ionic channels either for cations, which results in membrane depolarization, i.e., in cell excitation, or for chloride ions, which causes membrane hyperpolarization, i.e., cell inhibition. The cationic channels have been found in both vertebrates and invertebrates, while the anionic channels are only in invertebrates.

In a normal ionic environment, open cationic acetylcholine receptor channels are permeable for sodium and potassium ions (the P_{Na} to P_K ratio ranges from 0.6 to 0.9) and, to a lesser extent, for calcium ions. The permeant ions, especially divalent ones, may bind within the channel. The energy profile of the cationic channel is supposed to have two energy maxima (barriers) separated by an energy minimum (well), the ion binding site.

The mean conductance γ of a single cationic acetylcholine receptor channel in different types of neurons ranges from 8 to 42 pS; in sympathetic ganglion neurons, there is an additional population of channels, with γ equal to 50 pS. The mean values of γ for neuronal anionic acetylcholine receptor channels are much smaller: they range from 0.6 to 16 pS. The conductance for cationic channels is decreased with increased concentration of divalent cations because of their binding to the open channel and, as a rule, it does not depend on the membrane potential level.

The mean apparent open time in cationic acetylcholine receptor channels has been estimated by three different methods: analysis of the e.p.s.c. decay time course, and acetylcholine-induced membrane current fluctuations and relaxations. When measured in mammalian sympathetic ganglion neurons at resting membrane potential level (–60 to –50 mV) and at body temperature (36–37°C), this time is 3–4 ms. Similar values of the mean apparent open time have been found in molluscan and amphibian autonomic neurons at room temperature. In some neurons, a much faster kinetic component is present.

The mean apparent open time depends on temperature and membrane potential. It is increased by decreasing temperature, with Q_{10} values ranging from 3 to 6, or by membrane hyperpolarization, with an e-fold increase corresponding to a 100–400 mV voltage drop in mammalian neurons or higher voltage drops in amphibian neurons. The voltage dependence of the channel open time is increased by decreasing the temperature.

The mean apparent open time in anionic acetylcholine receptor channels is similar to that in cationic channels, although it may either increase, decrease, or remain unchanged with membrane hyperpolarization.

The kinetics of single cationic acetylcholine receptor channels, studied in mammalian sympathetic ganglion neurons with the use of the patch-clamp method, reveals two kinds of openings: isolated short openings and longer openings appearing in bursts. It follows from a comparison of the e.p.s.c. decay time constant with the duration of channel openings and the duration of bursts of openings that burst duration, rather than a single opening duration, corresponds to the mean apparent open time found by other methods. The binding of acetylcholine to a receptor molecule results in a burst, on the average, of six channel openings, each 1.3 ms long (which is a true channel open time), separated by channel closings, each 0.15 ms long. The decay of summed individual bursts imitates the e.p.s.c. decay.

From the above and some other kinetic components revealed in single-channel activity, the rate constants k_{+1} and k_{-1} of acetylcholine binding to a receptor and its dissociation from a receptor, respectively, as well as the rate constant of channel opening β and that of channel closing α, have been calculated. The values of these rate constants in neurons are markedly lower than in muscle.

Using a deduced amino acid sequence and the theoretically predicted secondary structure of the acetylcholine receptor molecules, a model of its ionic channel incorporated into a lipid membrane has been built. According to the model, the internal surface of the channel is lined by α-helical portions of each of five subunits, with their charged side chains exposed to the channel. Recent findings suggest, however, that δ subunit is most important for channel formation.

Chapter 4

GENERAL CHARACTERISTICS OF THE EFFECTS PRODUCED BY BLOCKING AGENTS IN NEURONAL ACETYLCHOLINE RECEPTORS

4.1. RELATIONSHIP BETWEEN CHEMICAL STRUCTURE AND GANGLION-BLOCKING ACTIVITY

Nicotinic receptors of autonomic ganglion neurons are best studied among neuronal nicotinic receptors in relation to their blocking mechanisms. Historically, their studies have been mainly initiated and stimulated by demands for hypotensive drugs. In this chapter, only one formal characteristic of ganglion-blocking agents, their structure–activity relationship, will be briefly summarized, considering only those compounds whose effects are most accountable for the main ganglion-blocking mechanisms (for a more detailed description see special reviews: Barlow, 1964; Kharkevich, 1967; Triggle, 1965; Triggle and Triggle, 1976; Trčka, 1980a, b).

4.1.1. Mono-Quaternary Ammonium Compounds

The simplest in structure and the best studied as a ganglion-blocking drug is tetraethylammonium (TEA, see **2.6**). The principle realized in TEA structure compared with the structure of the ganglionic stimulant TMA – the substitution of ethyl groups for methyl groups attached to quaternary nitrogen – has been commonly used to obtain many other ganglion-blocking agents from ganglionic stimulants (Table 4.1). Table 4.2 illustrates the combination of substituents resulting in the highest blocking activity: two ethyls and two isopropyls (compound 6 in Table 4.2, diethyldiisopropylammonium). The compound possessing this combination is 12.6 times more effective than TEA. Larger or more polar substituents decrease blocking activity.

A similar tendency is observed if the cationic group is included in the cyclic structure: the increase in the size of the group enhances blocking activity, whereas

TABLE 4.1. Effect of N-Substitution on Activity at Ganglionic Junctions (Ing, 1957)

No.	Activator	Antagonist
1	$CH_3CO_2CH_2CH_2N^+Me_3$	$CH_3CO_2CH_2CH_2N^+Et_3$
2	Me_4N^+	Et_4N^+
3	$C_6H_5OCH_2CH_2N^+Me_3$	$C_6H_5OCH_2CH_2N^+Et_3$
4	Me_3S^+	Et_3S^+
5	$C_2H_5OCH_2CH_2N^+Me_3$	$C_2H_5OCH_2CH_2N^+Et_3$

the presence of a hydroxyl group decreases blocking activity. Maximum activity is observed in 2,6-dimethyl-1,1-diethylpiperidinium (**4.1**):

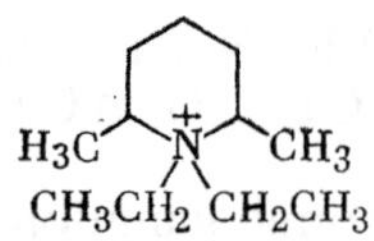

2,6-Dimethyl-1,1-diethylpiperidinium, **4.1**

Other highly active heterocyclic mono-quaternary compounds are tropan derivatives with a benzyl substituent at the nitrogen atom (Nador, 1960); compound **4.2** is about 40 times more active than TEA:

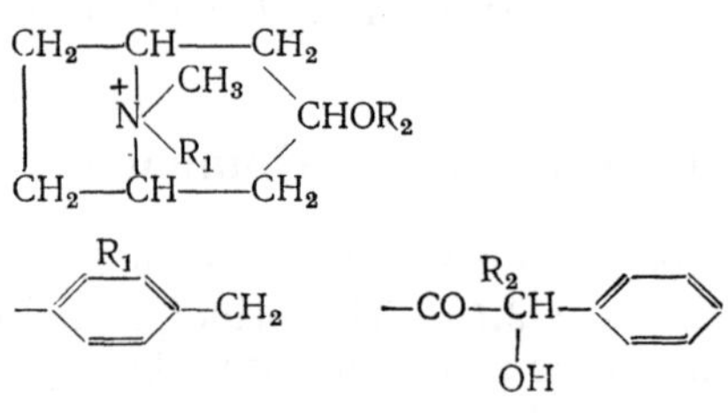

4.2

It can be concluded that most active mono-quaternary blocking agents possess the following structural component (Trčka, 1980a):

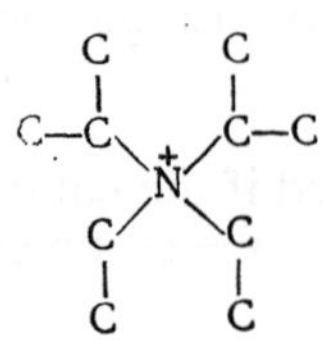

4.3

TABLE 4.2. Ganglion-Blocking Activities of Simple Quaternary Ammonium Ions (Kharkevich, 1967)

Activity No.	$R_1R_2R_3R_4N^+$ R_1	R_2	R_3	R_4	Equipotent molar activity (TEA = 1)
1	CH_2CH_3	CH_2CH_3	CH_2CH_3	CH_2CH_3	1.0
2	CH_2CH_3	CH_2CH_3	CH_2CH_3	CH_2CH_2OH	0.24
3	CH_2CH_2OH	CH_2CH_2OH	CH_2CH_2OH	CH_2CH_2OH	inactive
4	CH_2CH_3	CH_2CH_3	$(CH_2)_2CH_3$	$(CH_2)_2CH_3$	0.92
5	CH_2CH_3	CH_2CH_3	CH_2CH_3	$CH(CH_3)_2$	2.6
6	CH_2CH_3	CH_2CH_3	$CH(CH_3)_2$	$CH(CH_3)_2$	12.6
7	CH_2CH_3	CH_3	$CH(CH_3)_2$	$CH(CH_3)_2$	5.9
8	CH_3	CH_3	$CH(CH_3)_2$	$CH(CH_3)_2$	1.9
9	CH_3	CH_3	$CH_2CH(CH_3)_2$	$CH_2CH(CH_3)_2$	3.0
10	CH_2CH_3	CH_3	$CH_2CH(CH_3)_2$	$CH_2CH(CH_3)_2$	1.1
11	CH_2CH_3	CH_3	$(CH_2)_2CH(CH_3)_2$	$(CH_2)_2CH(CH_3)_2$	0.36
12	CH_2CH_3	CH_2CH_3	CH_2CH_3	$CH_2C_6H_5$	0.5
13	CH_2CH_3	CH_2CH_3	$CH_2C_6H_5$	$CH_2C_6H_4$	0.4
14	CH_2CH_3	CH_2CH_3	CH_2CH_3	1-$CH_2C_{10}H_7$	0.4

This component or some of its elements are present in all the blocking agents mentioned above that are the most active in each corresponding series.

4.1.2. Tertiary and Secondary Amines

A high ganglion-blocking activity has also been observed in tertiary and secondary amines. Examples are 3-methylaminoisocamphane (macamylamine, **4.4**), 1,2,2-trimethylcyclohexylamine (**4.5**), isopropylisobutylamine (**4.6**), and 1,2,2,6,6-pentamethylpiperidine (pempidine). Pempidine is a tartrate (**4.7**), while the toluenesulfonate with the same cation is another potent ganglion-blocking agent, pirilenum (**4.8**):

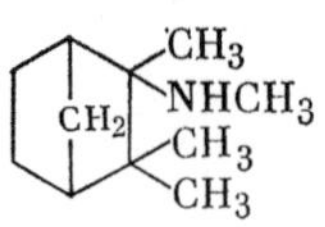

Mecamylamine, **4.4**

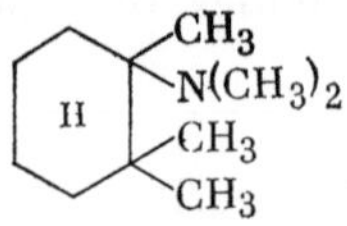

4.5

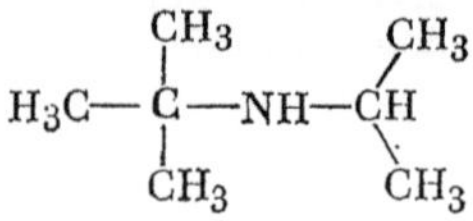

4.6

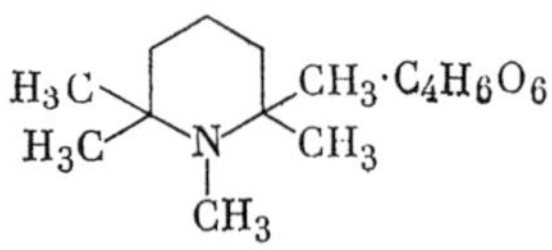

Pempidine, **4.7**

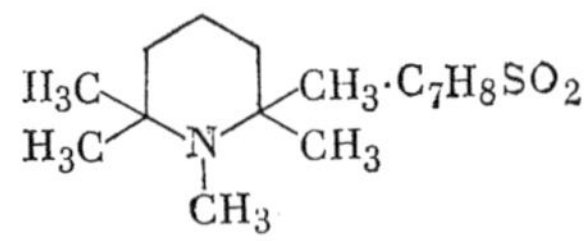

Pirilenum, **4.8**

Macamylamine is almost as active as hexamethonium (Van Rossum, 1962a, b), whereas the other compounds mentioned above are 1.5–2.5 times (**4.5**), 2–10 times (**4.6**), and 1.2–2.5 times (**4.7**) more active (see Trčka, 1980b).

Each of these compounds has either of the following two structural components (**4.9** or **4.10**):

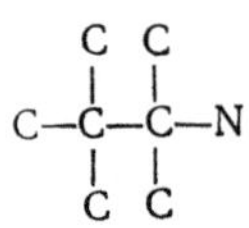

4.9

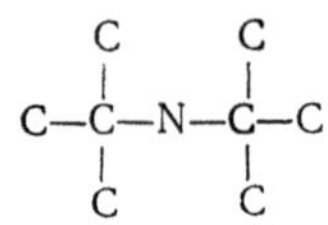

4.10

Component **4.9** is present in compounds **4.4** and **4.5**, while compounds **4.6**, **4.7**, and **4.8** contain component **4.10** (Trčka, 1980b). One can conclude from the above results that the main structural components responsible for high ganglion-blocking activity are actually similar in quaternary, tertiary, and secondary amines. Their most common element is a positively charged nitrogen atom in connection with the alkyl chain branching at the α-carbon.

4.1.3 Bis-Quaternary Ammonium Compounds

In Table 4.3 (Trčka, 1980a), the activities of some symmetrical bis-quaternary ammonium compounds are compared with those of hexamethonium (**2.18**, $n = 6$) taken as 100%.

It is evident from Table 4.3 that the addition to TMA of a second trimethylammonium group turns stimulant TMA into a blocking agent. Furthermore, the presence of one or two methyls in the cationic group increases the blocking activity compared with that of the compounds possessing only ethyl groups. On the other hand, the compounds with two trimethylammonium groups exhibit two maxima in their blocking activities, at $n = 5$–6 and at $n = 16$–17. Another interesting point is that the blocking activity maximum shifts toward lower n values as two identical cationic groups become larger.

TABLE 4.3. Ganglion-Blocking Activities of Symmetrical Bis-Quaternary Ammonium Compounds 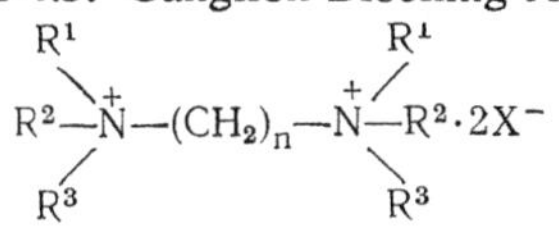

No.	R^1	R^2	R^3	*n*	Percentage activity (hexamethonium = 100%)	References
1	CH_3	CH_3	CH_3	4	2 1	Paton and Zaimis, 1949 Wien et al., 1952
2	CH_2CH_3	CH_3	CH_3	4	10	Wien et al., 1952
3	CH_2CH_3	CH_2CH_3	CH_3	4	100	Wien et al., 1952
4	CH_2CH_3	CH_2CH_3	CH_2CH_3	4	5	Wien et al., 1952
5	CH_3	CH_3	CH_3	5	65 80	Wien et al., 1952 Paton and Zaimis, 1949
6	CH_2CH_3	CH_3	CH_3	5	150	Wien et al., 1952
7	CH_2CH_3	CH_2CH_3	CH_3	5	125	Wien et al., 1952
8	CH_2CH_3	CH_2CH_3	CH_2CH_3	5	7 1	Wien et al., 1952 Chou and de Elio, 1948
9	CH_3 hexamethonium	CH_3	CH_3	6	100	Paton and Zaimis, 1949
10	CH_2CH_3	CH_3	CH_3	6	150	Wien et al., 1952
11	CH_2CH_3	CH_2CH_3	CH_3	6	75	Wien et al., 1952
12	CH_2CH_3	CH_2CH_3	CH_2CH_3	6	5	Wien et al., 1952
13	CH_3	CH_3	CH_3	7	10	Paton and Zaimis, 1949
14	CH_3	CH_3	CH_3	12	25	Barlow and Zoller, 1964
15	CH_3	CH_3	CH_3	14	80	Barlow and Zoller, 1964
16	CH_3	CH_3	CH_3	16	400	Barlow and Zoller, 1964
17	CH_3	CH_3	CH_3	18	400	Barlow and Zoller, 1964
18	CH_3	CH_3	CH_3	20	80	Barlow and Zoller, 1964

TABLE 4.4. Ganglion-Blocking Activities of Phenylalkane-bis-onium Compounds (Wien and Mason, 1953) $4R_3\overset{+}{N}C_6H_4(CH_2)_n\overset{+}{N}R_3$

	Equiactive molar ratios (hexamethonium = 1)			
n	Me_3	Me_2Et	$MeEt_2$	Et_3
1	160	–	–	–
2	0.25	0.22	0.38	33
3	2.9	1.4	4.4	17
4	–	6	12	12

TABLE 4.5. Ganglion-Blocking Activity of Bis-Pyrollidinium Series (see Triggle and Triggle, 1976)

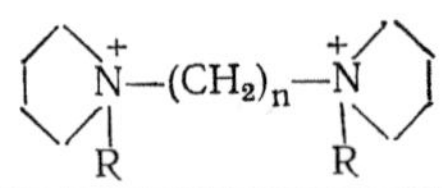

n	R	Equieffective molar ratio (hexamethonium = 1)
3	Me	20.0
4	Me	2.0
5	Me	0.2
6	Me	0.33
6	C_2H_5	1.8

Estimates for comparative blocking activities of ganglion-blocking drugs obtained by different authors vary greatly, probably because of differences in the methods and objects used in the experiments. For example, in early experiments hexamethonium was found to be 20 times more potent as a blocking agent in sympathetic ganglia than TEA (Paton and Zaimis, 1949), whereas in later works this estimate dropped to only 2 times (see Trčka, 1980b). Some authors found that hexamethonium had 0.84 of the mecamylamine potency in cat superior cervical ganglion (see Trčka, 1980b), while others found it 17 times less potent than mecamylamine (Table 4.10).

There is no doubt that the dependence of the blocking activity on the number of methylene groups *n* in bis-quaternary ammonium compounds is actually a dependence on the intercharge distance, since a similar dependence is observed in compounds with a rigid carbon skeleton (Table 4.4, except compounds with R = Et) and in compounds whose charged nitrogen atoms are included in the cyclic ring (Table 4.5).

The maximum in ganglion-blocking activities observed in Table 4.3 at $n = 5$–6 (see also Fig. 6.1) was initially accounted for by simultaneous interaction of the blocking agent with two "receptor sites" (Barlow and Ing, 1948; Paton and Zaimis, 1949); the receptor sites were believed to normally bind two acetylcholine molecules and to be separated by a distance of 0.67–0.8 nm (Gill, 1959). This interpretation was favored by the anticipation that doubled ionic interaction would result in much stronger binding of a ligand to a receptor than other interactions, since the energy of ionic interaction is higher than the energy of any other ligand–protein interaction (see above).

Further analysis has revealed that the two "receptor sites" are not identical to one another. This is indicated by the higher blocking activities of nonsymmetrical bis-quaternary ammonium compounds of type **4.11** (Fakstorp et al., 1957) compared with those of the symmetrical ones:

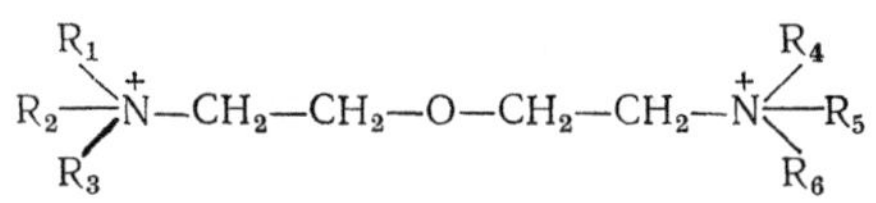

4.11

With $R_1, R_2 = CH_3$ at R_3–$R_6 = CH_2CH_3$, or with $R_1 = CH_3$ at R_2–$R_6 = CH_2CH_3$, the activity is higher than with R_1–$R_6 = CH_2CH_3$ and is equal to 1.4, 1.7, and 0.06 of hexamethonium activity, respectively. This conclusion has been confirmed in recent works with other compounds (Lukomskaya and Gmiro, 1982).

One of the nonsymmetrical compounds in this series, the bis-quaternary ether dicolinium (**4.12**), is 4 times more active than hexamethonium (see Kharkevich, 1967; Trčka, 1980a):

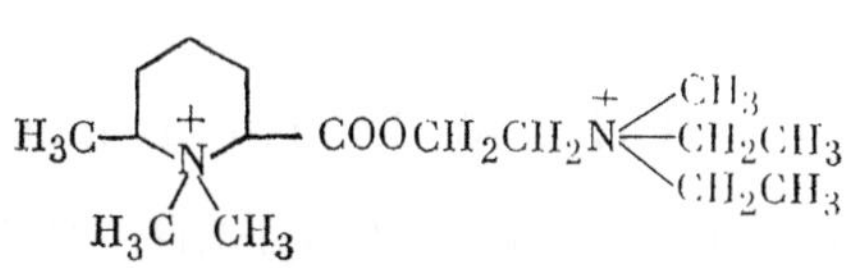

Dicolinium, **4.12**

The related compound, dimecolinium, which has two methyl groups attached to the nitrogen instead of the ethyl groups in dicolinium, is 6.4 times more active than hexamethonium (N. Ya. Lukomskaya, personal communication).

In each group of related nonsymmetrical bis-quaternary ammonium compounds, the most active compound usually possesses three methyl groups attached to one of its two quaternary nitrogens:

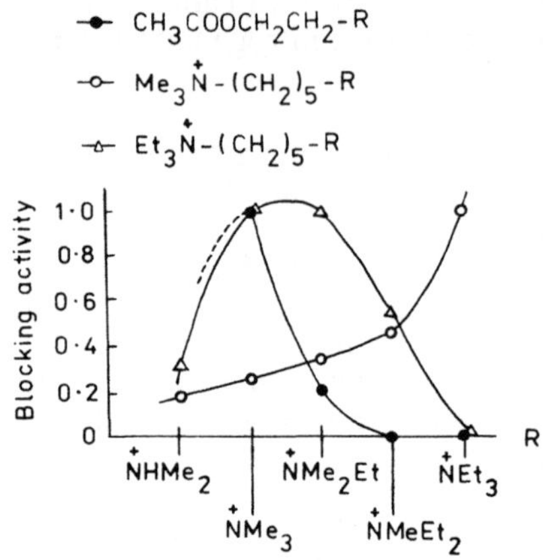

Fig. 4.1. Relationship between the structure of the cationic group in bis-cationic ganglion-blocking compounds and in acetylcholine, on the one hand, and their potencies, on the other hand. The ED_{50} for the most potent analog in each of three groups was taken as the corresponding potency unit. For further details see text (Lukomskaya and Gmiro, 1982).

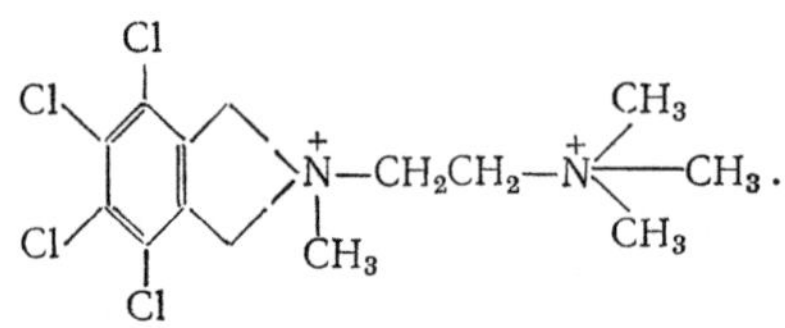

Chlorisondamine, **4.13**

Chlorisondamine (**4.13**) is 6.5 times more active as a ganglion-blocking agent than hexamethonium. Other isondoline derivatives with two ethyl groups substituted for two methyl groups of chlorisondamine are about 6 times less active than chlorisondamine (Plummer et al., 1956). More examples of structure–activity relationships that result from methyl substitutions in cationic groups are illustrated in Fig. 4.1. The activities of pentamethylene bis-trialkylammoniums $(CH_2CH_3)_3\overset{+}{N}–(CH_2)_5–R$ decrease as their cationic groups R are modified in the direction $–\overset{+}{N}(CH_3)_3$ [or $–\overset{+}{N}(CH_3)_2CH_2CH_3$] $\rightarrow \overset{+}{N}CH_3(CH_2CH_3)_2 \rightarrow$ $–\overset{+}{N}H(CH_3)_2 \rightarrow –\overset{+}{N}–(CH_2CH_3)_3$. It is evident that to maintain a high blocking activity all groups attached to one of two quaternary nitrogens should be methyls, similarly to what is needed for the high stimulant activity of agonists. Based on these observations, it was suggested that one of the two anionic sites that bind a bis-onium blocking compound belongs to a recognition center that normally binds acetylcholine (see Triggle and Triggle, 1976; Lukomskaya and Gmiro, 1982).

However, some other data obtained are not consistent with this suggestion. Loss of blocking activity in a series of bis-onium compounds that have one trimethylammonium group, resulting from successive substitution of methyls by ethyls in another trimethylammonium group, is not parallel to the loss of stimulant activity observed in acetylcholine with the equivalent substitution. Figure 4.1 shows that substitution of only one methyl in acetylcholine is followed by a fivefold decrease in its stimulant activity, whereas almost no change in blocking activity is observed following a similar substitution in a bis-onium compound. Furthermore, when two methyls are substituted, the stimulant activity disappears while only about a half-decrease in blocking activity is observed. Thus, the presence of only one methyl in the cationic group of a bis-onium compound ensures half of its blocking activity compared with that observed with a trimethylammonium group, which is not seen with acetylcholine. This result indicates that the anionic sites binding a bis-onium blocking compound are probably not common with that binding acetylcholine in a recognition center.

Whereas the hypothesis that one of the two anionic sites binding a bis-onium ganglion-blocking compound may normally bind acetylcholine remains debatable, it is agreed that the other anionic site binding a bis-onium ganglion-blocking compound essentially differs in structure from that which may bind acetylcholine. The evidence for this has been obtained using a series of bis-onium compounds possessing one trimethylammonium group and another group of variable structure separated by a variable distance (Lukomskaya and Gmiro, 1982), as demonstrated in Table 4.6.

The data shown in Table 4.6 indicate that maximum blocking activity is attained if one of the two cationic groups is not quaternary but tertiary. This is evident from a comparison of the activities within each of three pairs of compounds, 1–2, 4–5, and 8–9, and indicates that the hydrogen bond is important for binding a ligand to a peripheral anionic site. Compound 5 is the most active (almost 5 times more active than hexamethonium). This is obviously attributed not to the large size of the cationic group in compound 5, but rather to the presence of a hydrogen atom, since the bis-onium compounds that contain sulfur instead of nitrogen are similarly active to those containing nitrogen. On the other hand, the presence of only hydrogen bonds is not sufficient for high blocking activity. This follows from the fact that the activity increases more than 200-fold resulting from the substitution of one hydrogen in compound 7 by an alkyl in compound 6, a procedure that should strengthen hydrophobic interaction. The highest activity, 9.4 times higher than in hexamethonium, is achieved if two hydrogen atoms in the tertiary amino group are substituted by isopropyls (Table 4.7). Therefore, most effective is the structural component **4.10**, found in mono-onium tertiary and secondary amines, or half of the structural component **4.3**, found in quaternary mono-onium compounds, as described above.

The next point of interest is the distance between the two anionic sites binding a bis-onium ganglion-blocking compound. To examine the possible functional role of this distance, a series of bis-onium compounds with variable intercharged distances and with constant total length of the aminoalkyl chain shown in Table 4.7 has been synthesized. The most active compounds are those with 4 or 5 methyl groups between their charged nitrogen atoms. A decrease in this number to 3 groups causes about a tenfold loss in blocking activity (compounds 6 and 7), making a bis-onium compound as active as a mono-onium one with the same length of the aliphatic chain (compound 8).

TABLE 4.6. Ganglion-Blocking Activity of Pentamethonium Series Containing Tertiary or Secondary Amino Group (Lukomskaya and Gmiro, 1982)

	$Me_3\overset{+}{N}-(CH_2)_5-R$	Relative potency (hexamethonium = 1)
(1)	$-\overset{+}{N}(Me)(Me)(Pr-i)$	3.9
(2)	$-\overset{+}{N}H(Me)(Pr-i)$	8.0
(3)	$-\overset{+}{N}H_2-Pr-i$	7.7
(4)	$-\overset{+}{N}(Me)(Pr-i)(Pr-i)$	5.0
(5)	$-\overset{+}{N}H(Pr-i)(Pr-i)$	9.4
(6)	$-\overset{+}{N}H_2-Pr-i$	7.7
(7)	$-\overset{+}{N}H_3$	0.03
(8)	$-\overset{+}{N}H(Me)(Pr-i)$	8.0
(9)	$-\overset{+}{S}(Me)(Pr-i)$	4.1

Based on these results, the hypothetical scheme of binding sites for biscationic ganglion-blocking compounds shown in Table 4.8 has been suggested. The distance between the two anionic sites in this scheme corresponds to the length of the tetramethylene chain. There is a hydrophobic area adjacent to the second (peripheral) anionic site, which begins at a distance equivalent to 7 intercarbon distances away from the main anionic site or 2 intercarbon distances from the second anionic site and extends for a distance equivalent to a chain of three methylene groups.

TABLE 4.7. Effect of Chain Length on the Ganglion-Blocking Activity of Bis-Cations (Lukomskaya and Gmiro, 1982) (for simplicity, hydrogens at C-atoms are omitted in structural formulas)

		Relative potency (hexamethonium = 1)
(1)	$Me_3\overset{+}{N}$—C—C—C—C—C—$\overset{+}{N}H_2$—C$<^{C}_{C}$	7.7
(2)	$Me_3\overset{+}{N}$—C—C—C—C—$\overset{+}{N}H_2$—C—C$<^{C}_{C}$	7.7
(3)	$Me_3\overset{+}{N}$—C—C—C—$\overset{+}{N}H_2$—C—C—C$<^{C}_{C}$	0.3
(4)	$Me_3\overset{+}{N}$—C—C—C—C—C—C—$\overset{+}{N}H_2$—C—C	2.3
(5)	$Me_3\overset{+}{N}$—C—C—C—C—C—$\overset{+}{N}H_2$—C—C—C	3.9
(6)	$Me_3\overset{+}{N}$—C—C—C—C—$\overset{+}{N}H_2$—C—C—C—C	4.2
(7)	$Me_3\overset{+}{N}$—C—C—C—$\overset{+}{N}H_2$—C—C—C—C—C	0.38
(8)	$Me_3\overset{+}{N}$—C—C—C—C—C—C—C—C—C	0.53
(9)	$Me_3\overset{+}{N}$—C—C—C—C—C—$\overset{+}{N}H_2$—C	0.2
(10)	$Me_3\overset{+}{N}$—C—C—C—C—C—C—C	stimulant

The results shown in Table 4.8 provide further support for this scheme. In tetramethonium (compound 1) methyls of the second cationic group do not fit the hydrophobic area, and therefore compound 1 is of low activity notwithstanding the suitable position of positively charged nitrogen. The activity increases 140 times as the methyls start to fit the hydrophobic area (compound 2), and a further increase in activity is achieved by a fit of the positively charged nitrogen to the second anionic

TABLE 4.8. Hypothetical Scheme of Binding Site for Bis-Cationic Ganglion-Blocking Drugs (Lukomskaya and Gmiro, 1982, modified)

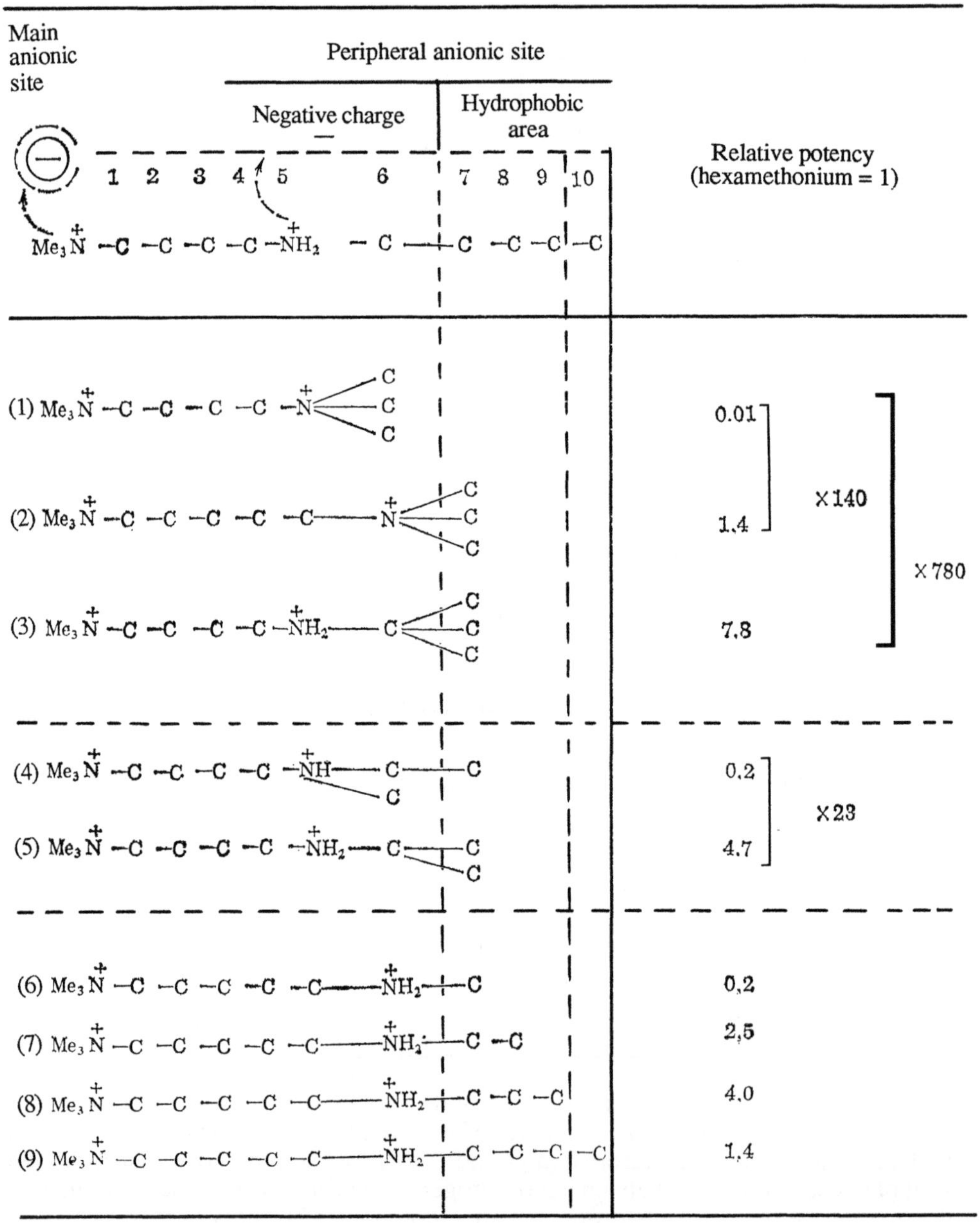

site (compound 3). A decrease in the number of methyls that reach the hydrophobic area (compounds 4 and 5), as well as the lengthening of the aliphatic chain that interacts with the hydrophobic area to more than three carbons (compounds 6–9), results in a loss of activity. In contrast to the hydrophobic area attached to the first anionic site and extending no more than one intercarbon distance away from the negative charge, the hydrophobic area attached to the second anionic site begins two intercarbon distances away from the negative charge. Although the latter area extends farther than that related to the first anionic site, its most functionally important part is not longer than one intercarbon distance, i.e., is actually similar to that related to the first anionic site (Table 4.8). The same configuration is apparently typical, in general, of the sites binding mono-onium blockers, as it follows from the structures of their most functionally important components **4.3** and **4.10**.

Of additional interest is the fact that the two most probable chemical structures for the anionic sites, aspartic and glutamic residues, differ by one intercarbon distance. This allows one to speculate that the most simple structure for a binding site which corresponds to the scheme in Table 4.8 is that composed of aspartic and glutamic residues (which differ by exactly one intercarbon distance) separated by a distance equivalent to five intercarbon distances, i.e., about 0.7 nm (see Bernhard, 1968). Additional support for this suggestion will be given later.

From an examination of structure–activity relationships, the principal structure for the ganglion-blocking agents ensuring their high blocking activities can be summarized as follows. In mono-onium compounds, it is a positively charged nitrogen atom incorporated into a carbon skeleton branching at the α-carbon. It has been assumed that these mono-onium compounds, as well as one of the two cationic groups in bis-onium compounds, may bind to a recognition center (Triggle and Triggle, 1976). This would predict the high blocking activity in such bis-onium compounds that possess at least one cationic group of the above principal structure. However, this is not what has been observed. The highest blocking activity has been found in bis-onium compounds that have one trimethylammonium group, i.e., the group similar to that in acetylcholine, with another cationic group differing from the first one. Insertion, instead of methyls, of one or two isopropyls, which are known to increase the blocking potency in mono-onium compounds, results in the opposite effect. For example, compounds **4.14** and **4.15**

$$\text{i-Pr(CH}_3)_2\overset{+}{\text{N}}\text{-(CH}_2)_5\text{–}\overset{+}{\text{N}}\text{HPr}_2\text{-i}$$

4.14

$$\text{i-Pr}_2\text{CH}_3\overset{+}{\text{N}}\text{–(CH}_2)_5\text{–}\overset{+}{\text{N}}\text{CH}_3\text{Pr}_2\text{-i}$$

4.15

are only 0.8 and 0.7 as active as hexamethonium (Lukomskaya and Gmiro, 1982).

This discrepancy makes doubtful the above suggestion that the mono-onium compounds of principal structures **4.3**, **4.9**, and **4.10** may bind to the anionic site that belongs to a recognition center. The alternative explanation is that these compounds may bind only to a second (peripheral) anionic site, most probably nonrelated to a recognition center, i.e., that they all are actually noncompetitive ganglion-blocking agents.

As has been mentioned above, only trimetaphan and some neurotoxins are purely competitive ganglion-blocking agents. Another compound with a possible competitive component in its blocking effect in rabbit superior cervical ganglion is tubocurarine. However, as will be shown below, in rat sympathetic ganglion tubocurarine exhibits only a noncompetitive effect.

It is apparent from a comparison of the structures of trimetaphan (**2.19**) and tubocurarine (**2.20**) with those of the mono- or bis-onium compounds that there is a structural component common to trimetaphan and tubocurarine and absent in the other compounds. This component is an oxygen located five intercarbon distances away from the positive charge, i.e., at a distance exactly similar to that for the carbonyl oxygen in acetylcholine. This structural component is probably complementary to the most functionally important points contained in the scheme of a recognition center discussed in Chapter 2. A similar component is possessed by bis-quaternary ether **4.12** and by a tropan derivative **4.2**, although in the latter the distance between an oxygen and positive charge is somewhat smaller. Both compounds **4.12** and **4.2** are very potent ganglion-blocking agents, although their blocking mechanisms are not yet clear.

One can thus suggest that compounds lacking either the trimethylammonium group or an oxygen equivalent to the carbonyl oxygen of acetylcholine are not able to interact with a recognition center. This means that the ganglion-blocking compounds of principal structures **4.3**, **4.9**, and **4.10** should be exclusively noncompetitive in their blocking mechanisms.

This suggestion is supported by the finding that compounds containing components **4.3**, **4.9**, or **4.10** as well as some bis-quaternary ammonium compounds bind more deeply than acetylcholine in the transmembrane electric field (see Chapter 7). Binding of bis-onium compounds to a site different from the recognition center will answer the puzzling question why hexamethonium possessing a cationic group similar to that in acetylcholine is completely devoid of any excitatory activity on the nicotinic acetylcholine receptor.

As mentioned above, there is one more peak in the blocking activity of bis-quaternary ammonium compounds, when the number of methylene groups $n = 15$–18, in addition to that observed when $n = 5$–6 (Table 4.3 and Fig. 6.1). This additional peak is illustrated in Table 4.9 with a more gradual change of n values. One can see from Table 4.9 that this peak is observed regardless of whether the two cationic groups are trimethylammonium or triethylammonium groups. This implies that compounds with $n = 15$–18 may interact with the second (peripheral) anionic site mentioned above and with an additional, third peripheral anionic site which, in its structure, is similar to the second anionic site rather than to the first (main) one normally interacting with acetylcholine (Triggle and Triggle, 1976). This suggestion has not yet been verified.

TABLE 4.9. Ganglion-Blocking Activity of Polymethylene-bis-onium Salts (Barlow and Zoller, 1964)*

$Me_3\overset{+}{N}(CH_2)_n\overset{+}{N}Me_3$	Equipotent molar activity	$Et_3\overset{+}{N}(CH_2)_n\overset{+}{N}Et_3$	Equipotent molar activity
n =10	11.5	*n* =10	20.0
11	22.0	11	20.0
12	18.5	12	10.0
13	11.5	13	4.5
14	6.1	14	2.3
15	1.9	15	1.5
16	1.3	16	1.0
17	1.3	17	0.8
18	1.2	18	2.5
19	2.7	19	10.0
20	5.7	20	18.0
21	14.5	21	80.0
(6	5.0)		

*Cat superior cervical ganglion. Activities relative to $Et_3\overset{+}{N}(CH_2)_{16}\overset{+}{N}Et_3 = 1$.

4.2. CLASSIFICATION OF GANGLION-BLOCKING AGENTS ACCORDING TO CLASSICAL CHARACTERISTICS OF THEIR BLOCKING EFFECTS

The ganglion-blocking agents were initially classified as 1) depolarizing, 2) competitive, 3) noncompetitive, and 4) mixed (competitive + noncompetitive) (Paton and Perry, 1953; Van Rossum, 1962a, b). This classification was based on measurements of the responses produced in autonomic ganglia or in their target organs by direct application of agonists and antagonists, e.g., the agonist-induced contraction of gut muscle. Depolarizing agents first activate nicotinic acetylcholine receptors (see Chapter 2) and then cause their blockade. For this reason, the compounds classified as depolarizing blocking agents are those already described in Chapter 2 as agonists: nicotine (**2.3**) and DMPP (**2.2**). The effect of nicotine and

DMPP on autonomic ganglia is somewhat similar to that produced by decamethonium in the end-plate, which has been studied in more detail.

Decamethonium first opens the ionic channels in nicotinic acetylcholine receptors of the end-plate and then penetrates the open channels (Case et al., 1977; Creese et al., 1977), causing their blockade similarly to other organic cations (Adams and Sakmann, 1978; Adams et al., 1981). However, in contrast to multisite binding of decamethonium to the nicotinic receptor (Auerbach et al., 1983), binding of nicotine is thought to be one-sited (Eldefrawi et al., 1971). In parasympathetic ganglion, decamethonium blocks open channels of nicotinic acetylcholine receptors without their preliminary opening (Ascher et al., 1979).

Therefore, one of the possible mechanisms underlying the blocking effect of nicotine and DMPP is the open-channel blockade. An alternative possibility is desensitization, which basically differs from the open-channel blockade (Fiekers et al., 1980; Sakmann et al., 1980; Magleby and Pallotta, 1981).

Competitive and noncompetitive ganglion-blocking agents block synaptic transmission through the ganglia without causing a preliminary excitatory effect. Competitive blocking agents are thought to bind to a recognition center, thus preventing interaction between the recognition center and agonist. The most simple classical test to determine whether a blocking agent is competitive is to examine its effect on the dose–response relationship: increased concentrations of a competitive blocker would result in a parallel shift of the dose–response curve, whereas noncompetitive blockers change its slope and shape instead. According to this test, TEA (**2.6**), the bis-quaternary ammonium compounds hexa- and pentamethonium (**2.18**, $n = 6$ and $n = 5$, respectively), and also azamethonium (**4.16**) and trimetaphan (**2.19**), were classified as competitive ganglion-blocking agents:

$$(CH_3)_3\overset{+}{N} - CH_2CH_2 - \underset{\displaystyle CH_3}{\underset{|}{N}} - CH_2CH_2\overset{+}{N}(CH_3)_3$$

Azamethonium, **4.16**

Of these compounds, only hexa-, pentamethonium, and trimetaphan have been studied in relation to the effects they produce in single nerve cells of autonomic ganglia. The results obtained indicate that penta- and hexamethonium block the ionic channels opened by the agonist, and only trimetaphan appears to be a pure competitive blocker (Ascher et al., 1979; Selyanko et al., 1981, 1982; Skok et al., 1983, 1984; Rang, 1982; Gurney and Rang, 1984; Bobrishev and Skok, 1987). The mechanisms of these effects are discussed in more detail in Chapter 5. The effects produced by TEA have been examined at the cellular level only in the end-plate. Both competitive and open-channel blocking effects have been observed (Adler et al., 1979).

The third group, noncompetitive blockers, is represented by chlorisondamine (**4.13**), pentacyne (**4.17**), trimethidine (**4.18**), and some dioxolane derivatives (Triggle and Triggle, 1976).

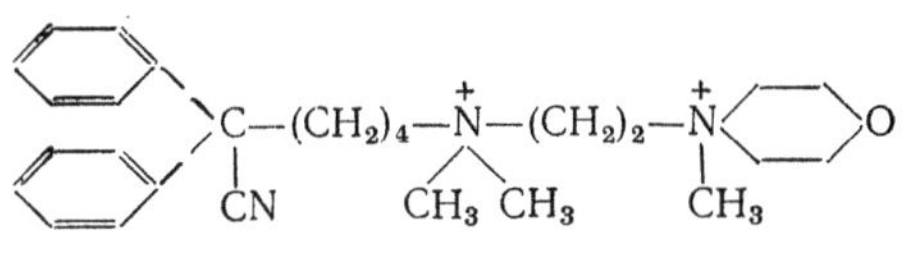

Pentacyne, **4.17**

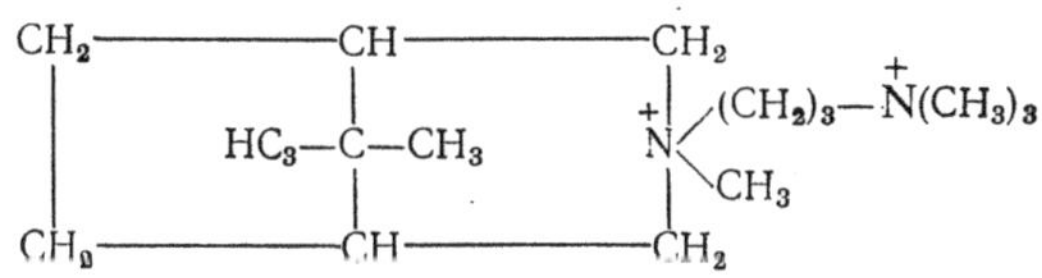

Trimethidine, **4.18**

The fourth group, mixed blockers, includes mecamylamine (**4.4**) and pempidine (**4.7**).

The effects of chlorisondamine (Lingle, 1983a), mecamylamine (Ascher et al., 1979; Lingle, 1983b; Varanda et al., 1985), pempidine (Lingle, 1983b), and IEM-1288, a compound which differs from pempidine (4.7) as a bromide instead of a bitartrate (Selyanko, Derkach, and Skok, unpublished), have been studied in single neurons of sympathetic and parasympathetic ganglia and also in single fibers of the lobster gastric mill, which have nicotinic receptors similar in their phamacology to those in autonomic ganglia. All these compounds have been found to produce a strong open-channel blocking effect, i.e., they should be considered as noncompetitive blockers (see Chapter 5).

4.3. PHARMACOLOGICAL SPECIFICITY OF NEURONAL NICOTINIC ACETYLCHOLINE RECEPTORS

The pharmacological specificity of receptors, i.e., their pharmacological features in which certain types of receptors differ from all other types, is of great interest in physiology and medicine, as it underlies the selectivity in drug-induced blocking effects on synaptic transmission.

An examination of receptors for their pharmacological specificity implies that the effects produced by the same drug on different objects should be compared. Therefore, the results obtained are usually contaminated by differences in experimental procedures. For example, a comparison of blocking activities of a drug in autonomic ganglia and in skeletal muscle requires that such basically different responses as blood pressure changes (or contraction of the gut muscle) and contraction of skeletal muscle, evoked either by nerve-released or exogenous acetylcholine, should be compared. In addition, the possibility that neuronal muscarinic receptors are involved in the responses cannot always be excluded. Because of these difficulties, knowledge of the pharmacological specificity of neuronal nicotinic receptors in different objects remains scarce despite the great number of studies performed.

4.3.1. Neurons of Autonomic Ganglia

The available data allow one to compare nicotinic acetylcholine receptors in autonomic ganglion neurons, with respect to their pharmacology, with those in skeletal muscle fibers and in parasympathetic ganglion neurons. In early experiments, the ganglion-blocking compound of the simplest structure, TEA, produced in skeletal muscle an excitatory effect, which was primarily considered as evidence for the substantial difference between the two types of nicotinic acetylcholine receptors (Barlow, 1955). However, subsequent studies have revealed that TEA, in addition to its direct effect on nicotinic acetylcholine receptors, increases the release of transmitter from presynaptic terminals, and the excitability of the postsynaptic membrane, by a selective blockade of potassium channels. The latter effect is characterized by a K_D value of about 0.4 mM (Hille, 1970), similar to that found in the TEA blocking effect on synaptic transmission through sympathetic ganglion (Riker, 1965). For this reason, the effect of TEA on nicotinic acetylcholine receptors should be studied separately from its effect on potassium channels. Using this approach, it has been found that TEA produces both competitive and open-channel blocking effects in the end-plate acetylcholine receptors (Adler et al., 1979). No results obtained with this approach from neuronal acetylcholine receptors have been reported as yet.

The relative potencies of some other compounds listed above, in their blocking effects on synaptic transmission through sympathetic ganglion and through neuromuscular junction, are compared in Table 4.10 (the tubocurarine potency is taken as 1). The potencies were estimated according to a drug-induced reduction in the contraction of cat nictitating membrane and cat soleus muscle. The potencies of each compound in the ganglion and in the muscle can be compared based on the observation that tubocurarine is, on the average, 4 times more potent in the muscle than in the ganglion (Guyton and Reeder, 1950).

It is apparent from Table 4.10 that the activities of different compounds are much more variable in the muscle than in the ganglion. Furthermore, only two compounds, mecamylamine (**4.4**) and hexamethonium, are selective blockers in the ganglion, whereas at least four are selective blockers in the muscle (in addition to tubocurarine): succinylcholine (**4.19**), decamethonium, pancuronium (**4.20**), and dimethyl-d-tubocurarine (**4.21**).

It is of interest that each selective neuromuscular blocking agent shown in Table 4.10 possesses two identical cationic groups that are either similar or very close in their structure to that of acetylcholine and are separated by a distance equivalent to 9–11 intercarbon distances (about 1.4 nm in decamethonium).

TABLE 4.10. Comparative Ganglion-Blocking and Neuromuscular Blocking Activities

No.	Compound	Relative neuromuscular blocking activity[a]	Relative ganglion-blocking activity[b]
1	Tubocurarine	1	1
2	Dimethyl-d-tubocurarine	4	0.2
3	Gallamine	0.2 (0.44)	0.01 (0.04)
4	Dihydro-β-erythroidine	0.5	0.33
5	Benzoquinonium	1 (2.4)	0.11 (0.12)
6	Decamethonium	8	0.08
7	Succinylcholine	2	0.01
8	Hexamethonium	0.003	0.19
9	Pancuronium	8.15	0.43
10	Mecamylamine	0.008	3.31

Activities of compounds 1–7 and 7–10 (and data in parentheses) are taken from Grob (1967) and Bowman and Webb (1972), respectively.
[a]Twitch depression, cat soleus muscle.
[b]Nictitating membrane contraction depression.

$$(CH_3)_3\overset{+}{N}CH_2CH_2OOC(CH_2)_2COOCH_2CH_2\overset{+}{N}(CH_3)_3$$

Succinylcholine, **4.19**

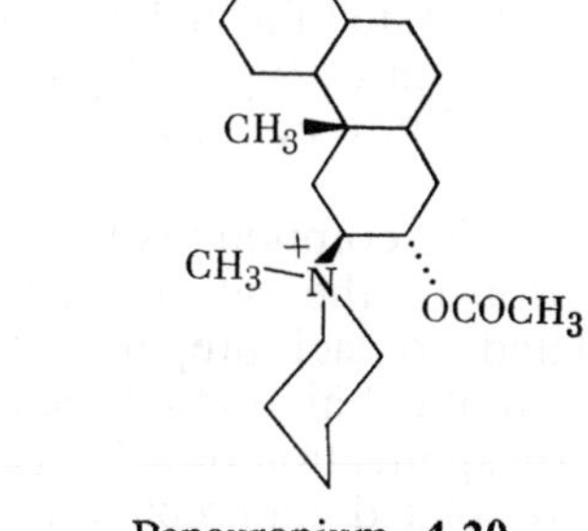

Pancuronium, **4.20**

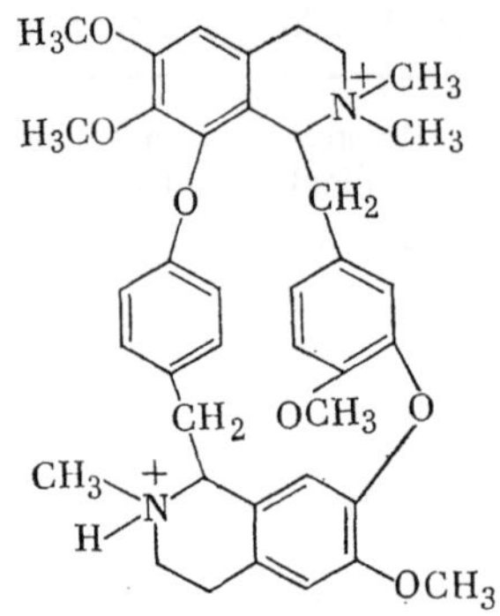

Dimethyl-d-tubocurarine, **4.21**

Succinylcholine, decamethonium, and pancuronium are known as depolarizing neuromuscular blocking agents. Recent results suggest that tubocurarine may activate nicotinic acetylcholine receptors as well (Trautmann, 1982). In addition, all neuromuscular blocking agents possess molecules of large size compared to those in ganglion-blocking agents. These properties suggest that selective neuromuscular blocking agents interact with a recognition center located at the surface of the acetylcholine receptor molecule, outside the anionic channel, and that there are two anionic sites, similar in their structure, about 1.4 nm apart, which are complementary to the trimethylammonium or dimethylammonium cationic group. By contrast, the selective ganglion-blocking agents (except for a few competitive ones) probably interact with a site different from a recognition center, which is localized deeply in the ionic channel and is composed of two structurally different anionic sites about 0.7 nm apart.

The next point of interest is to compare the activity of ganglion-blocking agents in sympathetic and in parasympathetic ganglia. It was revealed in early studies that parasympathetic ganglia in the head and heart of the cat are about 1.5 times less sensitive than cat skeletal muscles to the blocking effect produced by tubocurarine (see Kharkevich, 1967). Since this difference is smaller than the difference between sympathetic ganglia and skeletal muscles (4 times as small: Guyton and Reeder, 1950), one can conclude that tubocurarine is somewhat less potent in sympathetic than in parasympathetic ganglia.

More recently, the activities of many compounds of varying chemical structure as blockers of synaptic transmission through sympathetic and parasympathetic ganglia have been compared using three types of preparations: in situ, in isolated organs, and in isolated ganglion (Gmiro et al., 1984, 1987).

The blocking activities of the compounds tested have been estimated from their effects on the contraction of the rat stomach muscle elicited by vagal stimulation (in situ or in isolated stomach preparation), on the rat blood pressure, on the nerve-evoked contraction of cat nictitating membrane, on the contraction of guinea-pig gut evoked either by application of suberyldicholine or by mesenterical nerve stimulation (in situ or in isolated gut preparation), and on the contraction of

TABLE 4.11. Relative Activity of Ganglion-Blocking Agents Obtained in Experiments on Anesthetized Animals; Hexamethonium = 1 (Gmiro et al., 1987)

Compound	Parasympathetic ganglion		Sympathetic ganglion		
	Guinea-pig ileum (subecholine)	Rat stomach (electric stimulation of vagus)	Hypotensive action in doses equipotent to 3 μM/kg hexamethonium (rat)	Inhibition of nictitating membrane contraction (electric stimulation of sympathetic nerve)	Selectivity index (2 to 3* ratio)
	1	2	3	4	5
Hexamethonium	1	1	1	1	
ED_{50}	2.2 ± 0.32 μM/kg (n = 36)	4.4 ± 0.61 μM/kg (n = 33)		0.035 ± 0.001 μM/kg (n = 120)	
IEM-1460	10.2 ± 1.8 (n = 5)	12.0 ± 1.5 (n = 5)	1.0 ± 0.19 (n = 4)	0.3 ± 0.04 (n = 3)	12.0
IEM-1372	5.1 ± 0.76 (n = 6)	7.8 ± 0.9 (n = 5)	0.37 ± 0.06 (n = 5)	0.04 ± 0.003 (n = 3)	21.0
IEM-1194	24.2 ± 3.4 (n = 7)	23.5 ± 2.3 (n = 6)	0.34 ± 0.04 (n = 6)	0.2 ± 0.07 (n = 4)	69.0
IEM-1559	41.6 ± 5.3 (n = 7)	32.7 ± 3.5 (n = 6)	0.59 ± 0.09 (n = 5)	0.13 ± 0.03 (n = 3)	55.4

*Assuming ED_{50}s for hexamethonium in cases 2 and 3 are equal to each other (in case 3, ED_{50} for hexamethonium was not determined).

TABLE 4.12. Relative Activity of Ganglion-Blocking Agents Obtained in Experiments on Isolated Organs; Hexamethonium = 1 (Gmiro et al., 1987)

Compound	Parasympathetic ganglion			Sympathetic ganglion	
	Guinea-pig ileum (subecholine)	Guinea-pig ileum (electric stimulation of mesentery)	Rat stomach (electric stimulation of vagus)	Guinea-pig vas deferens (electric stimulation of hypogastric nerve)	Selectivity index (2 to 4* ratio)
	1	2	3	4	5
Hexamethonium	1	1	1	1	
ED_{50}	4.5 ± 0.52 $\times 10^{-6}$ M ($n = 42$)	2.9 ± 0.36 $\times 10^{-5}$ M ($n = 33$)	1.2 ± 0.17 $\times 10^{-5}$ M ($n = 41$)	1.2 ± 0.19 $\times 10^{-5}$ M ($n = 35$)	
IEM-1460	19.8 ± 2.2 ($n = 6$)	14.5 ± 2.3 ($n = 5$)	14.8 ± 2.6 ($n = 5$)	0.9 ± 0.18 ($n = 5$)	6.4
IEM-1372	19.7 ± 2.6 ($n = 6$)	10.8 ± 1.6 ($n = 5$)	9.3 ± 1.2 ($n = 6$)	1.2 ± 0.17 ($n = 5$)	3.6
IEM-1194	70.2 ± 7.8 ($n = 8$)	48.7 ± 5.7 ($n = 6$)	31.2 ± 4.6 ($n = 7$)	1.1 ± 0.13 ($n = 5$)	17.7
IEM-1559	150.3 ± 22.9 ($n = 7$)	91.4 ± 9.7 ($n = 6$)	59.9 ± 7.8 ($n = 6$)	0.9 ± 0.11 ($n = 5$)	41.0

*Corrected for the difference in EC_{50}s for hexamethonium in these particular test preparations ($1.2 \cdot 10^{-5}/2.9 \cdot 10^{-5} \approx 0.4$). Corresponding activity ratios are therefore multiplied by 0.4.

TABLE 4.13. Activities of Ganglion-Blocking Agents at Sympathetic and Parasympathetic Ganglia (L. V. Melnichenko, unpublished data). EC_{50} Values are Indicated (M ± m; *n* is the number of observations) for Depression of Postganglionic Compound Action Potential in Cat Superior Cervical and Ciliary Ganglia (Gmiro et al., 1987)

No.		Hexa-methonium	IEM-1316 (4.25)	IEM-1372 (4.23)	IEM-1194 (4.22)	IEM-1559 (4.26)	IEM-1460 (4.27)	Tubocurarine
1	Superior cervical ganglion	(11.6 ± 1.7) × 10^{-5} M $n = 10$	(4.6 ± 0.3) × 10^{-5} M $n = 10$	(5.3 ± 0.25) × 10^{-5} M $n = 4$	(3.2 ± 0.2) × 10^{-5} M $n = 3$	(1.3 ± 0.6) × 10^{-5} M $n = 3$	(3.43) ± 0.8) × 10^{-5} M $n = 4$	(0.44 ± 0.05) × 10^{-5} M $n = 3$
2	Ciliary ganglion	(38.5 ± 2.2) × 10^{-5} M $n = 4$	(8.93 ± 0.83) × 10^{-5} M $n = 5$	(3.6 ± 0.4) × 10^{-5} M $n = 4$	(1.15 ± 0.34) × 10^{-5} M $n = 3$	(0.39 ± 0.03) × 10^{-5} M $n = 3$	(7.85 ± 2.3) × 10^{-5} M $n = 3$	(0.33 ± 0.02) × 10^{-5} M $n = 3$
3	Selectivity index (1 to 2 ratio)	0.3	0.5	1.5	2.8	3.3	0.4	1.0

the guinea pig isolated vas deferens elicited by hypogastric nerve stimulation. In the experiments on isolated ganglia, the effects of blocking agents on post-ganglionic action potentials evoked in cat superior cervical and ciliary ganglia by a single stimulus applied to their preganglionic nerve fibers have been examined. To estimate the sympathetic to parasympathetic blocking selectivity of the compounds tested, the ratios of their EC_{50} or ED_{50} obtained in sympathetic ganglia to those obtained in parasympathetic ganglia (the "selectivity index") have been estimated.

The blocking activities found in these experiments are shown in Tables 4.11–4.13 with respect to that of hexamethonium. The highest blocking activity of bis-cationic compounds in sympathetic ganglia was observed for hexamethonium (Table 4.13). In parasympathetic ganglia, the highest blocking activity was found in IEM-1316, a pentamethonium derivative (**4.22**), which is 64 times more active than hexamethonium in blocking the guinea-pig gut responses to nerve stimulation.

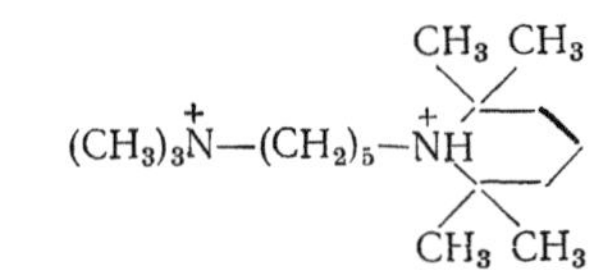

IEM-1316, **4.22**

A difference in the same direction, although much smaller (4.3 times), was observed when both substances were applied to another parasympathetic ganglion, cat ciliary ganglion (see Table 4.13). High parasympathetic blocking activity has also been found in a related compound, IEM-1460 (**4.23**), possessing an adamantane group (Ad):

$$(CH_3)_3\overset{+}{N}{-}(CH_2)_5{-}\overset{+}{N}H_2CH_2Ad$$

IEM-1460, **4.23**

The differences in activity between IEM-1460 and hexamethonium were 10.2–12.0 times, 14.5–19.8 times, and 4.8 times, when measured on the gut and stomach muscles in situ, on the isolated gut, and on the isolated ciliary ganglion, respectively (Tables 4.11–4.13).

Like hexamethonium, both compounds are more active in sympathetic than in parasympathetic isolated ganglia (the difference was 1.9 times in IEM-1316, 2.3 times in IEM-1460, and 3.3 times in hexamethonium; see Table 4.13). However, applied to isolated organs or in situ, IEM-1460 appears to be more active in parasympathetic than in sympathetic ganglia (Tables 4.11 and 4.12). Mono-onium compounds exhibit somewhat lower activity than that in bis-cationic compounds when tested in sympathetic ganglia in situ, and almost the same activity as in bis-cationic compounds when tested in isolated sympathetic ganglia. But, by contrast, bis-cationic compounds show a clear preference for blocking parasympathetic ganglia relative to sympathetic ganglia. A piperidine derivative possessing a decyl group, IEM-1559 (**4.24**), was 32.7–41.6 and 91.4–150.3 times more potent than hexamethonium in parasympathetic ganglia of the gastrointestinal tract when tested in situ and on the isolated organs, respectively. A somewhat lower parasympathetic blocking activity, although still much higher than in hexamethonium, was observed in decyldiisopropylamine (IEM-1194, for structure see **4.25**) and in decylisopropylamine (IEM-1372, for structure see **4.24**), as demonstrated by Tables 4.11 and 4.12.

$$C_{10}H_{21}\text{-}\overset{+}{N}H_2CH(CH_3)_2$$

IEM-1372, **4.24**

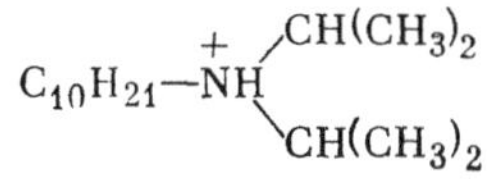

IEM-1194, **4.25**

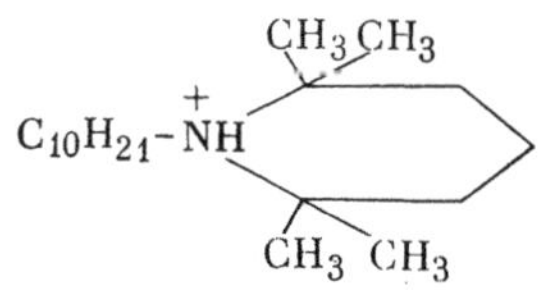

IEM-1559, **4.26**

Compounds **4.26**, **4.25**, and **4.24** are 3.3, 2.8, and 1.5 times, respectively, more active in isolated parasympathetic ganglion than in isolated sympathetic ganglion.

In further experiments, the contribution of the aliphatic group to the blocking activities of mono-onium compounds was investigated. The activities of triethylammonium derivatives **4.27** in their blocking effects on suberyldicholine-induced contraction of gut muscle increased with the increased length of the aliphatic group. This is illustrated by the following series:

$$(C_2H_5)_3\overset{+}{N}\text{–}R$$

4.27

R:	C_4H_9	C_5H_{11}	C_6H_{13}	C_7H_{15}	C_8H_{17}	C_9H_{19}	$C_{10}H_{21}$	$C_{10}H_{20}\text{–}\overset{+}{N}(C_2H_5)_3$
Activities:	0.05	0.17	0.54	2.1	2.5	8.2	11.7	1.0

The increased activity can be related either to binding of the aliphatic group to the hydrophobic area in the receptor molecule or to incorporation of this group into the lipid membrane (see Changeux et al., 1984). The introduction of a second cationic group in triethyldecylammonium markedly reduces its blocking activity, probably because of reduced hydrophobicity of the blocking molecule.

One can see from Table 4.13 that tubocurarine, although it is more active than hexamethonium, actually does not differentiate between isolated sympathetic and parasympathetic ganglia.

Furthermore, it has been shown using the same approach demonstrated in Table 4.7 that the most active bis-cationic compounds in intestinal parasympathetic ganglia are those with their intercharge distance equivalent to a chain of 4–5 methylene groups and whose propylic CH_3 in one cationic group is located 2 intercarbon distances away from the nitrogen atom (S. Serdyuk, personal communication). This implies that the two anionic sites binding such compounds should be located about 0.7–0.8 nm apart from each other, while the hydrophobic area should begin less than two intercarbon intervals away from one of the anionic sites. These distances are exactly similar to those found in the site that binds blocking agents in the acetylcholine receptor of sympathetic ganglion neuron (see Table 4.8).

One can thus summarize the characteristic structural differences between the blocking agents selective for sympathetic and parasympathetic ganglia and the corresponding differences in their binding sites in acetylcholine receptors:

1) The bis-cationic compounds with two trimethylammonium groups (e.g., hexamethonium) are less active in parasympathetic than in sympathetic ganglia. A somewhat lesser difference, although in the same direction, is observed if only one cationic group is trimethylammonium.

2) Most active in parasympathetic ganglia relative to hexamethonium, on the one hand, and relative to their activity in sympathetic ganglia, on the other hand, are aliphatic monocationic compounds whose cationic group is similar to structures **4.3**, **4.9**, or **4.10**, e.g., IEM-1372, IEM-1194, and IEM-1559. This implies that there should be a hydrophobic area attached to the only negative charge in the blocker-binding site of the parasympathetic neuronal acetylcholine receptor, in contrast to that of the sympathetic neuronal acetylcholine receptors, which should have two negative charges complementary to trimethylammonium groups, while the above hydrophobic area is probably less important.

4.3.2. Molluscan Neurons

As has been mentioned in Chapter 1, several types of responses can be observed in molluscan neurons treated with bath-applied acetylcholine. Primarily, two types of responses, depolarization (excitation) and hyperpolarization (inhibition), were found in *Helix* and in *Aplysia*. The excitatory response was due to an increase in sodium or chloride permeability. The inhibitory response was marked by an increase in chloride or potassium permeability. Similar responses were observed in molluscan neurons as evoked by presynaptic stimulation (Tauc and Gerschenfeld, 1960; see Gerschenfeld, 1973). The chloride equilibrium potential in these neurons is close to the resting membrane potential level. For this reason, some neurons reveal chloride-dependent excitation, whereas in others chloride-dependent inhibition occurs. Not in all molluscs can all four types of responses be observed. For example, only chloride-dependent depolarization and potassium-dependent hyperpolarization have been found in *Planorbarius* neurons (Ger et al., 1979a). The sodium- and chloride-dependent responses arise with short latency and can be blocked by tubocurarine. By contrast, the potassium-dependent response has a much longer latency and can be evoked by the muscarinic agonist arecoline (Kehoe, 1972). Therefore, only the first two types of responses can be considered as being related to nicotinic acetylcholine receptors.

Two types of tubocurarine-sensitive receptors also differ in their other pharmacological properties: in *Aplysia* an increase in sodium conductance is much more sensitive to hexamethonium than to decamethonium (Ascher et al., 1978b), while an increase in chloride conductance is not sensitive to the blocking effect of hexamethonium and can be elicited by decamethonium (Kehoe, 1972; Kehoe et al., 1976). Atropine selectively blocks sodium but not chloride conductance increase, although in much higher concentrations (10^{-3}–10^{-4} M) than that needed to block muscarinic responses (Kehoe, 1972).

Two types of nicotinic receptors also differ in their sensitivities to enzymes and to some compounds that modify the receptor protein. The effects produced by these compounds indicate that aspartic or glutamic residues in the receptor protein are important to evoke an increase in sodium conductance and that C-terminal lysine is probably responsible for an increase in chloride conductance (Sato et al., 1983). All the reversible blocking agents listed above interact with open ionic channels of acetylcholine receptors (Marty et al., 1976; Ascher et al., 1978a, b).

It is interesting that TEA, in contrast to tubocurarine, blocks only cation-selective acetylcholine receptor channels in *Aplysia* neurons. The potassium-selective channels can be blocked through a competitive mechanism, whereas the sodium-selective channels are blocked noncompetitively (Sasaki, 1985). Thus, as in autonomic ganglia, TEA blocks only the channels sensitive to hexamethonium.

In *Aplysia* neurons, carbachol and TMA open the acetylcholine-sensitive ionic channels for a shorter time than acetylcholine does (Ascher et al., 1978a), although in other molluscs (*Lymnaea*) the TMA- and carbachol-induced membrane conductance increases are higher and lower, respectively, than that induced by acetylcholine (Kurchikov et al., 1985).

The voltage dependence of the channel open time is linear in molluscan neurons (Ascher et al., 1978a; Slater et al., 1983) but exponential in autonomic ganglion neurons (Connor et al., 1983; Derkach et al., 1983) and skeletal muscle fibers (Magleby and Stevens, 1972a; Peper et al., 1982). On the other hand, nicotinic acetylcholine receptors in molluscan neurons, as in autonomic ganglion neurons and skeletal muscle fibers (Paton and Zaimis, 1949; Barlow and Zoller, 1964), are sensitive to the blocking action of bis-quaternary ammonium compounds with their cationic groups separated by a chain of 16 methylene groups. It has been suggested, based on this observation, that there should be two anionic sites separated by a distance of about 0.20–0.22 nm in molluscan neuronal acetylcholine receptor (Vulfius et al., 1969, 1980).

Vipoxin, a polypeptide obtained from viper toxin, at low concentrations blocks acetylcholine receptors of all three types in *Aplysia* neurons, while at higher concentrations it can activate them as a partial agonist (Slater et al., 1983, 1985). No effect of vipoxin at similar low concentrations has been observed in nicotinic acetylcholine receptors in vertebrates (Slater et al., 1985).

4.4. CONCLUSIONS

The most important structural component in monocationic compounds responsible for their high blocking activity in sympathetic ganglia is the positively charged nitrogen with attached alkyl groups that have branching in the α position. In bis-cationic compounds such components are: 1) one cationic group similar to that in acetylcholine and another cationic group similar to the one above; 2) the intercharge distance corresponding to a chain of 4–5 or 16–18 methylene groups. The ganglion-blocking effect of all these compounds (except those with 16–18 methylene groups, which have not been studied in detail) is due exclusively (or predominantly) to their blocking effect on the open ionic channels of neuronal nicotinic acetylcholine receptors. By contrast, molecules of a larger size and similar to the acetylcholine molecule in possessing an oxygen located as far from the positive charge as in acetylcholine may exert their ganglion-blocking effect through competitive binding to a recognition center.

As a rule, the ganglion-blocking agents are of smaller molecular size than neuromuscular blocking agents; they are less structurally similar to acetylcholine, and, in the case of bis-cationic compounds, possess an intercharge distance of about 0.7 nm. These features of ganglion-blocking agents suggest that they may interact with the open channel more easily than neuromuscular blocking agents.

Regarding the blocking selectivity between sympathetic and parasympathetic ganglia, the selective parasympathetic blockers are monocationic compounds that have a cationic group similar to that in the most active monocationic sympathetic blocking agents with a very long alkyl group attached.

The selective blocking agents for nicotinic acetylcholine receptors in molluscan neurons are similar to those for vertebrate autonomic ganglia.

Chapter 5

MOLECULAR MECHANISMS UNDERLYING THE EFFECTS OF GANGLION-BLOCKING AGENTS

5.1. OPEN-CHANNEL BLOCK

5.1.1. Effect of Ganglion-Blocking Agents on Integrated Activity of Nicotinic Acetylcholine-Sensitive Channels Recorded from the Whole Cell

Figure 5.1 illustrates the effect produced by a specific ganglion-blocking agent hexamethonium ($1 \cdot 10^{-5}$ M) on fast excitatory postsynaptic current (fast e.p.s.c.) evoked in the neuron of rabbit isolated superior cervical ganglion by a single preganglionic stimulus at 35–36°C. In Fig. 5.1A, the e.p.s.c. recorded in normal solution (left) and in a hexamethonium-containing solution (right) are shown. One can see that at membrane potential –70 mV, which is close to resting membrane potential level (about –50 mV), hexamethonium markedly shortens the e.p.s.c. decay time. This effect can be clearly seen in Fig. 5.1B, where the e.p.s.c. decay, as shown in semilogarithmic scale, is single-exponential. The hexamethonium-induced shortening of the e.p.s.c. decay is strongly enhanced by membrane hyperpolarization (Fig. 5.1C).

It should be emphasized that in its minimum effective concentration hexamethonium shortens the e.p.s.c. decay without reducing the e.p.s.c. amplitude. This is evident from Fig. 5.1D, where a small decrease in the e.p.s.c. amplitude can be observed only about 10 min after the e.p.s.c. decay shortening is first indicated. For a more significant and faster decrease of the e.p.s.c. amplitude at low concentrations of hexamethonium, the neuronal membrane should be strongly hyperpolarized (Fig. 5.1E).

As has been shown in Chapter 3, the time constant τ_{syn} of the e.p.s.c. decay is equal to the mean apparent time the channels stay open. Thus one would expect that hexamethonium should also shorten the channel apparent open time determined by

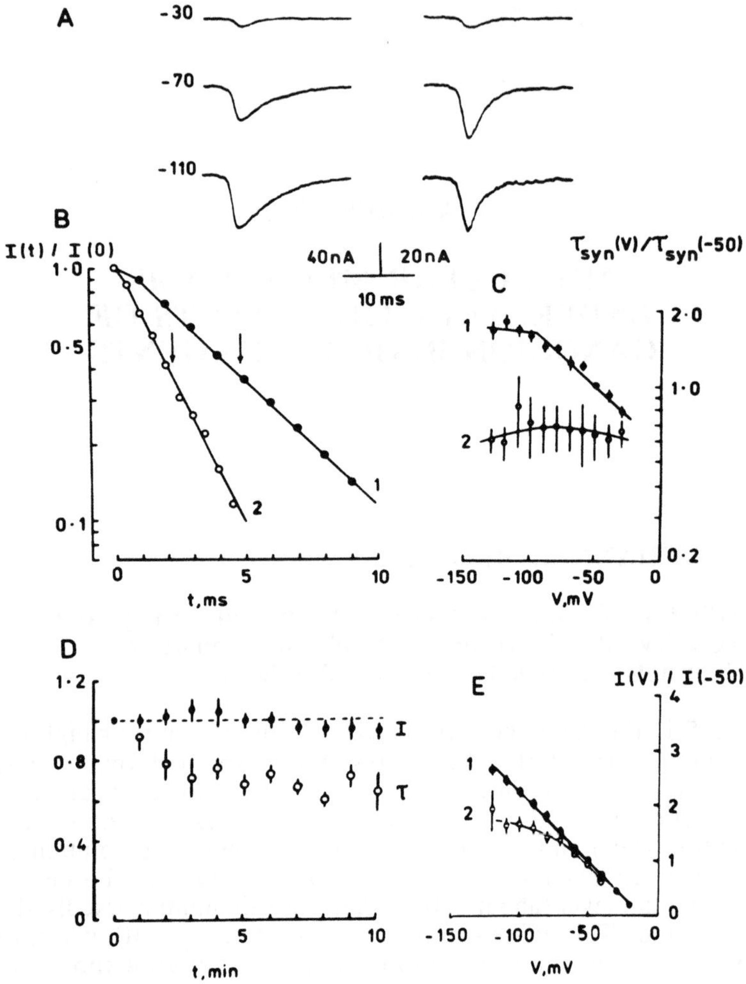

Fig. 5.1. Effect of hexamethonium (10 μM) on the e.p.s.c.'s recorded from the rabbit superior cervical ganglion neurons at 36°C. A) E.p.s.c.'s recorded at different membrane potentials (indicated in mV beside each trace) in normal solution (left) and in the presence of hexamethonium (right). B) Semilogarithmic plot of the normalized amplitude of the e.p.s.c.'s shown in A as a function of time, in normal solution (1) and in the presence of hexamethonium (2). Membrane potential –70 mV. Arrows indicate the decay time constants (τ_{syn}); control was 4.8 ms; in hexamethonium it was 2.3 ms. C) Voltage dependence of τ_{syn}. Semilogarithmic plot of τ_{syn} (normalized with respect to its value at –50 mV) vs. membrane potential in normal solution (1) and in the presence of hexamethonium (2). The absolute normal τ_{syn} value was 4.5 ± 0.3 ms at –80 mV. Points are means and SE of mean for 3 observations in C and for 6 observations in D or E. D) Amplitude (I) and τ_{syn} (τ) of the e.p.s.c. recorded at –50 mV as a function of time after the onset of perfusion with hexamethonium. Amplitudes and τ_{syn} are normalized with respect to those at zero time. E) E.p.s.c. amplitude as a function of membrane potential in normal solution (1) and in presence of hexamethonium (2). Amplitudes are normalized with respect to those in normal solution at –50 mV (Skok, 1986).

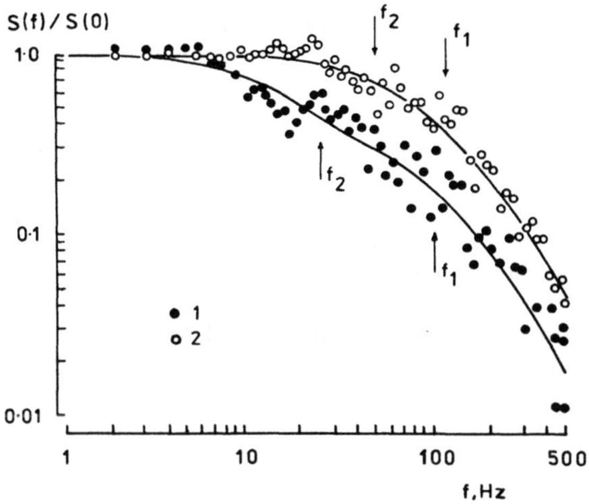

Fig. 5.2. Effect of hexamethonium (10 μM) on the two-Lorentzian power spectra of the acetylcholine-induced membrane current fluctuations recorded from rabbit superior cervical ganglion neuron at –80 mV and 36°C. Spectral density as a function of frequency was measured, as in Fig. 3.4, in control solution (1) and in hexamethonium-containing solution (2). Mean apparent channel lifetimes τ_{f1} and τ_{f2} as calculated from the cutoff frequencies f_1 and f_2 (indicated by arrows) were 1.5 and 10.6 ms in control solution and 1.2 and 3.2 ms in hexamethonium-containing solution (Skok et al., 1983).

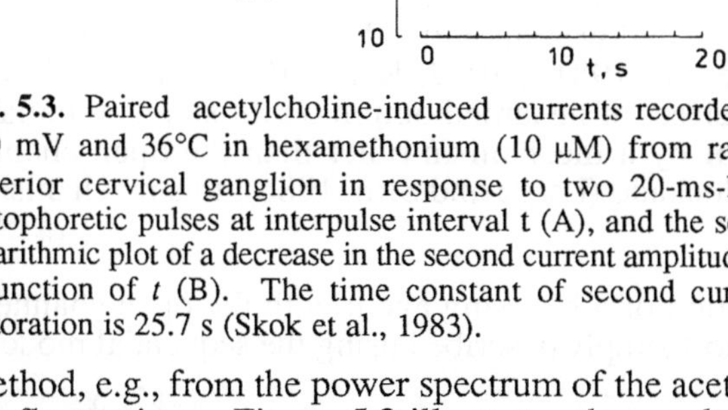

Fig. 5.3. Paired acetylcholine-induced currents recorded at –80 mV and 36°C in hexamethonium (10 μM) from rabbit superior cervical ganglion in response to two 20-ms-long iontophoretic pulses at interpulse interval t (A), and the semi-logarithmic plot of a decrease in the second current amplitude as a function of t (B). The time constant of second current restoration is 25.7 s (Skok et al., 1983).

an independent method, e.g., from the power spectrum of the acetylcholine-induced membrane current fluctuations. Figure 5.2 illustrates the results obtained in such experiments. The membrane current fluctuations have been evoked by iontophoretic application of acetylcholine from a micropipette on the surface of rabbit isolated superior cervical ganglion neuron. The power spectrum of the current fluctuations is fitted by two Lorentzian functions, as described in Chapter 3. It is evident from Fig. 5.2 that both the slow and fast kinetic components of the spectrum are shortened by hexamethonium. The mean shortening observed at this concentration of hexamethonium ($1 \cdot 10^{-5}$ M) is to 51% and 79% of the initial duration of the component, for the slow and fast components, respectively (Skok et al., 1983).

The voltage-dependent shortening of the chemosensitive channel open time without a decrease in the probability of the channel being opened by the agonist is one of the indications of the open-channel blocking mechanism (Adams, 1976, 1977; Adams and Feltz, 1980a, b; Neher and Steinbach, 1978; Ascher et al., 1978b; Colquhoun et al., 1979; Ogden et al., 1981). An alternative interpretation is that the open time shortening (e.g., caused by local anesthetics) is due to a decrease in membrane viscosity (Gage and Hamill, 1975).

The next indication of the open-channel block is a decrease in the number of channels that can be opened by the agonist following their conditional opening in the presence of a blocking agent (Ascher et al., 1978b, 1979). This phenomenon, called "use-dependence," is thought to underlie a commonly observed frequency dependence in the effects of many blocking agents (Courtney, 1975). Figure 5.3A shows membrane currents evoked in sympathetic ganglion neuron by paired iontophoretic applications of acetylcholine in the presence of $1 \cdot 10^{-5}$ M hexamethonium. The second current is reduced compared with the first one, which has not been observed in normal Krebs solution. The effect disappears as the interval between two acetylcholine applications increases (Fig. 5.3B). Another indication of the open-channel blocking mechanism is an enhancement of the blockade caused by increased agonist concentration in the presence of increased concentration of antagonist, which is usually responsible for the unparalleled fashion of the dose–response curve shift. This result has been observed in the hexamethonium effects in parasympathetic (Ascher et al., 1979) and sympathetic (Skok, Selyanko, and Derkach, unpublished results) ganglia.

Thus it can be concluded that hexamethonium blocks the open channel of the acetylcholine receptor. This mechanism had been first suggested by Blackman (1970) from his observation of the voltage dependence of hexamethonium-induced ganglionic blockade and was supported by later experimental results (Ascher et al., 1979; Selyanko et al., 1981b, 1982; Rang, 1982). It should be noted, however, that recently the effect of hexamethonium on rat submandibular ganglion neurons was accounted for by a mechanism other than the open-channel blockade – the closed-channel blockade (Large and Sim, 1986); this mechanism will be described below.

The interaction of a blocking agent with the open channel of a postsynaptic receptor can be most simply described using the sequential model:

$$\underset{4}{nA + R} \underset{k_{-1}}{\overset{k_1}{\rightleftharpoons}} \underset{3}{A_nR} \underset{\alpha}{\overset{\beta}{\rightleftharpoons}} \underset{1}{A_nR^* + B} \underset{k^*_{-B}}{\overset{k^*_{+B} X_B}{\rightleftharpoons}} \underset{2}{A_nR^*B} \qquad (5.1)$$

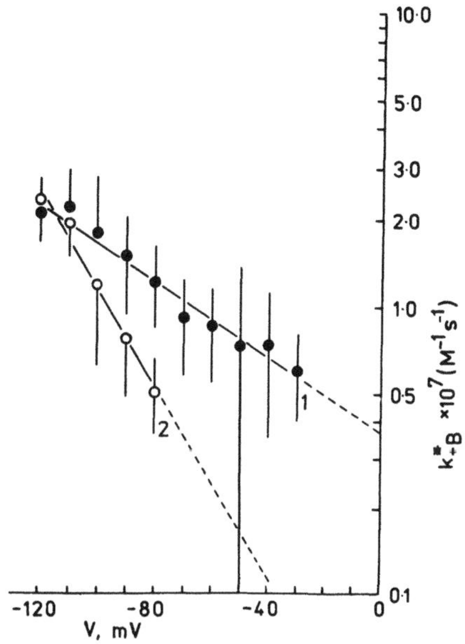

Fig. 5.4. Voltage dependence of the rate constants k_{+B}* that characterize binding of hexamethonium (1) and tubocurarine (2) to open ionic channels of nicotinic acetylcholine receptors in rabbit superior cervical ganglion neurons at 36°C. Slopes (H_{+B}* values) correspond to –67 (1) and –26 mV (2). Mean values (n = 3) and SE are shown (Selyanko et al., 1981b).

The left part of this equation is similar to Eq. (3.16), while the right part describes the interaction of agent B at concentration X_B with the open channel of the receptor–agonist complex A_nR*. The interaction is characterized by rate constants k_{+B}* and k_{-B}* and results in the appearance of complex A_nR*B (with an open-blocked channel). The numbers indicate corresponding states (see below in this chapter).

As mentioned above, the e.p.s.c. decay, although shortened by a blocking agent, remains single-exponential. This implies that k_{+B}*·X_B » k_{-B}* (otherwise the e.p.s.c. decay would exhibit two exponentials). Therefore, the rate k_{+B}*·X_B can be found as the difference between the apparent rate constant of e.p.s.c. decay estimated in the presence (α') and absence (α) of a blocking agent [see Eq. (5.1)]. Therefore,

$$k^*_{+B} = \left[(\tau')^{-1}_{syn} - \tau^{-1}_{syn}\right] X_B^{-1}, \qquad (5.2)$$

which is convenient for obtaining k_{+B}* values from experimental results.

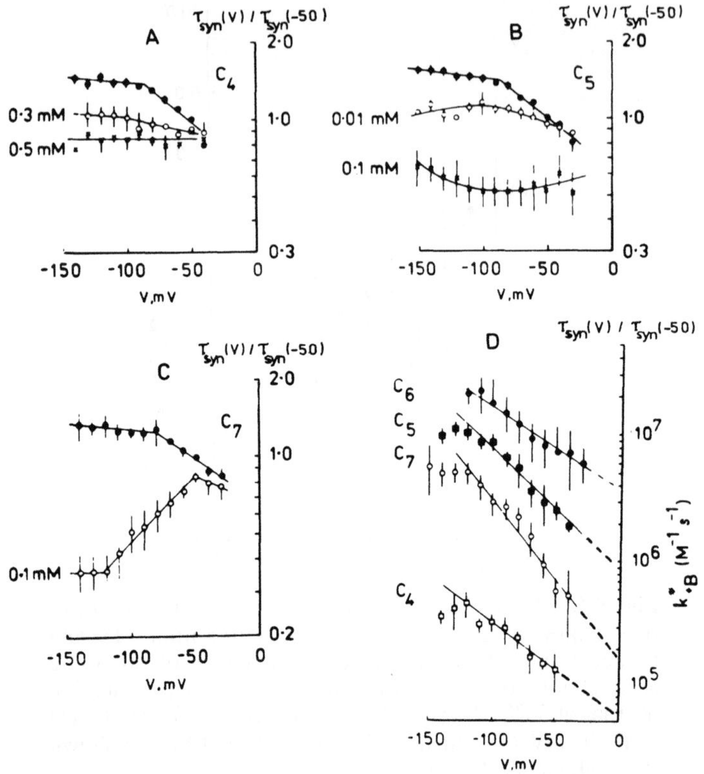

Fig. 5.5. Effects of ganglion-blocking agents on voltage dependence of the e.p.s.c. decay time constant τ_{syn} as measured in rabbit superior cervical ganglion neurons at 36°C. Effects produced by tetramethonium, C_4 (A), pentamethonium, C_5 (B), and heptamethonium, C_7 (C) in the concentrations as indicated are shown. Dark circles correspond to the results obtained in normal solution (without ganglion-blocking agents). Mean τ_{syn} values normalized relative to those obtained at –50 mV and SE values are shown. D) Values of k_{+B}^* for the above compounds and, for comparison, for hexamethonium (see Fig. 5.4), are plotted logarithmically as a function of membrane potential (Selyanko et al., 1982).

The k_{-B}^* value can be found as an inverse to the time constant that characterizes restoration of the acetylcholine-induced current when interpulse intervals are increased in the double-pulse experiment illustrated by Fig. 5.3.

For the hexamethonium effect which has been examined in rabbit superior cervical ganglion at a membrane potential –50 mV and temperature 36°C, $k_{+B}^* = (7.30 \pm 2.10) \cdot 10^6\ M^{-1} \cdot s^{-1}$, $k_{-B}^* = 0.06 \pm 0.01\ s^{-1}$, and $K_D^* = 8.2 \cdot 10^{-9}$ M (mean ± S.E., see Table 5.1).

TABLE 5.1. Characteristics of Channel-Blocking, Competitive, and Ganglion-Blocking Actions of Antagonists

Compound	$k_{+B}^* \cdot 10^6$ M^{-1}·s^{-1} (–50 mV)	H_{+B}^*, mV	k_{-B}^*, s^{-1} (at –50 mV)	k_{-B}^* (–50 mV) / k_{-B}^* (–80 mV)	Relative ganglion-blocking activity, %	Relative competitive activity, % (at –50 mV)	H_B, mV
	1	2	3	4	5	6	7
IEM-1119	27.0 ± 16.1 (*n* = 4)	–65.0 ± 13.4 (*n* = 4)	0.22 ± 0.06 (*n* = 8)	1.45	470	253 ± 37 (*n* = 4)	–264.7
C_4	0.13 ± 0.04 (*n* = 3)	–59.0 ± 12.4 (*n* = 3)	1.8 ± 0.7 (*n* = 3)	1.29	1 (2)	33 ± 5 (*n* = 5)	–219.7
C_5	2.60 ± 0.45 (*n* = 7)	–48.0 ± 9.4 (*n* = 4)	0.62 ± 0.14 (*n* = 4)	1.84	140 (80)	115 ± 8 (*n* = 4)	–236.4
C_6	7.30 ± 2.10 (*n* = 6)	–67.0 ± 25.0 (*n* = 3)	0.06 ± 0.01 (*n* = 5)	1.13	100 (100)	100 ± 17 (*n* = 6)	–61.3
C_7	0.56 ± 0.16 (*n* = 4)	–32.0 ± 5.6 (*n* = 4)	0.52 ± 0.09 (*n* = 6)	2.73	15 (10)	39 ± 5 (*n* = 4)	–52.6
IEM-1288	16.67 ± 6.6 (*n* = 3)	–	–	–	280	–	–
Tubocurarine	≅2.0 (*n* = 3)	–	–	–	537	922 ± 3 (*n* = 11)	–

Note. C_4, C_5, C_6, and C_7 are tetra-, penta-, hexa-, and heptamethonium, respectively. In columns 1–4 the values of rate constants k_{+B}^* and k_{-B}^* (Selyanko et al., 1981b; Skok et al., 1983, 1984), as well as of their voltage-dependences (Derkach et al., 1984), are indicated. In column 5 relative ganglion-blocking activities (hexamethonium = 100%) are given in parentheses – according to Paton and Zaimis (1949), without parentheses – according to Skok et al. (1984); tubocurarine activity is taken from Bowmann and Webb (1972). In column 6 the effectiveness of compounds, relative to that of C_6, in depression of e.p.s.c. amplitude, in concentrations $1 \cdot 10^{-4}$ M (C_4, C_5, and C_7) and $1 \cdot 10^{-5}$ M (IEM-1119 and tubocurarine; A. A. Selyanko and B. A. Derkach, unpublished data) are shown. In column 7 the voltage dependence of the effect given in column 6 is indicated (Skok et al., 1984).

The k_{+B}* value obtained from rabbit superior cervical ganglion is intermediate between those obtained from rat submandibular ganglion and from *Aplysia* neurons ($2.2 \cdot 10^7$ $M^{-1} \cdot s^{-1}$ and $7.5 \cdot 10^5$ $M^{-1} \cdot s^{-1}$, respectively, as found at –80 mV and room temperature). On the other hand, the k_{-B}* and K_D* values for hexamethonium in rabbit superior cervical ganglion are lower than in rat and molluscan neurons where they are 0.48 s^{-1} and 0.5 s^{-1}, and $2.2 \cdot 10^{-8}$ M and $6.7 \cdot 10^{-7}$ M, respectively (Ascher et al., 1978b, 1979).

The rate constants k_{+B}* and k_{-B}* strongly depend on membrane potential level, which is illustrated by Fig. 5.4 (for comparison, the results obtained with another antagonist, tubocurarine, are shown in Fig. 5.4). The voltage dependence of k_{+B}* is given by

$$k^*_{+B}(V) = k^*_{+B}(0) \exp(V/H^*_{+B}), \tag{5.3}$$

where k_{+B}*(V) and k_{+B}*(0) are the k_{+B}* values measured at potentials V and 0, and H_{+B}* is the shift in membrane potential that corresponds to an e-fold increase of k_{+B}*. For hexamethonium, the mean values of k_{+B}*(0) and H_{+B}* are $5.76 \cdot 10^6$ $M^{-1} \cdot s^{-1}$ and –67.0 mV (minus means that the k_{+B}* value increases with membrane hyperpolarization), respectively; the voltage dependence of k_{-B}* is characterized by a 1.13-fold decrease in the k_{-B}* value following a membrane potential shift from –50 to –80 mV (Table 5.1). Therefore, membrane hyperpolarizaton results in an increase in k_{+B}* and a decrease in k_{-B}*, thus causing a marked decrease in K_D*. It is evident from Fig. 5.4 that the voltage dependence of k_{+B}* for tubocurarine is higher than for hexamethonium: H_{+B}* for tubocurarine is –26 mV and k_{+B}*(0) is $2.0 \cdot 10^6$ $M^{-1} \cdot s^{-1}$ (Selyanko et al., 1981b).

Figure 5.5 illustrates the voltage dependence of the effects produced by bis-quaternary ammonium compounds tetra-, penta-, and heptamethonium (**2.18**) on the e.p.s.c. decay time constant τ_{syn} and on the rate constant k_{+B}* in rabbit superior cervical ganglion. To compare the effects produced by these compounds on e.p.s.c., in their possible relation to synaptic transmission blockade, the compounds should be applied in concentrations inversely proportional to their ganglion-blocking activities. Their activity ratio is 50:1.25:1:10 for tetra-, penta-, hexa-, and heptamethonium, respectively (Paton and Zaimis, 1949). Therefore, they have been used in concentrations $5 \cdot 10^{-4}$ M, $1 \cdot 10^{-5}$ M, $1 \cdot 10^{-5}$ M, and $1 \cdot 10^{-4}$ M, respectively (the hexamethonium concentration used is just high enough to cause the open-channel blocking effect shown in Fig. 5.1).

All of these compounds, in addition to decreasing the e.p.s.c. amplitude, also shorten the e.p.s.c. decay, leaving it single-exponential. As Fig. 5.5 shows, the compounds in question weaken or even reverse the voltage dependence of τ_{syn}, the effect being increased by their increased concentration. These features are consistent with the open-channel blocking mechanism, with the channel blocking potencies increasing from tetra- to heptamethonium.

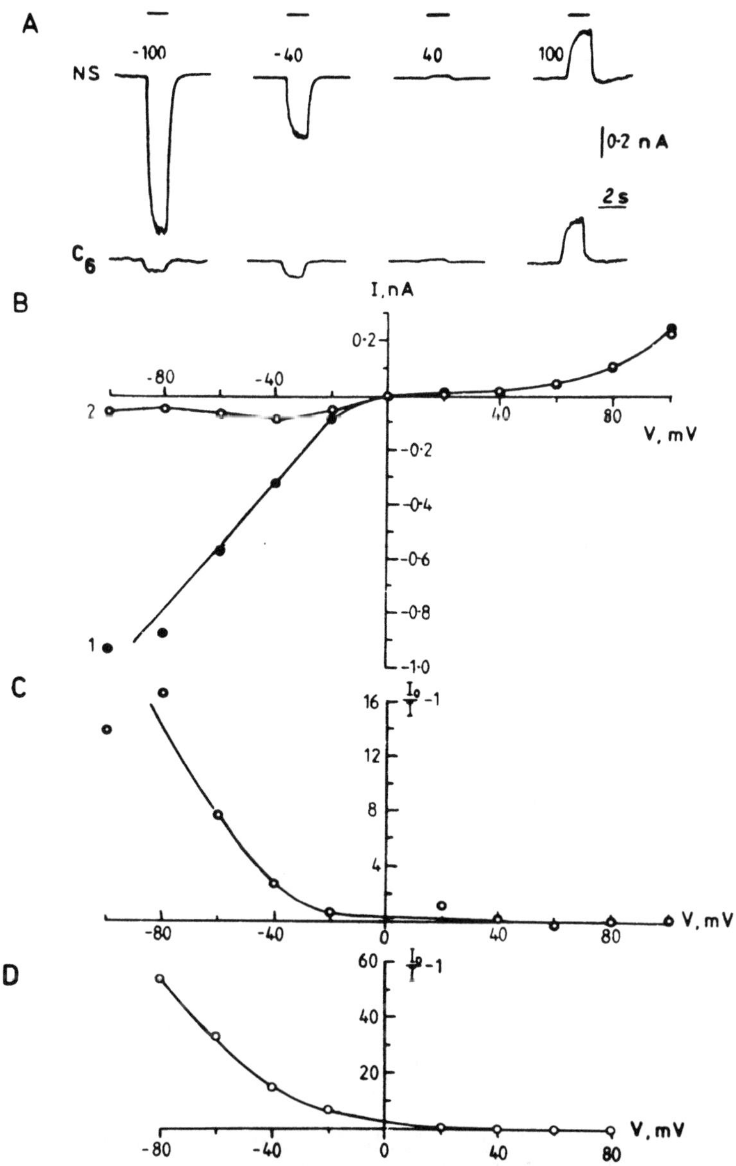

Fig. 5.6. Voltage dependence of the blocking effects produced by 50 μM hexamethonium (A–C) and 10 μM IEM-1288 (D) on the acetylcholine-induced responses in rat superior cervical ganglion neurons recorded with the whole-cell patch-clamp method at 23–25°C. A) Membrane currents recorded in normal solution (top records) and in the presence of hexamethonium (bottom records) at membrane potential levels indicated near the records. B) Peak amplitude of the currents, some of which are shown in A, as plotted against membrane potential level, recorded in the absence (1) and in the presence of hexamethonium (2). C) Blocking activity of hexamethonium calculated as $(I_0/I - 1)$, where I_0 and I are the amplitude of normal and blocked responses, respectively, plotted against membrane potential level. D) Blocking activity of IEM-1288 obtained with the same approach as that of hexamethonium and plotted against membrane potential level.

The k_{+B}* values and their voltage dependences H_{+B}* are shown in Table 5.1 (columns 1 and 2). These values have been estimated at –50 mV, which is close to a resting membrane potential level. Besides the effects of the compounds listed above, the effects of tubocurarine (**2.20**) and other ganglion-blocking agents trimethylammonium–tetramethylene–isopropylammonium (IEM-1119, **5.1**) and IEM-1288, a pempidine derivative (see **4.7**) with a bromide instead of a tartrate anion, are also demonstrated in Table 5.1. The latter three compounds have been used in similar concentrations, $1 \cdot 10^{-5}$ M.

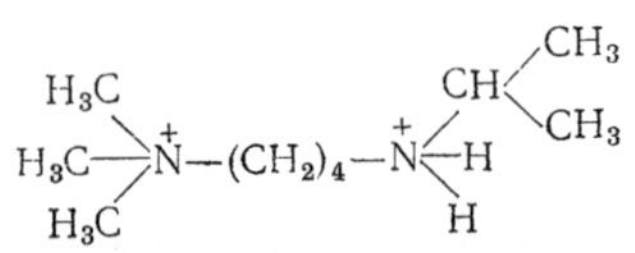

IEM-1119, **5.1**

It follows from Fig. 5.5 and Table 5.1 that the increase in the length of the polymethylene chain results in a somewhat increased voltage dependence of k_{+B}* values, although the k_{+B}* values themselves do not change.

5.1.2. Effect of Ganglion-Blocking Agents on the Acetylcholine-Induced Current in Strongly Depolarized Membrane

The high voltage dependence of k_{+B}* and k_{-B}* implies that, using strong depolarization of neuronal membrane, one should observe no reduction of the acetylcholine-induced current by the open-channel blocking agent, unless a non-voltage-dependent blocking effect (e.g., the competitive blockade of the recognition center) is involved. Figure 5.6 illustrates the effects produced under these conditions by two ganglion-blocking agents, hexamethonium and IEM-1288, which, according to the results described above, are potent open-channel blockers. The experiments illustrated in Fig. 5.6 have been performed on rat superior cervical ganglion neurons at room temperature, using the whole cell patch-clamp technique (see Hamill et al., 1981; to decrease the delayed rectification caused by strong membrane depolarization, the patch pipette was filled with 140 mM of NaCl-containing solution). Other experimental procedures were the same as for the single-channel current recordings described in Chapter 3 (Derkach et al., 1987).

One can see from Fig. 5.6 that both hexamethonium and IEM-1288 strongly reduce the acetylcholine-induced current in hyperpolarized membrane but are ineffective at positive membrane potentials. Their blocking activity expressed as $(I_0/I) - 1$, where I_0/I is the ratio of normal to blocked current amplitudes, drops actually to zero at membrane potentials more positive than +20 mV. Similar effects have been observed in four neurons studied (A. A. Selyanko, V. A. Derkach, and D. Kurenni, in press).

Thus, one can conclude that neither hexamethonium nor IEM-1288 has any competitive (non-voltage-dependent) component in their blocking effects on nicotinic acetylcholine receptors in rat sympathetic ganglion neurons.

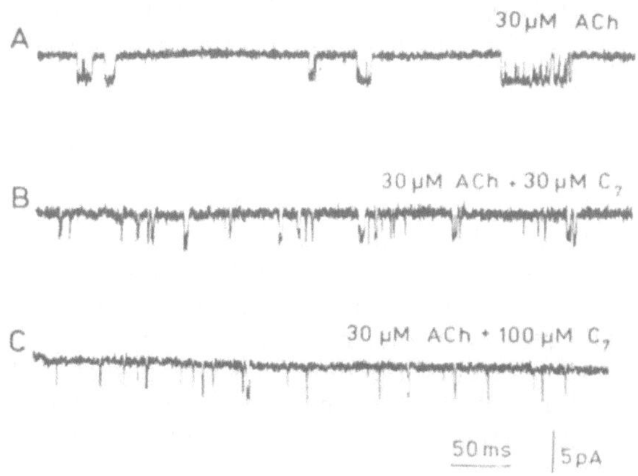

Fig. 5.7. Effect of heptamethonium (C_7) on the acetylcholine (ACh, 30 μM)-induced activity recorded with the cell-attached patch-clamp method from rat superior cervical ganglion neuron (–110 mV calculated membrane potential, 23–25°C). Records were obtained in normal solution (A) and in the presence of 30 μM (B) and 100 μM (C) heptamethonium (Derkach et al., 1987).

5.1.3. Effect of Ganglion-Blocking Agents on the Activity of a Single Acetylcholine-Sensitive Channel

As has been mentioned above, e.p.s.c. arises as a result of synchronous activation with the nerve-released acetylcholine of a great number of acetylcholine receptors, and the e.p.s.c. decay time constant corresponds to the mean duration of a burst of single-channel openings (see Chapter 3.3).

One may anticipate from this result that the open-channel blocking agents should decrease the mean duration of the bursts of openings recorded from a single channel. On the other hand, the model [Eq. (5.1)] predicts that each single-channel opening should be shortened by the open-channel blocker as well. To determine whether this is the case, the effect of heptamethonium on a single acetylcholine-sensitive channel activity was studied in the neurons of rat superior cervical ganglion (Selyanko et al., 1985b; Derkach et al., 1987). Among the many ganglion-blocking agents, heptamethonium is most appropriate for this purpose because of the comparatively high voltage sensitivity of its open-channel blocking effect (see Table 5.1). The high voltage sensitivity allows one to check whether the observed reduction of single channel activity is due to a heptamethonium effect, or whether it is an artifact, simply by changing the membrane potential level.

As expected, heptamethonium shortened the mean duration of a burst of single-channel openings τ_b, the effect being increased by increased heptamethonium concentration and also by membrane hyperpolarization (Fig. 5.7). The shortening of τ_b was due to both a decrease in the mean number of openings per burst and in the duration of each single opening τ_o. At –110 mV, heptamethonium ($3 \cdot 10^{-5}$ M) decreased τ_b from 8.5 ± 1.2 ms to 1.5 ± 0.1 ms and τ_o from 1.3 ± 0.2 ms to 0.4 ± 0.1 ms, while reducing 1.9-fold the mean number of openings per burst. The mean duration of each single-channel closing within a burst remained constant (0.15 ± 0.02 ms and 0.12 ± 0.02 ms, respectively), as well as the amplitude of a single-channel current (2.2 ± 0.01 pA and 2.0 ± 0.07 pA, respectively).

Only one kind of single-channel blockade, the blockade produced by fast-dissociating blockers (e.g., local anesthetics), has been analyzed in detail so far (Neher and Steinbach, 1978; Ogden et al., 1981; Colquhoun and Hawkes, 1983). This kind of blockade results in the transformation of end-plate channel single openings into bursts of shortened openings. It was assumed that termination of each blocker-induced burst of channel openings is due to channel transition into a long (nonblocked) state R [see Eq. (5.1)]. In contrast to the action of local anesthetics on the skeletal muscle end-plate, the action of hexamethonium and its analogs on nicotinic receptors of autonomic ganglion neurons is characterized by the slow dissociation of a blocker from the open channel (Ascher et al., 1979; Selyanko et al., 1981, 1982; Skok et al., 1983, 1984). This implies that there should be long blocked states in single-channel activity. It should be noted that the burstlike single-channel activity has been observed in both the absence and presence of a blocker (see above). This observation suggests that, in general, a burst in the presence of a slow-dissociating blocker can terminate due to a transition either to the long closed nonblocked state or to the long closed blocked state. Let us consider now this more general situation, with the burst observed in the presence of ganglion-blocking agents, like those produced by local anesthetics from single end-plate channel openings using the approach suggested by Colquhoun and Hawkes (1983). The probability $P(\mathrm{m}')$ that a burst will consist of m' successive channel openings is equal to the sum of the probabilities of two random events [see Eq. (5.1)] – the probability $P_4(\mathrm{m}')$ that a burst will be interrupted by channel transition to the closed state 4 and the probability $P_2(\mathrm{m}')$ that it will be interrupted by channel transition to the closed blocked state 2:

$$P(\mathrm{m}') = P_4(\mathrm{m}') + P_2(\mathrm{m}'). \tag{5.4}$$

On the other hand, the transition to state 4 occurs after the following random events: m' transitions from 1 to 3, (m' – 1) transitions from 3 to 1, and one transition from 3 to 4. The probabilities of these transitions are: $P_{13} = \alpha/(\alpha + k_{+B}^* \cdot X_B)$, $P_{31} = \beta/(\beta + k_{-1})$, and $P_{34} = k_{-1}/(\beta + k_{-1})$. Thus, the probability $P_4(\mathrm{m}')$ will be given by

$$P_4(\mathrm{m}') = P_{13}^{\mathrm{m}'} \cdot P_{31}^{\mathrm{m}'-1} \cdot P_{34}. \tag{5.5}$$

Similarly,

$$P_2(\mathrm{m}') = P_{13}^{\mathrm{m}'-1} \cdot P_{31}^{\mathrm{m}'-1} \cdot P_{12}, \tag{5.6}$$

where $P_{12} = k^*_{+B} \cdot X_B / (\alpha + k^*_{+B} \cdot X_B)$, and P_{13} and P_{31} are given above.

It follows from probability theory that the number of channel openings per burst m' can be deduced from the expression

$$\overline{\mathrm{m}}' = \sum_{\mathrm{m}'=1}^{\infty} \mathrm{m}' \cdot P(\mathrm{m}'). \tag{5.7}$$

Taking into account Eqs. (5.4)–(5.6) and transforming Eq. (5.7), one can obtain

$$\overline{\mathrm{m}}' = \overline{\mathrm{m}} \cdot (\alpha + k^*_{+B} \cdot X_B)/(\alpha + \overline{\mathrm{m}} \cdot k^*_{+B} \cdot X_B), \tag{5.8}$$

where $\overline{\mathrm{m}}$ is the mean number of channel openings per burst in the absence of a blocker ($\mathrm{m} = 1 + \beta/k_{-1}$). Equation (5.8) means that $\overline{\mathrm{m}}' \leq \overline{\mathrm{m}}$ at $X_B \geq 0$. Thus, the sequential blocking scheme [Eq. (5.1)] predicts that the number of channel openings per burst will decrease in the presence of a blocker compared with that in the absence of a blocker.

Moreover, it follows from Eq. (5.1) and from the rule (3.16) that the mean duration of openings within a burst in the presence of a blocker should decrease from $1/\alpha$ to $1/(\alpha + k_{+B}^* \cdot X_B)$. As a result, the mean overall time the channel stays open during a burst will be shortened from $\overline{\mathrm{m}}/\alpha$ to $\mathrm{m}'/(\alpha + k_{+B}^* \cdot X_B)$. The mean total closed state within a burst, defined as $1/(k_{+1} + \beta)$, is unaffected by a blocker, since the blocker does not interact with the channel in this state, according to Eq. (5.1). Then, using expression (5.8), one can obtain that the mean burst duration τ_b' in the presence of a blocker† is

$$\tau_b' = \frac{\overline{\mathrm{m}}'}{\alpha + k^*_{+B} \cdot X_B} + \frac{\overline{\mathrm{m}}' - 1}{k_{-1} + \beta} = \frac{\alpha}{\alpha + \overline{\mathrm{m}} k^*_{+B}} \times \left(\frac{\overline{\mathrm{m}}}{\alpha} + \frac{\overline{\mathrm{m}} - 1}{k_{-1} + \beta} \right) = \frac{\alpha}{\alpha + \overline{\mathrm{m}} k^*_{+B} \cdot X_B} \cdot \tau_b, \tag{5.9}$$

where τ_b, the mean burst duration in the absence of a blocker, is given by

$$\tau_b = \frac{\overline{\mathrm{m}}}{\alpha} + \frac{\overline{\mathrm{m}} - 1}{k_{-1} + \beta}. \tag{5.10}$$

A comparison of Eqs. (5.9) and (5.10) yields $\tau_b' \leq \tau_b$ at all $X_B \geq 0$.

† τ_b' indicates a burst duration in general terms, irrespective of any particular component in the experimentally found burst duration distribution described in Chapter 3.

It seemed interesting to know the number of blocked channel states per single activation of a receptor by an agonist. This required finding the proportion between the number of bursts terminating with closed blocked states and the number of bursts terminating with long closed nonblocked states. One can find this proportion provided that the proportion $r = P_2(\mathrm{m'})/P_4(\mathrm{m'})$ between the probabilities of transitions 1–2 and transitions 1–4 is known [see scheme (5.1)]. Using Eqs. (5.5) and (5.6), one can obtain

$$r = \frac{P_2(\mathrm{m'})}{P_4(\mathrm{m'})} = \frac{P_{12}}{P_{13}P_{34}} = \overline{\mathrm{m}}\frac{k^*_{+\mathrm{B}}\cdot X_\mathrm{B}}{\alpha}. \tag{5.11}$$

Equation (5.11) postulates that $r > 0$ at all values of $X_\mathrm{B} > 0$. Therefore, a single activation of the receptor in the presence of a blocker will result in a cluster of mean number $r + 1$ bursts instead of the single burst observed in the absence of a blocker. From the mean number of bursts per cluster and the mean single blocked state duration [$1/k_{-\mathrm{B}}*$, according to the rule (3.16)], one can find the mean cluster duration τ_{clu}:

$$\tau_{clu} = (r + 1)\cdot\tau_b' + r/k^*_{-\mathrm{B}}.$$

Using Eqs. (5.9) and (5.11), we obtain

$$\tau_{clu} = \tau_b + \overline{\mathrm{m}}\frac{k^*_{+\mathrm{B}}\cdot X_\mathrm{B}}{k^*_{-\mathrm{B}}}. \tag{5.12}$$

The mean overall channel open time within a cluster is apparently determined by the product

$$\frac{1}{\alpha + k^*_{+\mathrm{B}}\cdot X_\mathrm{B}}\cdot \mathrm{m'}(r + 1) = \frac{\overline{\mathrm{m}}}{\alpha}. \tag{5.13}$$

Therefore, the mean time the channel stays open during a cluster of bursts in the presence of a blocker is equal to the mean time the channel stays open during a burst in the absence of a blocker. This conclusion is consistent with the observation that the mean total time that the end-plate channel stays open during the burst of openings which results from a single opening under the influence of local anesthetics is exactly equal to the mean duration of the initial single opening observed in the absence of local anesthetics (Neher and Steinbach, 1978; Ogden et al., 1981; Colquhoun and Hawkes, 1983; Neher, 1983).

It is evident that when the dissociation rate of a blocker is relatively high [$k_{-\mathrm{B}}* \geq (k_{-1} + \beta)$], in contrast to the case considered above, the interburst time intervals will be indistinguishable from intraburst time intervals, and a cluster will transform to a continuous succession of openings with its mean duration $\tau_{clu} \geq \tau_b$

(at $X_B \geq 0$). This is exactly what has been observed during the action of local anesthetics at $\overline{m} = 1$. Then, as follows from Eq. (5.12),

$$\tau_{clu} = \left(1 + k^*_{+B} \cdot X_B / k^*_{-B}\right) / \alpha$$

(see Neher and Steinbach, 1978; Ogden et al., 1981; Colquhoun and Hawkes, 1983).

The "true" value of k_{+B}^* can be found from the following expression:

$$k_{+B}^* = [(\tau_o')^{-1} - (\tau_o)^{-1}]/X_B, \qquad (5.14)$$

where $\tau_o' = 1/(\alpha + k_{+B}^* \cdot X_b)$ and $\tau_o = 1/\alpha$.

Earlier, the "apparent" rate constant ($k^*_{+B,e.p.s.c.}$) was determined from the blocker-induced shortening of the e.p.s.c. decay [see Eq. (5.2)]:

$$k^*_{+B,e.p.s.c.} = \left[(\tau_{syn}')^{-1} - (\tau_{syn})^{-1}\right]/X_B,$$

where τ_{syn} and τ_{syn}' are time constants of the e.p.s.c. decay in the absence and presence of a blocker, respectively. Taking into account that τ_{syn} corresponds to τ_{b2} (see Chapter 3), one can obtain

$$k^*_{+B,e.p.s.c.} = k^*_{+B,b} = \left[\left(\tau_{b2}'^{-1}\right) - \left(\tau_{b2}\right)^{-1}\right]/X_B,$$

where τ_{b2} and τ_{b2}' are mean burst durations in the absence and presence of a blocker, respectively. Using expressions (5.9) and (5.10), we have

$$k^*_{+B,b} = k^*_{+B} \cdot \frac{\overline{m}}{\alpha \cdot \tau_{b2}} = \frac{k^*_{+B}}{1 + \dfrac{(\overline{m} - 1)\alpha}{\overline{m}(k_{-1} + \beta)}}. \qquad (5.15)$$

Using the values of α, k_{-1}, and β obtained for nicotinic receptors in rat sympathetic ganglion neurons (see Chapter 3), one can obtain

$$k^*_{+B,b} \cong 0.9 k^*_{+B}.$$

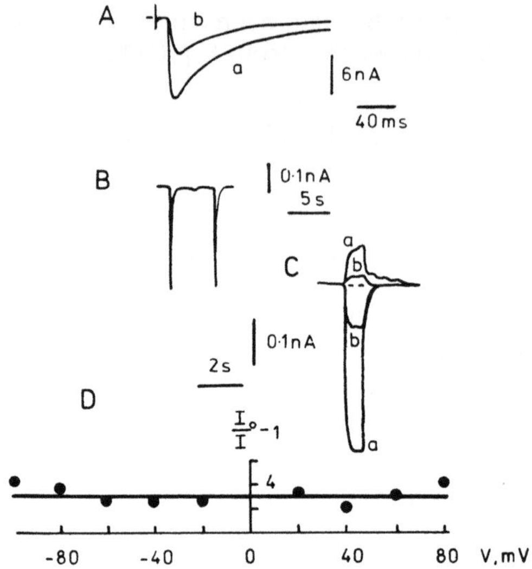

Fig. 5.8. Effect of trimetaphan on the responses of mammalian sympathetic ganglion neurons to nerve-released and iontophoretically applied acetylcholine. A) E.p.s.c.'s recorded from rabbit superior cervical ganglion neuron at –80 mV in normal solution (a) and in the presence of trimetaphan $1 \cdot 10^{-5}$ M (b); τ_{syn} remains unchanged by trimetaphan. B, C) Membrane currents obtained from rat superior cervical ganglion neurons and analyzed by methods similar to those in Fig. 5.6 (trimetaphan concentration $3 \cdot 10^{-6}$ M). No depression of second current is revealed in paired-pulse experiment (B). Superimposed currents shown in C were recorded at two distinct membrane potentials, –60 mV and +80 mV, in the absence (a) and in the presence of trimetaphan (b). No voltage dependence of the blocking effect was observed (D). All responses were obtained at 23–25°C. (A. Bobrishev, V. Derkach, D. Kurenny, A. Selyanko, and V. Skok, in press.)

The values of $k_{+B,b}{}^*$ and $k_{+B}{}^*$, found from Eqs. (5.14) and (5.15) using the parameters of single-channel activity, recorded in the presence of heptamethonium (see above), are $5.8 \cdot 10^7$ $M^{-1} \cdot s^{-1}$ and $1.8 \cdot 10^7$ $M^{-1} \cdot s^{-1}$, respectively. The threefold difference between these two values is higher than the predicted one (1.1-fold), which is probably due to the rather low accuracy of experimental determination of kinetic parameters from single-channel recordings. On the other hand, some basic characteristics of the heptamethonium effect on single-channel activity are consistent with those predicted by the above theory. These are: 1) the shortened mean channel open time, 2) the unchanged mean channel closed time within a burst, 3) the decreased number of channel openings within the burst, and 4) the shortened mean burst duration.

Two types of open-channel blocking actions, "fast," or high-frequency action, and "slow," or low-frequency action, can be distinguished, consistent with the above blocking model (see Colquhoun and Hawkes, 1983). The fast-blocking action is characterized by a high dissociation rate comparable to the rate of the blocker association with the open channel. An example of fast blockade is the interaction of local anesthetics with the acetylcholine-sensitive channels in the end-plate. In post-

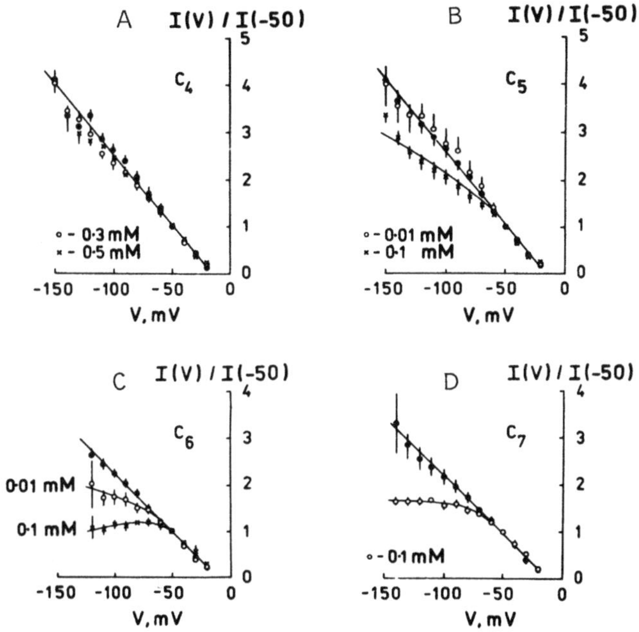

Fig. 5.9. Effect of ganglion-blocking agents tetra-, penta-, hexa-, and heptamethonium (C_4, C_5, C_6, and C_7, respectively) on the amplitude (*I*) of e.p.s.c. recorded from rabbit superior cervical ganglion neurons at 36°C. Symbols are the same as those used in Fig. 5.5 (Selyanko et al., 1982).

synaptic current evoked by a single presynaptic stimulus the fast-blocking agents produce bi-exponential decay instead of the normally observed single-exponential one (Adams, 1976; Neher and Steinbach, 1978). The fast-blocking agents also shorten single-channel openings, which yields a long-lasting burst of openings separated by relatively short interburst intervals that correspond to a channel blocked state (Neher and Steinbach, 1978; Colquhoun and Hawkes, 1983; Neher, 1983).

By contrast, the slow-blocking agents dissociate from the open channel at a rate much lower than their association rate (Ascher et al., 1978b). It is this type of blockade which is observed in the effects of the ganglion-blocking agents on the autonomic ganglion neurons, as evidenced by the single-exponential decay of e.p.s.c. observed in the presence of a blocking agent and also by the results of direct measurements of k_{-B}* values (Ascher et al., 1979; Selyanko et al., 1981b, 1982; Rang, 1982; Skok et al., 1983, 1984). The effects produced by heptamethonium on single-channel activity as described above, namely the shortenings of τ_o and τ_b without changing the short closings within the burst, are consistent with the slow-blocking mechanism.

5.2. BLOCKING OF RECOGNITION CENTER

It is commonly accepted that blocking of a recognition center, or a competitive blocking mechanism, is indicated by a parallel shift in the dose–response curve produced by increased concentration of the blocking agent. Another test for a competitive blocking mechanism is voltage-independent reduction by the blocking agent in the e.p.s.c. amplitude (although the same result can obviously be obtained with a presynaptic blockade). The classical example of competitive blockade, the effect of tubocurarine in low concentrations on neuromuscular transmission, is characterized by a voltage-independent reduction in the amplitude of the end-plate current (Albuquerque et al., 1978; Colquhoun et al., 1979). The effect of tubocurarine in autonomic ganglia exhibits several different mechanisms depending on the animal species. For this reason, this effect is considered in detail in a special section below.

A pure competitive blocking mechanism has been found in the effects produced in autonomic ganglia by trimetaphan (**2.19**) and by some neurotoxins, in particular by surugatoxin (Ascher et al., 1979; see Brown, 1980). This is evidenced by the fact that the blocking effects produced by these compounds in rat submandibular ganglion are not accompanied by a decrease in channel open time, they are voltage-independent, and are not enhanced by increased concentrations of the agonist. Similar results have been obtained in rabbit superior cervical ganglion (Bobrishev and Skok, 1987): trimetaphan ($1 \cdot 10^{-5}$ M) reduced the amplitude of e.p.s.c. without decreasing the e.p.s.c. decay time constant. On the other hand, as demonstrated in Fig. 5.8, trimetaphan reduces the acetylcholine-induced current in rat superior cervical ganglion neuron in a voltage-independent manner, which is in sharp contrast to what has been observed in the case of the open-channel blockers hexamethonium and IEM-1288 (cf. Fig. 5.6). A similar effect of trimetaphan on the acetylcholine-induced current excludes the possible interpretation of its effect upon the e.p.s.c. amplitude as being due to a presynaptic blockage.

5.3. BLOCKING OF CLOSED IONIC CHANNEL

Figure 5.9 illustrates the voltage dependence of the effects produced by a series of symmetrical bis-quaternary ammonium compounds on the amplitude of e.p.s.c. recorded from rabbit sympathetic ganglion neurons. One can see that hexa- and heptamethonium effects are strongly voltage dependent, in contrast to those of tetra- and pentamethonium. This is also evident from the values of H_B, the membrane hyperpolarization corresponding to an e-fold increase in the blocker-induced reduction of the e.p.s.c. amplitude, as shown in Table 5.1: the voltage dependence suddenly increases about fourfold with an increase in the polymethylene chain length between penta- and hexamethonium.

A similar observation was made with the voltage dependence of the blocking effect produced by symmetrical bis-quaternary ammonium compounds in potassium channels of sarcoplasmic reticulum (Miller, 1982). However, the voltage-dependence jump observed in potassium channels is less strong and occurs at greater chain length than that observed in sympathetic ganglion. One of the possible interpretations for this jump is that the compounds may interact with different sites

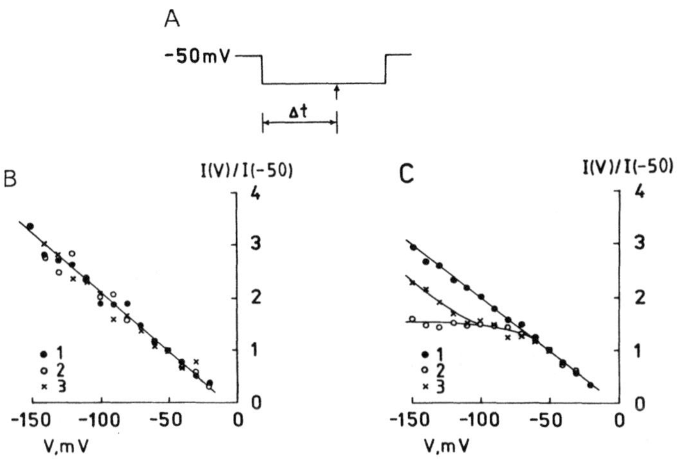

Fig. 5.10. Effect of time interval (Δt) between the voltage step and the appearance of e.p.s.c. on the amplitude (I) of the e.p.s.c. (see diagram of experimental procedure in A), as recorded in the presence of 10 μM pentamethonium (B) and 100 μM heptamethonium (C) at different membrane potential levels. The e.p.s.c.'s obtained in normal solution at Δt = 5 ms (1) are compared with those obtained in the presence of a blocking agent at Δt = 5 ms (2) and Δt = 7 s (3). The I values are normalized to those at 50 mV (Selyanko et al., 1982).

in the channel (see Kordas, 1970; Woodhull, 1973; Ascher et al., 1978b; Miller, 1982): hexa- and heptamethonium might interact with the sites more deeply located in the acetylcholine-sensitive channel than tetra- and pentamethonium. The localization of binding sites for these compounds is discussed further in Chapter 6.

What is the mechanism for the voltage-dependent decrease in the e.p.s.c. amplitude produced by hexa- and heptamethonium? As mentioned above, it is unlikely that this mechanism is the open-channel blockade (see also Rang, 1982). To account for this kind of phenomena, another blocking mechanism, the block of the closed ionic channel, has been suggested (Adams, 1977; Adams and Feltz, 1980a, b; Masukawa and Albuquerque, 1978; Tiedt et al., 1979). This mechanism, which is actually a noncompetitive blockade, has been observed, in particular, in the effects of TEA (Adler et al., 1979) and of guanidine derivatives (Farley et al., 1981) on the end-plate acetylcholine receptors. One of the tests for the closed-channel block is that the number of channels which can be opened by orthodromic stimulus in the presence of a blocking agent appears to be strongly influenced by the time interval between the membrane potential jump and the orthodromic stimulus applied at a new membrane potential level.

Figure 5.10 illustrates the experimental procedure used to examine the closed-channel blocking mechanism (A) and the results obtained using two ganglion-blocking agents, penta- and heptamethonium (B and C). These two compounds have been chosen because of the great difference in the voltage dependence between their effects on the e.p.s.c. amplitude. The e.p.s.c. was evoked at two different in-

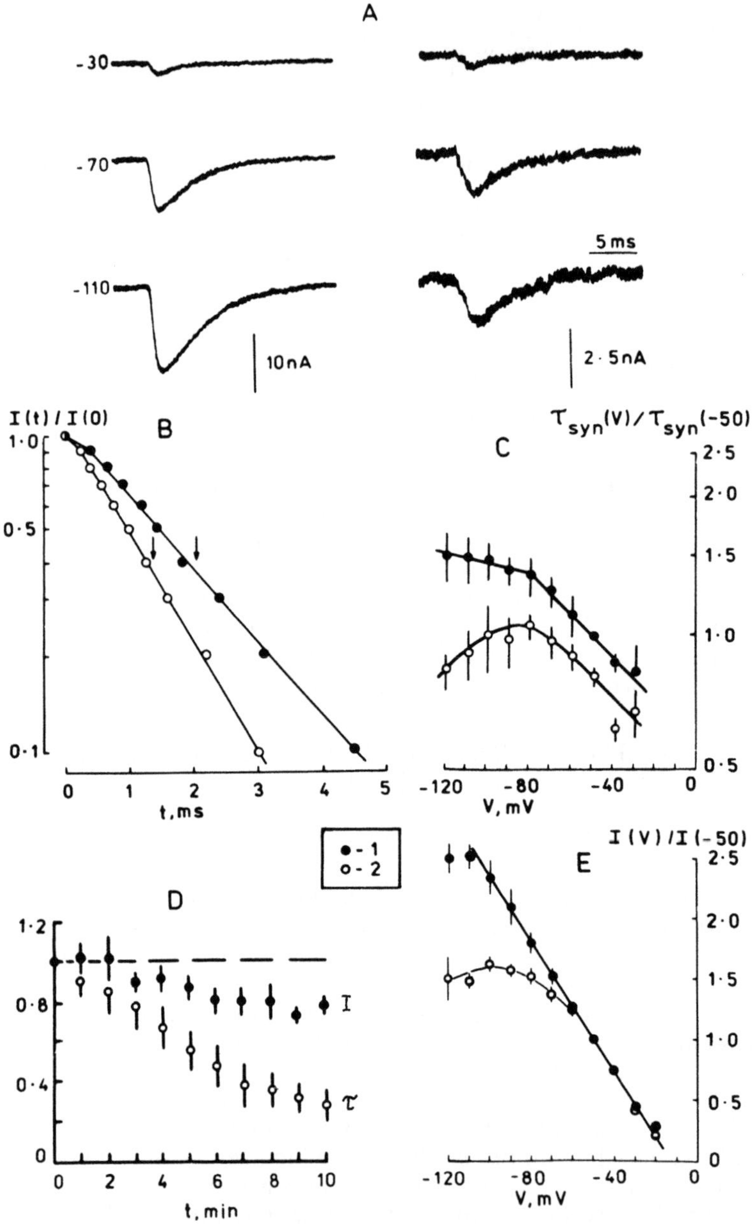

Fig. 5.11. Effect of tubocurarine (10 μM) on the e.p.s.c.'s recorded from rabbit cervical ganglion neurons at 36°C. Symbols A–E are the same as those used in Fig. 5.1. Note that the scales for e.p.s.c. amplitudes are different for left and right records in A (modified from Selyanko et al., 1981b).

tervals, $\Delta t = 5$ ms and $\Delta t = 7$ s, after the membrane had been hyperpolarized from -50 mV to various new levels. One can see from Fig. 5.10 that at $\Delta t = 5$ ms, heptamethonium markedly changes the slope in the current–voltage relationship, observed in the absence of a blocker, the effect being decreased at $\Delta t = 7$ s. By contrast, pentamethonium actually does not affect the current–voltage relationship observed without a blocker. Since hyperpolarization is applied before the acetylcholine-sensitive channels are opened, the heptamethonium effect cannot be explained as the open-channel block, but rather as the closed-channel block, the term used to differentiate this mechanism from the voltage-independent block of the recognition center. Recent studies performed on rat submandibular ganglion confirmed that hexamethonium may produce the closed-channel blocking effect (Large and Sim, 1986).

It should be noted that the closed-channel block can be as voltage dependent as the open-channel block or even more so (Farley et al., 1981; Vogel et al., 1984). This follows from a comparison of the values presented in columns 2 and 7 of Table 5.1 for hepta- and hexamethonium.

The molecular mechanism responsible for the closed-channel block is not yet clear. Compounds producing this kind of block probably interact with the acetylcholine receptor molecule at a site located closer to the channel "mouth" than the recognition center, as evidenced by the strong voltage dependence of their blocking effects.

The channel blocked-closed state described here should not be confused with its "closed-blocked" state found in recent experiments (Lingle, 1983a, b). The latter state has been suggested for nicotinic acetylcholine-sensitive channels of lobster gastric mill muscle treated by chlorizondamine, pempidine, mecamylamine, or decamethonium. The most essential detail in this closed-blocked mechanism is that the blocker that enters the open channel cannot leave it until the channel opens again with an agonist. No such mechanism has been described as yet for neuronal acetylcholine receptors (see, however, Gurney and Rang, 1984). Another trapping mechanism has recently been found for the action of some tertiary antagonists on nicotinic receptors in rat sympathetic ganglion neurons. This trapping effect was indicated by the fact that the acetylcholine-induced current in rat superior cervical ganglion neuron remained reduced, compared with its control value, after a blocker (IEM-1559) had been completely removed from the external solution with a prolonged washing-out procedure. Moreover, application of acetylcholine after the blocker had been washed out produced a marked additional reduction of the responses to subsequent acetylcholine applications, lasting as long as in the presence of the blocker. It seems that a blocker remains bound to a receptor, causing additional blockade after channel opening and slowly dissociating from the channel after its closing.

5.4. THE MECHANISM FOR A BLOCK BY TUBOCURARINE

In early studies, tubocurarine had been found to be four times less effective in blocking synaptic transmission through sympathetic ganglia than through a neuromuscular junction (Guyton and Reeder, 1950), although it was about five times more effective than hexamethonium in the ganglia (Bowman and Webb, 1972).

The effect produced by tubocurarine in rabbit superior cervical ganglion is shown in Fig. 5.11. Tubocurarine ($1 \cdot 10^{-5}$ M) strongly decreases the amplitude of e.p.s.c. without significant shortening of the e.p.s.c. decay τ_d (Fig. 5.11A, B, D). Within the range of membrane potential levels from –50 mV (the average resting membrane potential level) to about –80 mV, tubocurarine does not change the slope either in the current–voltage relationship or in the τ_d–voltage relationship (Fig. 5.11C, E). The latter effect is entirely different from that observed with hexamethonium in this ganglion (cf. Fig. 5.1). At the same time, the effect of tubocurarine on the e.p.s.c. amplitude is about as weakly voltage dependent as that observed for hexa-methonium (Fig. 5.11E). At membrane potentials more negative than –80 mV, the open channel blockade is obvious. The k_{+B}^* value, estimated (at –50 mV) from Eq. (5.2), is only $2.0 \cdot 10^6$ $M^{-1} \cdot s^{-1}$, which is three times less than in the case of hexamethonium. However, the H_{+B}^* is –26 mV, i.e., three times higher than in hexamethonium (see Fig. 5.4 and Table 5.1).

These results suggest that the tubocurarine effect on rabbit sympathetic acetylcholine receptors includes at least two components – the competitive component and the open-channel-blocking component. It is likely that it is not only the open-channel blockade which is responsible for the tubocurarine-induced ganglionic blockade observed in classical experiments. Support for this suggestion is given by the fact that the relative ganglion-blocking activity of tubocurarine is much higher than its relative k_{+B}^* value (Fig. 7.1). The relatively small voltage dependence of the tubocurarine-induced decrease in the e.p.s.c. amplitude which can be seen in Fig. 5.11 implies that the difference between the open-channel-blocking and the ganglion-blocking activities of tubocurarine shown in Fig. 7.1 is due to the contribution of a competitive blockade.

To examine further the possible contribution of a competitive mechanism in the blocking effect of tubocurarine, this effect has been studied in rat superior cervical ganglion neurons using the whole-cell recording technique already demonstrated in Figs. 5.6 and 5.8B, C. The results obtained are actually similar to those shown in Fig. 5.6. Surprisingly, no voltage-independent components were seen in the tubocurarine effect demonstrated by these results, which is not consistent with the above suggestion. One possible explanation for this discrepancy is that the mechanism of the tubocurarine effect in rat sympathetic ganglion differs from that in rabbit sympathetic ganglion.

A strong open-channel-blocking effect of tubocurarine was observed in rat submandibular ganglion (Ascher et al., 1979). Although k_{+B}^* for tubocurarine in this ganglion is lower than that for hexamethonium ($5 \cdot 10^6$ $M^{-1} \cdot s^{-1}$ and $2.2 \cdot 10^7$ $M^{-1} \cdot s^{-1}$, respectively), the dissociation constant K_B^* is somewhat higher ($3.4 \cdot 10^{-8}$ M vs $2.2 \cdot 10^{-8}$ M), due to the difference in k_{-B}^* (0.17 s^{-1} for tubocurarine and 0.48 s^{-1} for hexamethonium).

Only the open-channel blocking mechanism has been found in the effect of tubocurarine on molluscan neurons. In *Aplysia*, at –80 mV the rate constants are

$k_{+B}^* = 4.4 \cdot 10^5$ $M^{-1} \cdot s^{-1}$ and $k_{-B}^* = 0.1$ s^{-1}. Therefore, $K_B^* = 2.3 \cdot 10^{-7}$ M (Ascher et al., 1978b), which is an order lower than in rat parasympathetic ganglion neurons.

The fact that tubocurarine blocks not only e.p.s.c. but also the acetylcholine-induced current indicates that its effect on e.p.s.c. in autonomic ganglion cannot be entirely attributed to a presynaptic blockade. No presynaptic effect of tubocurarine has been observed in sympathetic ganglia by other authors (Ginsborg, 1971; Lees and Nishi, 1972).

Only a competitive blocking mechanism is responsible for the transmission blockade by tubocurarine at its low concentrations (not higher than $4 \cdot 10^{-7}$ M) in amphibian neuromuscular junction. At higher concentrations, the contribution of the open-channel-blocking component becomes more significant. At resting membrane potential level and at room temperature, the dissociation constant K_{+B}^* is about $1.2 \cdot 10^{-6}$ M, while the rate constants k_{+B}^* and k_{-B}^* are $9 \cdot 10^6$ $M^{-1} \cdot s^{-1}$ and 0.9 s^{-1} (Colquhoun et al., 1979; Colquhoun and Sheridan, 1982). In rat end-plate receptors k_{+B}^* is $13 \cdot 10^6$ $M^{-1} \cdot s^{-1}$, as measured at normal body temperature (Trautmann, 1982).

The k_{+B}^* and k_{-B}^* values for tubocurarine in the neuromuscular junction are very close to those in autonomic ganglia. Therefore, selectivity of tubocurarine as a myorelaxant agent is probably due to its competitive mechanism, which is much more significant in the neuromuscular junction than in autonomic ganglia.

5.5. THE MECHANISM FOR A BLOCK BY NEUROTOXINS

5.5.1. Variability of the Effects Produced by α-Toxins in Neuronal Nicotinic Acetylcholine Receptors

The α-toxins obtained from snake venom are highly specific and actually irreversible blockers in acetylcholine receptors of skeletal muscle and electric organ (Chang and Lee, 1963; Changeux et al., 1970; Banks et al., 1974; Kato et al., 1978). They are polypeptides consisting of 60–70 amino acids, with about 6–7 kD of total molecular mass. The most well known are toxins of the snakes *Bungarus* (α-bungarotoxin), *Naja* (α-cobratoxin), and *Denroaspis* (dendrotoxin), as well as of sea snakes *Laticauda* (erabutoxin *a* and erabutoxin *b*). All these toxins are very effective blockers for acetylcholine receptors in skeletal muscle and electric organ, although marked variability in their sensitivities has been reported in different species of amphibia (Lee et al., 1972a; Burden et al., 1975; Lee and Chen, 1976). Table 5.2 summarizes the effects produced by α-toxins in neuronal acetylcholine receptors. It is evident from Table 5.2 that receptors of numerous types, both in vertebrate and invertebrate neurons, are insensitive to α-toxins. Even in the few cases where sensitivity has been detected, it is different from that found in skeletal muscles in at least three respects. First, much higher concentrations of a toxin are needed (10^{-6}–10^{-5} M) to block interneuronal synaptic transmission than to block

TABLE 5.2. Effects of α-Toxins from Snake Venoms on Neuronal Nicotinic Receptors

Preparation	Ionic selectivity of nicotinic acetylcholine receptor channel: cationic (C) or anionic (A)	Presence (+) or absence (–) of blocking effect	Toxin origin (snake species)	References
Mammalian sympathetic ganglion	C	–	*Bungarus multicinctus, Naja naja siamensis, Laticauda semifasciata, Dendroaspis viridis*	Chou and Lee, 1969; Magazanik et al., 1974; Brown and Fumagalli, 1977; Patrick and Stallcup, 1977; Chiappinelli et al., 1981
Renshaw cells	C	–	*Bungarus multicinctus, Naja naja siamensis*	Duggan et al., 1976
Mammalian and avian sympathetic neurons in culture	C	–	*Bungarus multicinctus, Naja naja siamensis, Dendroaspis viridis*	Obata, 1974; Nurse et al., 1975; Kouvelas et al., 1978
Mammalian parasympathetic (pelvic and submandibular) ganglia	C	–	*Bungarus multicinctus, Naja naja siamensis, Laticauda semifasciata*	Ascher et al., 1979; Chiappinelli et al., 1981
Avian ciliary ganglion	C	+ –* –	*Naja naja siamensis* *Bungarus multicinctus* *Laticauda semifasciata*	Conti-Tronconi et al., 1979 Chiappinelli et al., 1981 Chiappinelli, 1983
Avian parasympathetic ganglion neurons in culture	C	+ –	*Bungarus multicinctus* *Dendroaspis viridis*	Ravdin and Berg, 1979 Patrick et al., 1980
Cochlea	C	+	*Bungarus multicinctus*	Fex and Adams, 1978

Goldfish and Toad retino-tectal tract	C	+	*Bungarus multicinctus*	Freeman et al., 1980
Amphibian spinal cord	C	+	*Dendroaspis viridis*	Miledi and Szepaniak, 1975; Quik et al., 1981
Myenteric plexus	C	–	*Bungarus multicinctus, Naja nivea*	Bursztajn and Gerson, 1977
Amphibian sympathetic ganglion	C	+ –	*Bungarus multicinctus, Laticauda semifasciata; Dendroaspis viridis*	Kato et al., 1980; Marshall, 1981; Quik et al., 1981
Amphibian cardiac ganglion	C	–	*Dendroaspis viridis*	Patrick et al., 1980
Molluscan neurons:				
Helix aspersa	C	–	*Bungarus multicinctus* *Dendroaspis viridis*	Szepaniak, 1974
	A	–	*Bungarus multicinctus*	
	A	+	*Dendroaspis viridis*	
Aplysia	C, A	+	*Bungarus multicinctus*	Shain et al., 1974
	C	–	*Bungarus multicinctus* *Bungarus caerules*	Kehoe et al., 1976
	A	+ –	*Bungarus multicinctus* *Bungarus caerules*	
	C	–	*Naja*	Ono and Salvaterra, 1981
	A	+	*Naja*	
	C, A	+	*Vipera russelli*	Slater et al., 1983, 1985

*Components, copurified with α-bungarotoxin, exhibit blocking activity (see text).

neuromuscular transmission (about 10^{-7} M); in addition, the toxin-induced blockade of interneuronal transmission is usually not complete. Second, the toxin-induced blockade of interneuronal transmission is usually reversible. Third, very high selectivity has been found in neurons toward different types of toxins.

One interesting peculiarity observed in molluscan neurons is that the acetylcholine-induced increase in anionic permeability appears to be sensitive to α-toxins more often than an increase in cationic permeability (see Table 5.2). Based on this fact and also on the fact that the anion-conducting acetylcholine-sensitive channels in molluscan neurons are more resistant to the blocking effect of hexamethonium than the cation-conducting channels (see Chapter 4), some authors considered the acetylcholine-sensitive anion-conducting channels in molluscan neurons to be more similar to end-plate channels than the acetylcholine-sensitive cation-conducting channels in neurons (Kehoe, 1972). However, one of the α-toxins, vipoxin, obtained from the snake *Vipera russelli* (molecular mass 13 kD), blocks both types of channels in molluscan neurons without blocking the end-plate channels (Slater et al., 1983, 1985). It should be noted that vipoxin is somewhat more effective in anion-conducting than in cation-conducting acetylcholine receptor channels. An interesting feature of vipoxin is that, in contrast to other toxins, it may stimulate the acetylcholine receptor channels if used at higher concentrations than those needed for its blocking effect.

The blocking effect produced by α-bungarotoxin on the acetylcholine-gated anion-conducting channels in molluscan neurons can be abolished by eserine (physostigmine) (Carpenter et al., 1976). The effect of eserine has been attributed to its interaction with the acetylcholine receptor (Carpenter et al., 1976) rather than to its direct action upon α-bungarotoxin (Lee et al., 1972b). By contrast, eserine does not reduce the blocking effect produced in the same receptors by vipoxin (Slater et al., 1985).

Another factor that may strongly modify the α-toxin effect is the mode of toxin application. For example, α-bungarotoxin, when injected into the external carotid artery, effectively blocks synaptic transmission through rat superior cervical ganglion, but it is ineffective if it is bath-applied (see Toldi et al., 1983; Table 5.2). To account for this difference, it has been suggested that there are enzymatic barriers in the ganglion tissue that make acetylcholine receptors unavailable for bath-applied α-bungarotoxin. This suggestion, however, is not consistent with the results of autoradiographic observations indicating that the bath-applied α-bungarotoxin binds to acetylcholine receptors in the ganglion neurons (see Chapter 1).

The most intriguing question is why α-bungarotoxin, although interacting with nicotinic acetylcholine receptors in autonomic ganglion neurons, does not influence ganglionic transmission. Some aspects of this problem are discussed below.

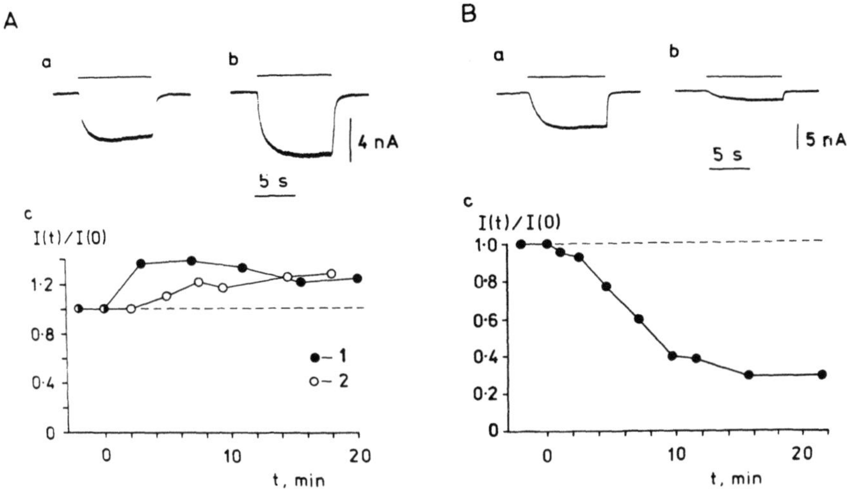

Fig. 5.12. Effects of α-neurotoxins on acetylcholine-induced membrane currents in the neurons of rabbit superior cervical ganglion. A) Potentiating effect of α-neurotoxins; a) a response to iontophoretic application of acetylcholine recorded in normal solution; b) the same but 7 min after application of α-bungarotoxin (10^{-6} M). Horizontal bars indicate the time when acetylcholine was applied; c) time courses of α-bungarotoxin (1) and α-carbatoxin (2) effects observed in two different neurons. B) Inhibitory effect of α-cobratoxin; symbols are the same as those in A (Selyanko, 1983).

5.5.2. Binding of Neurotoxins to Neuronal Nicotinic Acetylcholine Receptors without Causing Their Blocking

One may assume that this kind of binding, although not followed by receptor blocking, nevertheless may somehow modify the receptor functions. To test this assumption, the effects of α-toxins on the responses to acetylcholine application were studied in rabbit superior cervical ganglion neurons (Selyanko, 1983). As Table 5.2 indicates, the transmission through the ganglion is resistant to α-toxins.

It was found in this study that α-bungarotoxin, α-cobratoxin, and erabutoxin *b* when used at a concentration of $1 \cdot 10^{-6}$ M produced variable effects on acetylcholine-induced membrane currents (Fig. 5.12). In one part of the neurons, the amplitude of acetylcholine-induced current was increased (Fig. 5.12A), while in the rest either a decrease of the current (Fig. 5.12B) or its transient increase, followed by a subsequent decrease, were observed. Therefore, α-neurotoxins produced two opposite effects.

The amplitude of a steady-state current induced by acetylcholine (I_{ACh}) can be described by the following equation (see Stevens, 1976):

$$I_{ACh} = \left(\frac{\beta_a}{\alpha_a}\right)\frac{iN[ACh]^n}{K}, \qquad (5.16)$$

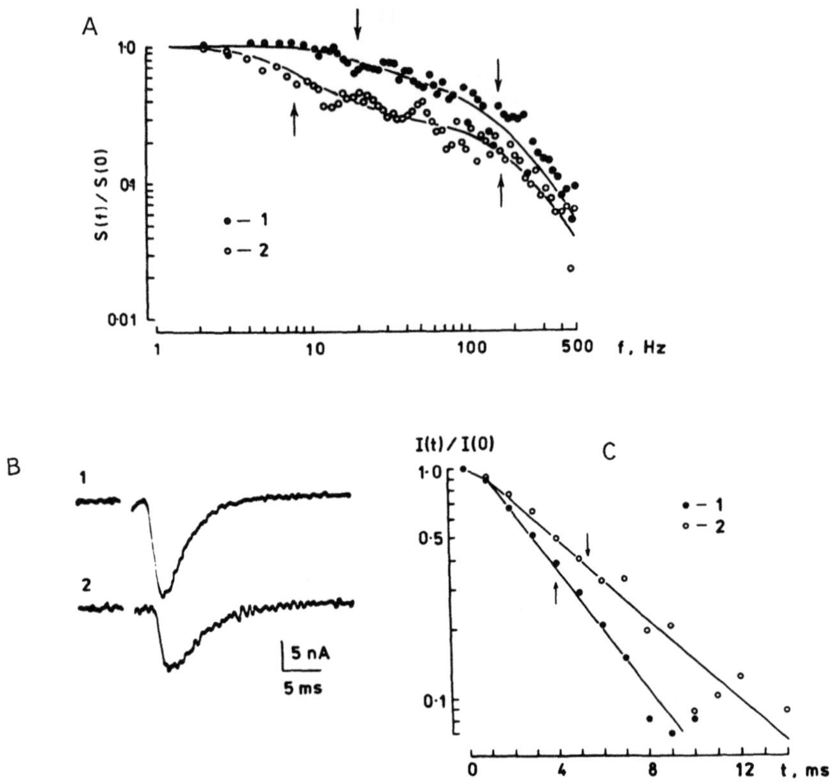

Fig. 5.13. Effects of α-neurotoxins on nicotinic acetylcholine receptor channel kinetics in rabbit superior cervical ganglion neurons. A) Effect of α-cobratoxin on the power spectrum of acetylcholine-induced membrane current fluctuations. 1, 2) Power spectra obtained in normal solution and in the solution containing 10^{-6} M α-cobratoxin, respectively. Each spectrum is fitted by the sum of two Lorentzian functions [see Eq. (3.14)] with f_1 = 155 Hz (τ_{f1} = 1 ms) and f_2 = 20.2 Hz (τ_{f2} = 7.9 ms) in 1 and f_1 = 183 Hz (τ_{f1} = 0.9 ms) and f_2 = 7.9 Hz (τ_{f2} = 20.2 ms) in 2. The f_1 and f_2 values are indicated by arrows. B) Effect of α-bungarotoxin on e.p.s.c.; 1, 2) e.p.s.c. recorded in normal solution and in the solution containing 10^{-6} M α-bungarotoxin, respectively. C) Semilogarithmic plots of the decays of e.p.s.c.'s shown in B. Arrows indicate the time constants 3.9 (1) and 5.4 ms (2) (Selyanko, 1983).

where i is the single-channel current, K is the dissociation constant for the interaction between acetylcholine and the acetylcholine receptor, β_a and α_a are the apparent rate constants of channel opening and channel closing, respectively, N is the number of receptors available for acetylcholine, and n is the number of acetylcholine molecules that interact with a single receptor to open its ionic channel.

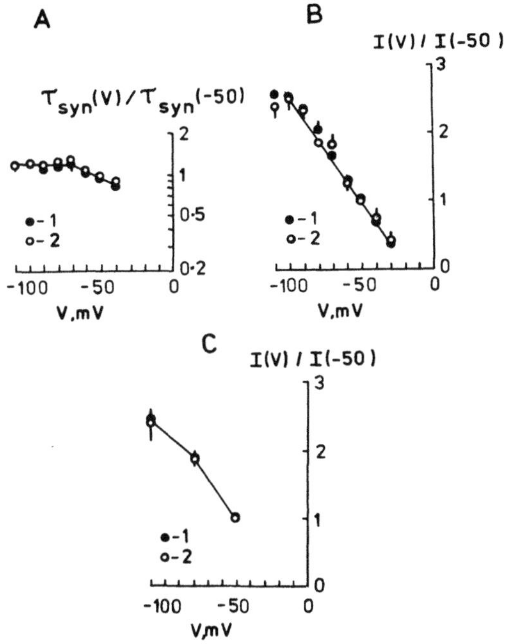

Fig. 5.14. Voltage dependence of the e.p.s.c. decay time constant (A), e.p.s.c. amplitude (B), and acetylcholine-induced membrane current amplitude (C) recorded from rabbit superior cervical ganglion in normal solution (1) and in the solution containing 10^{-6} M erabutoxin *b* (2) at 36°C. The corresponding values obtained at –50 mV are taken as unity. The averaged results are shown as obtained from four (A, B) and six (C) neurons (Selyanko, 1983).

It has been concluded from an analysis of the power spectrum of the acetylcholine-induced membrane current (Fig. 5.13) that α-toxins increase the α_a^{-1} value for population of channels with long open time (Selyanko, 1983). This effect may explain an increase in the acetylcholine-induced current caused by α-toxins. The opposite effect of α-toxins – a decrease in the acetylcholine-induced current – is probably due to a decrease in *N*, i.e., to a blockade of part of the receptors. Considering the rather large size of the α-toxin molecule, one could hardly suggest that it may directly interact with the ionic channel of the acetylcholine receptor. It should be noted, however, that one of the α-toxins, (±) dipentylperhydrohistrionicotoxin, may interact with the end-plate open channel, as evidenced by the high voltage dependence of its effect on the end-plate current (Maleque et al., 1984).

To know whether α-toxins may interact with the ionic channel in neuronal acetylcholine receptor, the voltage dependence of the effects produced by toxins on steady-state acetylcholine-induced current and on e.p.s.c. has been examined.

It has been found that erabutoxin *b*, which influences the acetylcholine-induced steady-state membrane current and e.p.s.c. similarly to α-bungarotoxin and α-cobratoxin, exerts its effect in a voltage-independent manner (Fig. 5.14). Therefore, both potentiating and depressing actions of α-toxins should result from their interaction with a site in the acetylcholine receptor remote from its ionic channel and probably located at the extracellular portion of the receptor.

5.5.3. Effects Produced on Neuronal Nicotinic Acetylcholine Receptors by Snake Venom Components Ineffective in Other Acetylcholine Receptors

Some components of the snake venom which are without blocking effect on acetylcholine receptors in skeletal muscle and electric organ are able to block neuronal nicotinic acetylcholine receptors (Ravdin and Berg, 1979). Four such components have been found in venom of the snake *Bungarus multicinctus*, in addition to the well-known α-bungarotoxin. All four components can block acetylcholine receptors in cultured cells of avian ciliary ganglion; one of them has blocking activity two orders higher than in α-bungarotoxin.

Some commercial species of α-bungarotoxin that block neuromuscular transmission are ineffective in ciliary ganglion (Chiappinelli et al., 1981). Their ineffectiveness is due to the lack of another component, ϰ-bungarotoxin, a polypeptide of about 6.5 kD molecular mass which is lower than in α-bungarotoxin (Chiappinelli, 1983). ϰ-Bungarotoxin blocks acetylcholine receptors in ciliary ganglion neurons at concentration as low as $7.5 \cdot 10^{-8}$ M, whereas α-bungarotoxin, although it binds in the ganglion with $K_D = 1 \cdot 10^{-9}$ M, does not block ganglionic transmission even at concentrations of $1.2 \cdot 10^{-6}$ M.

It has been suggested from the above studies of toxin effects that ciliary ganglion neurons possess two kinds of toxin-binding sites. The first kind, which are functionally important for the acetylcholine receptor, may bind ϰ-bungarotoxin but not α-bungarotoxin. The second kind may bind both ϰ- and α-bungarotoxins; however, these sites are not functionally important. The fact that only ϰ-bungarotoxin, but not α-bungarotoxin, can block ganglionic transmission and that tubocurarine, but not α-bungarotoxin, can protect the acetylcholine receptors against the blocking effect of ϰ-bungarotoxin is evidence for the existence of the former binding sites. On the other hand, ϰ-bungarotoxin can bind to its specific binding site even in the presence of α-bungarotoxin, but the binding is prevented by the irreversible agonist bromoacetylcholine.

That there is a second toxin-binding site, not important for functioning of a receptor molecule and not distinguishing between α- and ϰ-bungarotoxins, is evident from the fact that α-bungarotoxin, although it does not block ganglionic transmission, binds in the ganglion, and this binding partially reduces the ϰ-bungarotoxin binding. The interaction of both toxins with the second site can likewise be prevented by bromoacetylcholine.

Thus, there are probably two distinct binding sites for toxins in the acetylcholine receptor. However, it is not yet clear whether they are located on the same receptor or on different receptors.

The effects produced by different components of the *Bungarus multicinctus* toxin were also studied in the rat superior cervical ganglion (Quik and Lamarca, 1982). In early experiments, the ineffectiveness of α-bungarotoxin in mammalian sympathetic ganglion was shown (see Table 5.2). In addition to the toxin components purified by other authors, Quick and Lamarca purified a component with molecular mass about 15 kD. This component (indicated by the authors as II-S1) does not block neuromuscular transmission but instead exerts a strong blocking effect on transmission through sympathetic ganglion. The blocking effect of II-S1 is resistant to eserine and, therefore, cannot be attributed to modification of acetylcholinesterase activity. The influence of II-S1 on presynaptic acetylcholine release is negligible (see also Quik, 1985a). It has been concluded from these results that II-S1 blocks nicotinic acetylcholine receptors in sympathetic ganglion neurons.

Surprisingly, ϰ-bungarotoxin, which blocks acetylcholine receptors in avian ciliary ganglion, does not affect transmission through rat superior cervical ganglion (Quik and Lamarca, 1982), thus indicating that there are basic differences between the two corresponding receptor groups.

Although the II-S1 component blocks transmission through the sympathetic ganglion, it does not reduce the depolarizing responses produced in the ganglion neurons by carbachol (Quik and Lamarca, 1982). Initially this result had been explained by the lack of sensitivity to II-S1 in extrasynaptic acetylcholine receptors in contrast to synaptic ones (see Dun and Karczmar, 1980). However, it was later found that II-S1 blocks the carbachol-induced accumulation by ganglion neurons of channel-permeant ions (Quik, 1985a).

It is most interesting that binding of II-S1 in the ganglion is not prevented by nicotinic agonists or antagonists. This result implies that the binding site for II-S1 is not common with a recognition center. In addition, II-S1 possesses high phospholipase-like activity correlating with its ganglion-blocking activity as well as with its effect on the carbachol-induced ion flow through the neuronal membrane. Therefore, the effects of II-S1 are probably due to its interaction with the phospholipid environment of the acetylcholine receptors rather than with their recognition centers or with their ionic channels. A similar mechanism could be responsible for the II-S1 effect on nicotinic acetylcholine receptors in the brain (Quik, 1985b).

Another polypeptide purified from the *Bungarus multicinctus* venom is a P 15 component, which in its features is close to II-S1 and differs from α- and β-bungarotoxins (Gotti et al., 1985a). The P 15 is similar to II-S1 in its molecular mass (about 15 kD), in possessing the phospholipase-like activity, and in blocking ganglionic transmission without affecting neuromuscular transmission. However, unlike II-S1, P 15 does not bind to ganglionic receptors after they are treated with tubocurarine, and its ganglion-blocking activity does not correlate with its phospholipase-like activity. These features of P 15 imply that it is not identical to II-S1.

A more detailed comparison of the effects produced in sympathetic ganglion by α-bungarotoxin and P 15 (Gotti et al., 1985a, b) yielded the following results: 1) each polypeptide inhibited binding of another one in the ganglion homogenate; 2) although α-bungarotoxin does not block ganglionic transmission, it protects the transmission against the ganglion-blocking effect of P 15; 3) the levels of binding in the ganglion are similar for both toxins and are not affected by ganglionic denervation. These results suggest that both toxins interact with the same binding site in the receptor molecule, which is not identical to the acetylcholine-binding site. Because of its larger molecular size, P 15 probably can more effectively interfere with the interaction of acetylcholine with its binding site, which may account for a greater ganglion-blocking activity than that of α-bungarotoxin (Gotti et al., 1985b).

Both toxins probably interact with a receptor at a site located outside its ionic channel. This is suggested by the fact that hexamethonium, the open-channel blocking agent, does not essentially influence binding of P 15 and of α-bungarotoxin in sympathetic ganglion. At the same time, α-bungarotoxin protects ganglionic transmission against the blocking effect of hexamethonium, which can be explained by steric hindrance (Gotti et al., 1985b).

5.5.4. Possible Functional Significance of the Sites That Bind Neurotoxins without Affecting Synaptic Transmission

As mentioned previously, α-bungarotoxin does not block synaptic transmission in autonomic ganglia (see Table 5.2), although it binds to the neurons of the ganglia. One possible explanation is that the binding sites for α-bungarotoxin in the ganglion are not identical to acetylcholine receptors, which is supported by their immunological differences (Patrick and Stallcup, 1977). This is consistent with the observation that an increase in sensitivity to acetylcholine in cultured neurons, which is probably due to an increase in the number of their acetylcholine receptors, does not correlate with their binding of α-bungarotoxin (Mitsuka and Hatanaka, 1983; Smith et al., 1983).

The functional role of the toxin-binding sites that are not involved in synaptic transmission remained unclear for a long time. Some authors thought that these were needed to maintain normal functioning of synaptic contacts (Freeman, 1977; Schmidt, 1979). This hypothesis was supported by the observation that during development of the brain the toxin-binding sites appear earlier than other structural components needed for cholinergic transmission (Salvaterra and Moore, 1973; Ben-Barak and Dudai, 1979; Hunt and Schmidt, 1979). More recently, the endogenous substance of about 1 kD molecular mass that bound to the sites identical with the toxin-binding sites was found both in muscle fibers, where α-bungarotoxin blocks synaptic transmission, and in central neurons, where α-bungarotoxin is ineffective as a blocking agent (Quick, 1982). One possibility is that through interaction of this endogenous substance with the toxin-binding sites trophic communications between cells may occur.

5.5.5. The Effects of a Neurotoxin Obtained from Mollusc

Surugatoxin, a polypeptide obtained from sea mollusc *Babylonia japonica* (molecular mass 811), selectively blocks transmission through autonomic ganglia without blocking it in neuromuscular junction (Hayashi and Yamada, 1975; Brown et al., 1976). The effect of surugatoxin is reversible and is characterized by an apparent affinity constant about two orders higher than that for hexamethonium. Surugatoxin does not interact with receptors other than nicotinic in ganglion neurons and does not affect the release of acetylcholine from preganglionic nerve terminals, i.e., is a specific ganglion-blocking agent (Brown et al., 1976). In contrast to the effects of many other ganglion-blocking agents (except trimetaphan), the blocking effect of surugatoxin, as examined in rat parasympathetic ganglion, is voltage independent (Ascher et al., 1979). This suggests that surugatoxin, like trimetaphan, is a pure competitive blocking agent interacting with the recognition center of the acetylcholine receptor. In brain neurons surugatoxin binds to the sites that are not identical to the α-bungarotoxin-binding sites (Rapier et al., 1985).

5.6. CONCLUSIONS

The reversible effects of ganglion-blocking agents on neuronal acetylcholine receptors are marked by three mechanisms: blockade of the open ionic channel, blockade of the recognition center, and blockade of the closed ionic channel.

The open ionic channel blocking mechanism has been found in the effects produced by various ganglion-blocking agents, including mono- and bis-quaternary ammonium compounds and tertiary amines. In particular, hexamethonium, even at its minimum effective concentration, shortens the mean apparent open time of acetylcholine receptor channels in a voltage-dependent manner in rabbit superior cervical ganglion. The hexamethonium-induced receptor blockade is enhanced by preliminary receptor activation and by increased agonist concentration. A decrease in the channel mean apparent open time is due to both a decrease in true open time (single opening in the burst) and in the number of openings in the burst.

The blockage of a recognition center is produced by trimetaphan and surugatoxin at comparatively low concentrations. This mechanism is indicated by a voltage-independent decrease in the probability of acetylcholine-sensitive channels being opened by agonists, with the kinetics of their activity remaining unchanged.

The blockade of the closed ionic channel has been revealed in the effects produced by heptamethonium. One of the indications of this mechanism is a blocker-induced dependence of the probability of acetylcholine-sensitive channels being opened by agonists in the time interval between the membrane potential jump and the moment when the receptors are activated.

α-Toxins purified from snake venom exert variable effects on neuronal acetylcholine receptors. Some types of receptors can be easily blocked by α-toxins whereas others are resistant to their blocking effect, in contrast to the end-plate receptors. In acetylcholine receptors of sympathetic ganglion neurons α-toxins produce two effects that are opposite in modifying ganglion transmission: in one part of the receptors, the probability of their being activated with an agonist is reduced, while in another part the channel mean apparent open time is increased. Some components of snake venom (other than α-toxins), ϰ-bungarotoxin, and II-S1, as well as surugatoxin, a peptide purified from sea molluscs, and vipoxin, are selective in blocking neuronal acetylcholine receptors compared with other types of acetylcholine receptors.

Chapter 6

LOCALIZATION AND POSSIBLE FUNCTIONAL SIGNIFICANCE OF THE SITES IN NEURONAL NICOTINIC ACETYLCHOLINE RECEPTOR INTERACTING WITH BLOCKING AGENTS

6.1. LOCALIZATION OF THE SITES INTERACTING WITH BLOCKING AGENTS IN THE RECEPTOR MOLECULE

As follows from Table 4.3, the activities of symmetrical bis-quaternary ammonium compounds as ganglion-blocking agents depend on the length of the polymethylene chain that links their two cationic groups. This dependence is also illustrated by Fig. 6.1. For comparison, the blocking activities in neuromuscular synapse are indicated. One can see that as the number of methylene groups n grows from 4 to 21 (the corresponding symmetrical compounds will be further indicated as C_4–C_{21}), three peaks of activity are observed, at n equal to 5 or 6, 9 or 10, and 16–18. The activity peak at n = 16–18 is observed both in ganglionic and in neuromuscular synapses, whereas each of the activity peaks at n = 9–10 and n = 5–6 is observed in only one type of synapse – the former peak in neuromuscular synapses and the latter one in ganglionic synapses (Paton and Zaimis, 1949; Barlow and Zoller, 1964).

These activity peaks have been interpreted as being due to binding of the ligand molecule with two negatively charged sites in the postsynaptic membrane. According to the initial hypothesis, the sites binding one ligand molecule of n = 9–10 or n = 16–18 are two separate acetylcholine receptor molecules (Michelson and Zeimal, 1973), while those binding a ligand molecule of n = 5–6 may belong to the same acetylcholine receptor molecule (Gill, 1959; see Anichkov, 1974). Later, however, it became commonly accepted that the ligands of all three types are bound within one acetylcholine receptor molecule. This was evidenced by the fact that the diameter of a receptor molecule was much larger than the longest intercharge distance in the above compounds and also that the density of the receptor distribution at the postsynaptic membrane is too low to allow interaction of two receptors with the same ligand molecule.

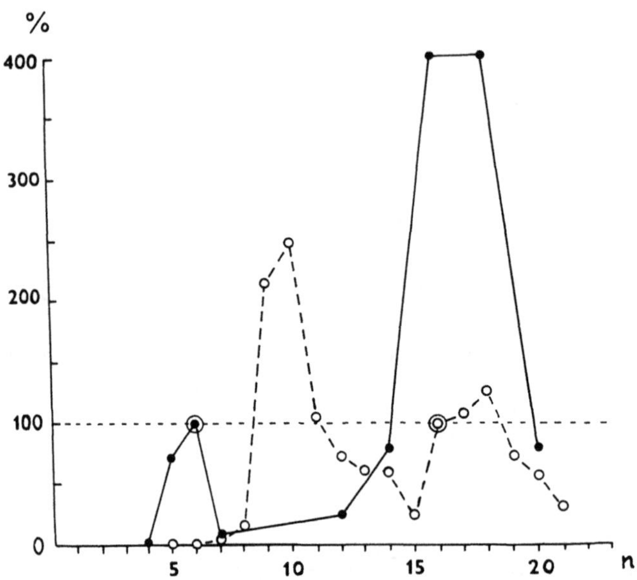

Fig. 6.1. Relationship between the number *n* of CH_2 groups in the compounds $(CH_3)_3\overset{+}{N}-(CH_2)_n-\overset{+}{N}(CH_3)_3$ and their synaptic transmission-blocking potencies. Smooth and dashed lines represent the results obtained from cat sympathetic ganglion and from cat skeletal muscle, respectively (Paton and Zaimis, 1949; Barlow and Zoller, 1964). Potencies are expressed in percent relative to those of hexamethonium (for sympathetic ganglion) and hexadecamethonium (for skeletal muscle).

If we assume that the better the fit between the complementary electric charges of ligand and receptor, the higher the blocking activity, we might localize the tentative binding sites in the receptor.

For this purpose, the most stable conformations in a series of symmetrical bis-quaternary ammonium compounds with variable intercharge distances were predicted based on their intramolecular nonvalent electrostatic and torsion interactions and also on the deformation energy of their angles of valences (Rosengart and Zhorov, 1983). Using this method it was found that in the medium with low dielectric constant ($\varepsilon = 4$), which characterizes the ligand–receptor interactions (see Triggle and Triggle, 1976), the most probable intercharge distances R for compounds C_5, C_6, and C_{10} are 0.8, 0.9, and 1.3 nm (with bin width $R = 0.05$ nm). According to other results obtained with a somewhat different approach (Dyadyusha et al., 1985), the intercharge distances for compounds C_5, C_6, C_9, C_{10}, C_{16}, and C_{18} are 0.75, 0.89, 1.26, 1.39, 1.91, and 2.11 nm ($\varepsilon = 3$). One may thus suppose that the blocking activity peaks for C_5–C_6, C_9–C_{10}, and C_{16}–C_{18} correspond to the intercharge distances 0.75–0.90, 1.26–1.39, and 1.91–2.11 nm, respectively.

In order to know the intercharge distances in the receptor molecule, the structure of nicotinic receptor protein was analyzed. As has been mentioned in Chapter 2.1, the most probable structure that may interact with a ligand cationic group is a carboxyl group of aspartate or glutamate.

If the secondary structure of a polypeptide chain is a β-strand, the neighboring amino residues that are exposed to the same side of the strand stay apart from each other at 0.67 nm (see Volkenstein, 1975) and are separated by one amino residue in the amino acid sequence. This means that the binding site for C_5–C_6 should contain two aspartic or glutamic residues separated by any other residue in the amino acid sequence. This structure will be further referred to as combination R_3. The same approach suggests that in the binding sites for C_9–C_{10} and C_{16}–C_{18} two aspartic or glutamic residues in the amino acid sequence are separated by three and five other residues, respectively (combinations R_5 and R_7).

Another possible secondary structure for a blocker-binding site in a receptor molecule is the α-helix. Assuming that the averaged α-helix is a cylinder 0.96 nm in diameter, with a 0.54-nm step and 3.6 residues per step (Schulz and Schirmer, 1979; Devillers-Thiery et al., 1983), one can find that the intercharge distance 0.75–0.90 nm (C_5–C_6) is close to a straight-line distance between each fourth or fifth residue in the sequence, while an intercharge distance 1.91–2.11 nm (C_{16}–C_{18}) corresponds to a nearly-straight-line distance between each fifteenth residue. No straight-line intercharge distance that would correspond to the 1.26–1.39 nm range (C_9–C_{10}) has been found in the α-helix.

The binding sites for C_5–C_6, C_9–C_{10}, and C_{16}–C_{18} that are identified using the above approach are indicated in Figs. 2.5 and 2.6 as C_6, C_{10}, and C_{16}, respectively. In the receptor molecule of $\alpha_2\beta\gamma\delta$ stoichiometry, there are 14 possible binding sites for C_5–C_6, 6 possible binding sites for C_9–C_{10}, and 8 possible binding sites for C_{16}–C_{18}. Using the model shown in Fig. 1.8B, one can see that the binding sites are located in the portions of the receptor that are exposed either in extracellular or intracellular space. Interestingly, in the extracellular portion the binding sites are concentrated in two relatively small areas within the ranges Glu 45–Glu 93 and Asp 157–Asp 186. [Figs. 2.5 and 2.6, see also Fig. 2.9, where the amino acid sequences of all subunits are aligned to reach their maximum homology]. One cannot exclude the possibility that binding of a blocker to the former area results in a competitive blockade, since the latter site is probably located closer to the ionic channel than the former one. However, evidence for this suggestion is lacking.

It is clear from Figs. 2.5, 2.6, and 2.9 that, although some possible binding sites for blockers are located in close vicinity to those for acetylcholine, they do not overlap, except site 6 in the β subunit that overlaps with a site for C_{16}. This correlates with the fact that all blockers listed are noncompetitive in relation to the acetylcholine receptor of the end-plate, which is very close in its primary structure to that of the electric organ shown in Figs. 2.5, 2.6, and 2.9 (Adams, 1976; Adams and Sakmann, 1978; Lukomskaya et al., 1981; Gurney and Rang, 1984). Figure 2.10 illustrates the tentative binding sites for blockers at the neuronal acetylcholine receptor α subunit localized using the above approach. One can see that there are four binding sites for C_6 in the neuronal α subunit instead of two in the electric organ receptor α subunit, which correlates with the difference in their sensitivities to hexamethonium. Another point of interest is that, as in the case of electric organ, neither one of the C_6-binding sites overlaps with the tentative acetylcholine-binding sites.

Another possibility is that the two cationic groups of a bis-quaternary ammonium compound may bind to two negatively charged amino residues that, although they are located close to each other in the protein tertiary or quaternary structure, are remote from each other in the amino acid sequence or even belong to different subunits. This possibility cannot be tested at present because the tertiary and quaternary structures of the receptor protein remain unknown.

As has been mentioned in Chapter 3, the ionic channel is probably lined by five α-helical portions L of the corresponding five subunits. This makes it possible that the binding sites responsible for the channel-blocking effect are located in the L portion of each subunit. Indeed, the glutamate residues which are thought to preferentially bind the cationic groups of channel-blocking drugs, compared with aspartate residues (see Chapter 4) are more concentrated in the L-portion than in any other portion of the receptor protein. In particular, in the L-portion of the neuronal acetylcholine receptor, they constitute 81% of the total content of aspartate and glutamate residues, whereas in the whole α subunit the contents of aspartate and glutamate residues are almost identical (see Boulter et al., 1986). The amino acid sequence in the L portion of the neuronal receptor differs from that of the electric organ receptor in the following peculiar respect: there is a combination R_4 in the neuronal receptor instead of the combination R_8 in the electric organ receptor. It is tempting to speculate that this difference is responsible for the increased sensitivities to hexamethonium in neuronal receptor and to decamethonium in the end-plate (see Fig. 6.1), although evidence supporting this speculation has not yet been obtained.

An independent way of localizing the binding site for the charged ligand in the membrane electric field is through the voltage dependence of the binding reaction. This approach, which was initially used to explain the open-channel blockade produced by protons (Woodhull, 1973), was later used to localize the binding sites for various charged ligands in the membrane channels (Ascher et al., 1978b; Miller, 1982). The principle used in this method is that the stronger the voltage sensitivity of the binding rate constants, the deeper in the membrane electric field the charged particle binds, since its movement toward the binding site becomes more dependent on the electric field.

Using this approach it has been found that the binding sites for penta-, hexa-, and heptamethonium and for IEM-1119 in open ionic channels of rabbit superior cervical ganglion neuronal receptors lie within 33–60% of the membrane electric field, i.e., in the middle one-third. A similar localization has been found for procaine and tubocurarine in other receptors (Adams, 1977; Ascher et al., 1978b).

However, some results obtained with this method are questionable. As mentioned in Chapter 5, the binding rate constant $k_{+B}{}^*$ is more voltage dependent in tubocurarine than in hexamethonium, which indicates that tubocurarine binds deeper than hexamethonium in the membrane electric field. This, however, is unlikely because the tubocurarine molecule is much larger than the hexamethonium molecule, and it is hardly believable that it can penetrate the ionic channel. More evidence against the localization of the binding sites by the above approach is that the voltage dependence of the open-channel blockade observed in the effect of procaine on the end-plate is identical for the charged and uncharged procaine forms obtained by changing the pH of the perfusion solution (Gage et al., 1983). It thus seems likely that some mechanisms other than the influence of electric field on the movement of charged ligand molecules toward binding sites are responsible for the voltage sensitivity of the binding reaction. One possible mechanism is the direct influence of the membrane electric field on the conformational transitions that occur in the activated receptor molecule, e.g., similar to that responsible for the voltage sensitivity of the channel open time (see Magleby and Stevens, 1972).

6.2. POSSIBLE FUNCTIONAL SIGNIFICANCE OF SITES IN THE IONIC CHANNEL OF THE NICOTINIC ACETYLCHOLINE RECEPTOR THAT BIND BLOCKING AGENTS

As has been mentioned above, the combinations R_3, R_5, and R_7 revealed in the β-structure of a polypeptide chain are probably able to bind the bis-ammonium compounds which are specific blocking agents for nicotinic receptors. The question arises, therefore, whether these combinations are specific for nicotinic receptor protein compared with other membrane proteins. This is mostly related to combination R_5, since decamethonium, which probably can selectively bind the R_5 combination, is a specific blocking agent for the particular protein whose structure is shown in Figs. 2.5 and 2.6.

To answer this question, the amino acid sequence of seven membrane proteins other than the nicotinic acetylcholine receptor protein was analyzed. The proteins chosen were retinal rhodopsin (Ovchinnikov et al., 1983a), bacteriorhodopsin (Ovchinnikov et al., 1979; Engelman et al., 1980), glycophorine (Capadli, 1982), protein I (porine) of *Escherichia coli* (Chen et al., 1979), Ca^{2+}-ATP-ase of sarcoplasmic reticulum (Allen et al., 1980), cytochrome-C-oxidase (Anderson et al., 1981; Gregor and Tsugita, 1982), and neuraminidase (Fields et al., 1981). It was found that the ratio of R_5 combinations to the total number of negatively charged amino residues in nicotinic receptor protein (see Noda et al., 1982, 1983a, b) was about twice as high as the averaged ratio in the other membrane proteins (0.13 and 0.07, respectively), although the ratios of their negatively charged amino residues to the total amino acid content were identical (0.12).

Therefore, the combination R_5 is indeed specific for nicotinic receptor. Only in two other membrane proteins does the ratio of R_5 to the total number of negatively charged amino acids approach that in nicotinic receptor. These are cytochrome-C-oxidase and Ca^{2+}-ATP-ase, with the above ratios equal to 0.13 and 0.10, respectively.

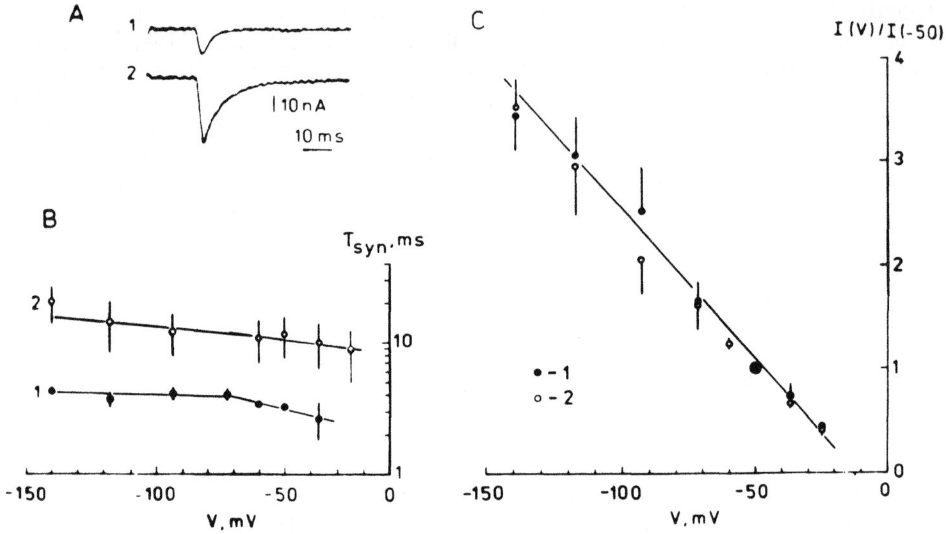

Fig. 6.2. Effect of increased external concentration of Ca^{2+} ions on e.p.s.c. recorded from rabbit superior cervical ganglion neurons at 36°C. A) E.p.s.c.'s recorded from the same neuron at –50 mV in normal solution containing 2.5 mM Ca^{2+} (1) and in the solution containing 25 mM Ca^{2+} (2). B, C) Voltage dependences of e.p.s.c. decay time constant (B) and e.p.s.c. amplitude (C) in normal solution (1) and in the solution containing 25 mM Ca^{2+} (2). In B and C the averaged results obtained from 5 neurons are represented (vertical bars indicated SE). In C the values are normalized relative to those obtained at –50 mV (Selyanko et al., 1985a).

Cytochrome-C-oxidase is functionally different from nicotinic receptor, since the latter does not bind iron ions. Ca^{2+}-ATP-ase has some functional similarity with nicotinic receptor, for nicotinic receptor can bind Ca^{2+} ions (see Chapter 1). Correlation analysis of the amino acid sequences, both of Ca^{2+}-ATP-ase and β, γ, and δ subunits of nicotinic receptor, reveals a peak in the distribution of the intercharge intervals that corresponds to the R_5 combination (V. I. Skok, unpublished).

The latter result encouraged comparison of other Ca^{2+}-binding proteins with the nicotinic receptor. Especially interesting were those whose conformation was known, such as Ca^{2+}-binding parvalbumin, the Ca^{2+} transmitter in mammalian muscles, and troponin (Kretsinger, 1976a, b). In their Ca^{2+}-binding site ("calcium loop"), the negatively charged amino acids play the role of ligands that coordinate the Ca^{2+} ion. The calcium loop is similar to the nicotinic receptor protein not only in the high content of negatively charged amino residues, but also in the high ratio of R_5 combinations to their total number. For eleven homologous portions of the amino acid sequence in the calcium loop (see Kretsinger, 1976a, b) this ratio is 0.14, which is close to the value found in the nicotinic receptor protein.

The secondary structure of calcium loop is a turn in the polypeptide chain located between two short β-antiparallel sheets. There are portions in each subunit of the nicotinic receptor that are similar to the calcium loop in their secondary structure and also in the following characteristics: 1) at least four amino acids in their sequence are identical to those found in the calcium loop; 2) the combinations of these amino acids in their sequence is identical to those in the calcium loop. Altogether, there are 20 portions that correspond to these requirements in nicotinic receptor of $\alpha_2\beta\gamma\delta$ stoichiometry (Skok, 1985). These portions are indicated in Figs. 2.5 and 2.6 by dark rectangles.

Since there is a turn in the calcium loop, the distances between the oxygen atoms of the neighboring carboxylic groups that coordinate the Ca ion are only 0.254–0.262 nm. This is much less than the predicted intercharge distances in bis-quaternary ammonium compounds that selectively block the nicotinic receptor. However, one may find the corresponding intercharge distances in the β structures adjacent to the turn in nicotinic receptor protein, which allows one to suggest that the hypothetical Ca^{2+} binding sites and blocker-binding sites may partially overlap. Interestingly, as one can see from Figs. 2.5 and 2.6, this overlap is typical for predicted binding sites for decamethonium (in γ and δ subunits) but not for hexa- or hexadecamethonium, which correlates with the specific blocking function of decamethonium in this type of nicotinic receptor compared with other bis-quaternary ammonium compounds.

It should be noted that the calcium loop is apparently not the only structure that may bind Ca^{2+}, for it has not been found in some Ca^{2+}-binding proteins, e.g., in Ca^{2+}-ATP-ase (Allen et al., 1980), which, nevertheless, has a ratio of R_5 combinations to the total number of negatively charged amino residues as high as that in nicotinic receptor protein.

If the suggestion that the blocker-binding sites in nicotinic receptor are somehow related to the Ca^{2+}-binding function is correct, one would expect that the increase in external Ca^{2+} concentration would produce an effect on nicotinic receptor channel features opposite to that produced by blocking drugs, and also that the increased external Ca^{2+} concentration would affect the blocker–receptor interaction. The results obtained in experiments indicate that this is indeed the case. Figure 6.2 illustrates that in rabbit sympathetic ganglion neurons the increased external concentration of Ca^{2+} ($[Ca]_o$) markedly prolongs the e.p.s.c. decay while at low $[Ca]_o$ the e.p.s.c. decay shortens (Selyanko et al., 1985a). Similar effects have been observed in rat submandibular ganglion (Rang, 1981), frog sympathetic ganglion (Kuba and Nishi, 1979), and molluscan neurons (Ascher et al., 1978a; Marty, 1980).

The increased $[Ca]_o$ also decreases the rate constant $k_{+B}*$ that characterizes binding of hexamethonium to an open channel of nicotinic receptor, without changing its voltage dependence (Fig. 6.3).

This effect of Ca^{2+} is not mediated by its possible shielding of negative charges that are known to exist normally on cell membrane surface. Evidence for this is that the increased $[Ca]_o$ decreases $k_{+B}*$ instead of increasing it. At the same time, the effect which one might expect from shielding of the membrane negative charges, which apparently would result in membrane hyperpolarization, is an increase, rather than a decrease, in $k_{+B}*$ (see Chapter 5). It can therefore be sug-

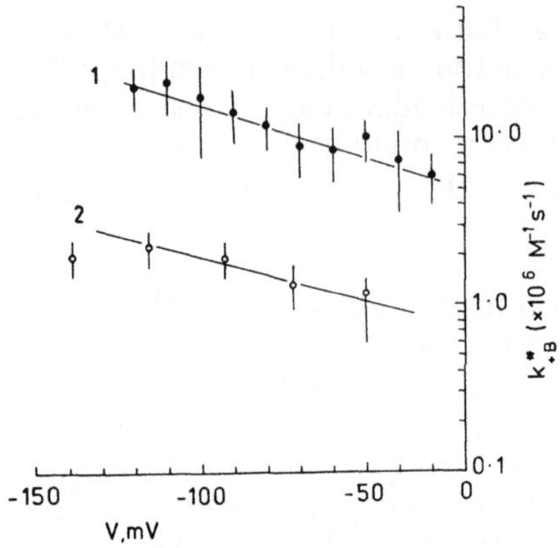

Fig. 6.3. Effect of increased external concentration of Ca^{2+} ions on the open channel-blocking activity of hexamethonium in nicotinic acetylcholine receptors of rabbit superior cervical ganglion. The rate constant k_{+B}* of hexamethonium binding to the open channel was calculated (see Chapter 5) at different membrane potential levels in normal solution containing 2.5 mM Ca^{2+} (1) and in the solution containing 25 mM Ca^{2+} (2) at 36°C. The averaged results shown (means ± SE) were obtained from 4 neurons (Selyanko et al., 1985a).

gested that hexamethonium somehow competes with Ca^{2+} for common binding sites in the receptor molecule.

No prolongation by increased $[Ca]_o$ has been observed in the end-plate current decay (Bregestovski et al., 1979; Kuba and Takeshita, 1983), in contrast to neuronal e.p.s.c. This difference correlates with the fact that hexamethonium actually does not effectively bind to open channels in the end-plate receptors (Lukomskaya et al., 1981; Rang and Rylett, 1984), in contrast to neuronal nicotinic receptors. The lack of the hexamethonium channel-blocking effect in the end-plate is probably due to a much higher k_{-B}* in the end-plate, while k_{+B}* is not much different from that in the neurons (L. G. Magazanik, personal communication). These results are consistent with the hypothesis that Ca^{2+} and the open-channel blocking drugs may compete for common binding sites in the neuronal nicotinic receptor channel. Further support for this hypothesis is provided by direct evidence that the ionic channel in molluscan neuronal nicotinic receptor can bind Ca^{2+} (Chemeris et al., 1982; Slater et al., 1984). The specific blocking agent for calcium channels D-600 can block the open channels in molluscan neuronal nicotinic receptors (Bregestovski and Iljin, 1980; Slater et al., 1984). Other specific organic blockers of calcium channels are not effective in nicotinic receptors of sympathetic ganglion neurons (Ito and Nishi, 1982). The ability to bind Ca^{2+} has also been reported for sodium (Yamamoto et al., 1984) and calcium (Kostyuk and Doroshenko, 1983) channels of electrically excitable membrane.

One might suggest, based on the above results, that hexamethonium and related compounds would also affect the interaction of Ca^{2+} with other Ca^{2+}-binding proteins. To check this hypothesis, the effects of a series of bis-quaternary ammonium compounds on Mg^{2+}–Ca^{2+}-ATP-ase, an enzyme involved in active transport of Ca^{2+} through the membrane of sarcoplasmic reticulum, were analyzed (Fedorov et al., 1985). This enzyme has three functionally distinct components: the active center which hydrolyzes ATP, the Ca^{2+}-binding component, and the transmembrane component, which is somewhat functionally similar to the ionic channel (MacLennan and Reitmeier, 1982). The results obtained indicate that hexa-, tetra-, and pentamethonium ($1 \cdot 10^{-4}$ M) do not affect ATP-ase activity both in native sarcoplasmic reticulum and in its solubilized form. However, all these compounds, including another compound, decamethonium, markedly inhibited the translocation of Ca^{2+} through liposomal membranes containing reconstructed Mg^{2+}–Ca^{2+}-ATP-ase. The activity of the compounds increased from tetra- to decamethonium. A similar tendency has been reported for the blocking effect produced by bis-quaternary ammonium compounds of variable chain length on potassium channels of sarcoplasmic reticulum as being due to their interaction with hydrophobic sites of the channel wall (Miller, 1982).

No activity peaks similar to those found at C_6 and C_{10} in nicotinic receptor (see Fig. 6.1) have been reported from the effects produced by bis-quaternary ammonium compounds in the potassium channels of sarcoplasmic reticulum; it is also not yet clear whether the peak at C_{10} exists in the case of Mg^{2+}–Ca^{2+}-ATP-ase. It therefore remains unknown whether the effects produced by bis-quaternary ammonium compounds on potassium channels of sarcoplasmic reticulum and on Mg^{2+}–Ca^{2+}-ATP-ase are specific, as in nicotinic receptors.

The above results suggest that the sites that bind bis-quaternary ammonium compounds in the nicotinic receptor may also bind Ca^{2+}. The functional significance of Ca^{2+}-binding in the receptor is not yet clear. In another type of receptor – glutamate receptors of mammalian brain and spinal cord neurons – Mg^{2+} ions in their physiological concentrations may block the receptor open channels, thus modifying the postsynaptic membrane conductance (Mayer et al., 1984; Nowak et al., 1984). Whether divalent ions may modify cholinergic synaptic transmission in the same way remains unclear.

It has been suggested that binding of divalent ions to open channels of nicotinic receptors may prevent their closing (Ascher et al., 1978a). The most convincing arguments in favor of this hypothesis have been obtained for Ca^{2+} ions (Marchais and Marty, 1979), as was mentioned in Chapter 3.

Another endogenous compound which may affect the nicotinic receptor channel kinetics is acetylcholine. According to recent results (Sakmann et al., 1980; Colquhoun and Sakmann, 1983; Sine and Steinbach, 1984b), acetylcholine may bind to the nicotinic receptor open channel, causing its blockade. This effect was observed at concentrations of acetylcholine $1 \cdot 10^{-4}$ M or higher (Colquhoun and Ogden, 1984), which corresponds to normal transient concentrations of acetylcholine in synaptic cleft (see Kuffler and Yoshikami, 1975). Therefore, the acetylcholine-induced open channel blockade may provide negative feedback in cholinergic transmission.

The channel-blocking effect of acetylcholine is thought to cause the burstlike kinetics of single-channel activity in the end-plate (Colquhoun and Sakmann, 1983; Colquhoun and Ogden, 1984). This mechanism may also exist in sympathetic neuronal nicotinic receptor channels, since they likewise exhibit the burstlike activity (Derkach et al., 1987). Calculations have shown a marked difference between neuronal and muscular nicotinic receptors in the affinity of acetylcholine to the open ionic channel, which is about 4 times higher in sympathetic ganglion neurons than in the end-plate. The possible significance of this difference for pharmacological specificity of nicotinic receptors will be discussed below.

6.3. CONCLUSIONS

A comparison between the structure of specific blocking agents for nicotinic receptor, the bis-quaternary ammonium compounds, and the predicted secondary structure of the receptor protein suggests that there are a few possible binding sites in the receptor molecule which are concentrated in two areas exposed at extracellular space as well as in the intrabilayer portion of the receptor protein and in the area of the receptor protein exposed at the cytoplasmic side.

The structure of tentative binding sites for specific blocking agents reveals some similarity with the sites that coordinate Ca^{2+} in calcium-binding proteins. On the other hand, the increased $[Ca]_o$ changes the mean apparent channel open time in neuronal nicotine receptors in the opposite direction compared with what is observed in the hexamethonium effect and strongly inhibits the interaction of hexamethonium with the receptor open ionic channel. No such effect of increased $[Ca]_o$ on mean apparent channel open time is observed in the hexamethonium-insensitive nicotinic receptors of the end-plate. These facts suggest that the sites in the acetylcholine receptor which bind selectively the open-channel blocking drugs, the specific ganglion-blockers, may normally bind Ca^{2+} ions.

Chapter 7

THE RELATION OF NICOTINIC RECEPTOR OPEN-CHANNEL BLOCKADE TO BLOCKADE OF SYNAPTIC TRANSMISSION

7.1. THEORETICAL PREDICTIONS OF SYNAPTIC TRANSMISSION BLOCKADE AS CAUSED BY BLOCKADE OF THE OPEN CHANNEL IN POSTSYNAPTIC RECEPTORS

It has been shown in Chapter 5 that the open-channel blockers decrease both true and apparent mean channel open time in nicotinic receptors, thus causing a decrease in τ_{syn}, the e.p.s.c. decay time constant. Let us show that a decrease in τ_{syn} alone may result in the blockade of synaptic transmission.

Both in nerve cell and muscle fiber the postsynaptic current possesses a much faster time course than that of the postsynaptic potential. This difference is accounted for by recharging of the membrane capacity during the e.p.s.p. development. The rate of recharging is limited by the membrane time constant, which is longer than τ_{syn}. On the other hand, the amplitude of e.p.s.p. depends on the electric charge $q_{e.p.s.c.}$ which is carried by e.p.s.c. and is given by

$$q_{e.p.s.c.} = \int_0^t I_{e.p.s.c.} dt, \tag{7.1}$$

where $I_{e.p.s.c.}$ is the e.p.s.c. amplitude and t is time. The time course of e.p.s.c. can be described as

$$I_{e.p.s.c.} = g_{max}(E_r - V) \times \exp(-t/\tau_{syn}), \tag{7.2}$$

where g_{max} is the peak in membrane conductance change observed at the e.p.s.c. maximum amplitude, E_r is the reversal potential for e.p.s.c., and V is the membrane potential.

It follows from Eqs. (7.1) and (7.2) that

$$q_{\text{e.p.s.c.}} = g_{\max}(E_r - V)\tau_{\text{syn}}. \tag{7.3}$$

Therefore, the higher the τ_{syn} value, the higher the maximum amplitude of e.p.s.p. One should thus expect that if τ_{syn} is too small, the e.p.s.p. will not reach the threshold amplitude for the postsynaptic spike initiation, and synaptic transmission will not occur.

7.2. MECHANISMS UNDERLYING SELECTIVITY IN DRUG-INDUCED BLOCKADE OF SYNAPTIC TRANSMISSION

The above calculations suggest that the blockade of the open ionic channel of the postsynaptic receptor may result in the blockade of synaptic transmission. The question arises whether this mechanism is related to a specific blockade of synaptic transmission or whether it causes only nonspecific transmission blockade. Although the open-channel blocking mechanism was suggested a rather long time ago (Blackman, 1970), because of the great variability in the chemical structures of the compounds that exhibit this kind of block (local anesthetics, barbiturates, alkaloids, quaternary ammonium compounds, etc.), the specificity of their effect has long been questioned (Ascher et al., 1978b) or even denied (see Peper et al., 1982; Karlin et al., 1983).

The first evidence that the open-channel blocking mechanism may be responsible for specific blockade of synaptic transmission was obtained from a comparison of the channel-blocking effect produced in nicotinic receptors of rabbit sympathetic ganglion neurons by two compounds hexamethonium and tubocurarine (Selyanko et al., 1981b; Skok et al., 1981). The rate constant of blocker binding to the open channel k_{+B}*, measured at the resting membrane potential level, was a few times higher in hexamethonium than in tubocurarine (see Table 5.1). Although tubocurarine is about 5 times more effective than hexamethonium in ganglionic blockade, the tubocurarine-induced ganglionic blockade is commonly thought to be nonspecific, in contrast to that produced by hexamethonium, since neuromuscular transmission blockade can be attained at a much lower concentration of tubocurarine than is needed to produce ganglionic blockade, just the opposite of what is observed for hexamethonium.

The suggestion that the open-channel blockade may be responsible for the specific blockade of ganglionic transmission was tested using a series of ganglion blockers of basically different chemical structures, bycomparing their channel- and transmission-blocking activities (Skok et al., 1983, 1984). If the above suggestion is correct, one should expect a close correlation between the open-channel-blocking and ganglion-blocking activities of the compounds in question. The drugs used were symmetrical bis-cationic compounds tetra-, penta-, hexa-, and heptamethonium (for structures see **2.18**, n = 4–7), a nonsymmetrical bis-cationic compound IEM-1119 (for structure see **5.1**), and IEM-1288, a monocationic com-

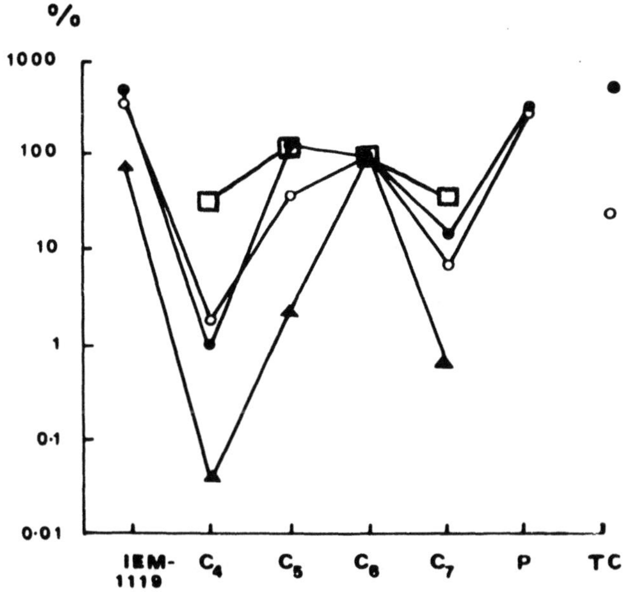

Fig. 7.1. Relationship between the ganglion-blocking potencies (•) of compounds IEM-1119, tetra- (C_4), penta- (C_5), hexa- (C_6), heptamethonium (C_7), IEM-1288 (pyrilene derivative, P), and tubocurarine (TC), and the rate constants k_{+B}* of their binding to open ionic channels of nicotinic acetylcholine receptors in sympathetic ganglion (o). For some compounds the values of $K_B^* = k_{+B}^*/k_{-B}^*$ are given (Δ) as well as their potencies to decrease the e.p.s.c. amplitude ( ). The values represented are taken from Table 5.1 and are expressed in percent relative to those for hexamethonium (Skok, 1986).

pound similar to the selective ganglion-blocker pempidine but possessing bromide instead of tartrate anion in pempidine (for structure see **4.7**).

The rate constants k_{+B}* and k_{-B}* that characterize channel-blocking activities of the compounds listed, and their ganglion-blocking activities, relative to that of hexamethonium, are indicated in Table 5.1 and in Fig. 7.1. The values for k_{+B}* and k_{-B}* were estimated in rabbit superior cervical ganglion neurons at –50 mV, which was close to resting membrane potential level. The ganglion-blocking activities of the compounds were estimated in cat superior cervical ganglion (Skok et al., 1984; Paton and Zaimis, 1949; the activities reported in the latter work are shown in parentheses in Table 5.1) at a frequency of preganglionic stimulation 10/s.

As Table 5.1 and Fig. 7.1 indicate, there is a statistically significant correlation ($p > 0.95$) between k_{+B}* and ganglion-blocking activities within the series of compounds tested. On the other hand, no statistically significant correlation exists between the affinity constants $K_B^* = k_{+B}^* / k_{-B}^*$ and the ganglion-blocking activities. This is probably due to the fact that the differences among the compounds in their k_{-B}* values are much smaller than the differences in their k_{+B}* values and, in

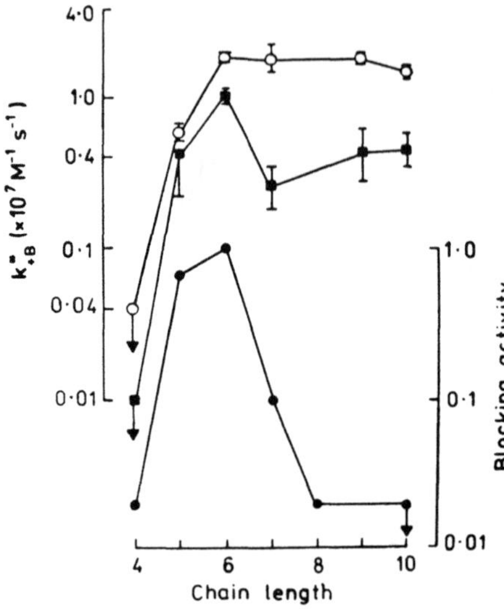

Fig. 7.2. Relationship between relative ganglion-blocking activities (•, right ordinate) of the compounds $(CH_3)_3\overset{+}{N}\text{–}(CH_2)_n\text{–}\overset{+}{N}(CH_3)_3$ in cat sympathetic ganglia (from Paton and Zaimis, 1949) and rate constant k_{+B}* of their binding to open "fast" (o) and "slow" (■) ionic channels in rat submandibular ganglion (left ordinate). Ganglion-blocking potency of hexamethonium taken as 1. Abscissa: the number *n* of CH_2 groups. Arrows indicate the values corresponding to the upper limit (Gurney and Rang, 1984).

addition, are not essential at the frequency of preganglionic stimulation used in transmission-testing experiments (10/s).

An alternative possibility is that the transmission blockade is produced by some mechanisms other than the open-channel blockade, but which are characterized by blocker–receptor interaction similar to that in the open-channel blocking mechanism. One of the possible mechanisms of this kind is a competitive blockade. This alternative is favored by the fact that each of the blockers listed, when used at a concentration of about EC_{50} needed for transmission blockade, in addition to the e.p.s.c. decay shortening also decreases the e.p.s.c. To determine whether this is the case, the e.p.s.c. amplitude-reducing activities of four drugs used (tetra-, penta-, hexa-, and heptamethonium) were estimated and compared with their ganglion-blocking activities, indicated in Table 5.1 and Fig. 7.1. No correlation has been found between these two activities, indicating that the competitive blocking effect is probably not responsible for the specific ganglionic blockade produced by the compounds tested. It should be noted that, as shown in Table 5.1, reduction of the e.p.s.c. amplitude and shortening of the e.p.s.c. decay produced by each of the compounds used are characterized by markedly different voltage sensitivities.

The correlation between the channel-blocking and transmission blocking effects of symmetrical bis-quaternary ammonium compounds with $n = 4$–7 can also be seen in their effects produced in "slow" nicotinic receptor channels in rat parasympathetic ganglion, as illustrated in Fig. 7.2, taken from the work of Gurney and Rang (1984). The correlation disappears at $n > 9$, i.e., with compounds that are selective blockers of neuromuscular rather than ganglionic transmission (see Fig. 6.1). "Slow" channels correspond in their kinetics to transmitter-activated nicotinic receptor channels in sympathetic ganglion neurons.

For comparison, in Table 5.1 and Fig. 7.1 the results obtained for tubocurarine, the selective neuromuscular blocker, are shown. It can be seen that the ganglion-blocking activity of tubocurarine (Bowmann and Webb, 1972) is much higher than one would expect from its $k_{+B}{}^*$ value (Selyanko et al., 1981b). The mechanism responsible for this discrepancy is not yet clear.

If the open-channel blockade were responsible for the selective ganglionic blockade in the above series of specific ganglion-blocking drugs, one might expect that their channel-blocking activity in neuromuscular junctions would be less strong than in autonomic ganglia. That this is the case has been found in both frog (Lukomskaya et al., 1981) and rat (Rang and Rylett, 1984) muscles. In the former work, no significant channel-blocking effect has been revealed in tetra-, penta-, and hexamethonium. According to L. G. Magazanik (personal communication) this result is due to the relatively high $k_{-B}{}^*$ values, while the values for $k_{+B}{}^*$ are not much different from those found in rabbit sympathetic ganglion. In the latter work, no channel-blocking effect has been observed with hexamethonium at concentrations lower than $4 \cdot 10^{-4}$ M.

Thus, it can be concluded that the open-channel blockade produced by some specific ganglion-blocking drugs, rather than the competitive blockade, i.e., the blockade of the receptor recognition center, is responsible for their specific blocking effect on ganglionic transmission. Other antagonists, such as trimethapan or α-toxins, as described above, exert their ganglion-blocking effects through a pure blockade of the receptor recognition center.

7.3. SELECTIVITY OF OPEN-CHANNEL BLOCKADE IN POSTSYNAPTIC RECEPTORS OTHER THAN NICOTINIC RECEPTORS IN AUTONOMIC GANGLIA

The question arises whether the selective blockade of the open channel can be observed in synapses other than the synapses in autonomic ganglia, and whether it can determine there the selective blockade of synaptic transmission.

Although the effect on the end-plate of tubocurarine, a selective blocker of neuromuscular transmission in skeletal muscles, is mixed (competitive and open-channel blocking), its predominant effect at concentrations lower than $4 \cdot 10^{-7}$ M (according to other authors, lower than $5 \cdot 10^{-6}$ M: Albuquerque et al., 1978) is competitive (Colquhoun et al., 1979; Colquhoun and Sheridan, 1982). Steroid anesthetics, barbiturates, and antimalarial drugs exhibit here an open-channel blockade, but their blocking effects can be observed only at comparatively high concentrations and are not selective (Adams, 1976; Adams and Feltz, 1980a, b; Gillo and Lass, 1984). Histamine, an endogenous modulator of synaptic transmission, can block neuromuscular transmission both through blocking of the recognition center of nicotinic receptor and through the open-channel blockade (Ariyoshi et al., 1985). Nicotinic interneuronal and neuromuscular transmission in invertebrates can likewise be selectively blocked through the open-channel blocking mechanism. Hexamethonium is a selective blocker in one type and tubocurarine in another type of interneuronal transmission in molluscs achieved via the open-channel blockade (Ascher et al., 1978b). The same mechanism has been found in the effect of chlorizondamine, pempidine, mecamilamine, and decamethonium in lobster neuromuscular transmission (Lingle, 1983a, b).

Glutamate receptors in the insect neuromuscular junction can be selectively blocked by δ-phylanthotoxin (Clark et al., 1982) and by large-size derivatives of pentamethonium (Magazanik et al., 1984). Both groups of compounds are powerful open-channel blockers. The same mechanism was suggested for the effects of picrotoxin in GABA receptors, the effects of strychnine in glycine receptors, and the effect of nembutal on glutamate receptors in mammalian cultured spinal neurons (Barker, 1975), as well as for the apamine-induced blockade of noncholinergic and nonadrenergic transmission in smooth muscle (Vladimirova and Shuba, 1978; Banks et al., 1974; Burnstock and Hoyle, 1985). Similar calcium-dependent potassium channels in other types of cells need other compounds for their selective blockade (Cherubini and North, 1984; Nohmi and Kuba, 1984).

Thus, the open-channel blockade occurs in different types of synapses. However, the contribution of the open-channel blockade to the selective blockade of synaptic transmission differs markedly from one type of synapse to another. The reasons for this variability are not yet clear. The end-plate channels which are least susceptible to open-channel blockade possess the shortest mean apparent lifetime. This time, when measured as the end-plate current decay time constant, is in the mouse 0.3 ms at 39°C and 1.0 ms at 23°C and at holding membrane potential between –65 mV and –80 mV (Dreyer et al., 1976; Head, 1983) and in frog about 1.0 ms (see Colquhoun, 1979).

This is much less than in those channels where the open-channel blockade is more effective. The mean channel open time is 12 (see Lingle, 1983a) and 17.7 ms (Finger, 1983) in the crustacean neuromuscular junction, 2.6–2.9 ms in the insect neuromuscular junction (Cull-Candy et al., 1981; Cull-Candy and Miledi, 1982), and 30 ms in cultured neurons of mammalian spinal cord (Sakmann et al., 1983a), measured at animal normal body temperature (room temperature for poikilotherms except the crustacean neuromuscular junction, which has been studied at 12°C) and at resting membrane potential level. On the other hand, the channels responsible for the faster $\tau_{e.p.s.c.}$ component of biexponential e.p.s.c. decay in rat parasympathetic ganglion (6.2 ms at –80 mV and 30°C) are blocked by hexamethonium at higher k_{+B}* ($17.9 \cdot 10^6$ $M^{-1} \cdot s^{-1}$) than those responsible for its slower component (20.0 ms), which are blocked at k_{+B}* = $10.3 \cdot 10^6$ $M^{-1} \cdot s^{-1}$ (Gurney and Rang, 1984). The k_{+B}* value for hexamethonium in rabbit sympathetic ganglion ($7.3 \cdot 10^6$ $M^{-1} \cdot s^{-1}$ at –80 mV and 34–37°C: Skok et al., 1984) is similar to that in molluscan ganglion ($7.5 \cdot 10^6$ $M^{-1} \cdot s^{-1}$ at –80 mV and 12°C: Ascher et al., 1978b), although the mean apparent channel lifetime in the former case is much shorter (4.5–5.0 ms: Derkach et al., 1983) than in the latter (27 ms: Ascher et al., 1978a). These results suggest that although the lowest sensitivity of the open channel to the blocking drug may somehow be related to the very short channel lifetime (0.3–1.0 ms), there is no clear correlation between these channel properties at longer lifetimes.

As mentioned above, the mean channel open time is probably determined by the interaction of open channels with divalent cations (Ascher et al., 1978a; Marchais and Marty, 1979; Mayer et al., 1984; Nowak et al., 1984) or with transmitter ions (Adams, 1976). This allows one to suggest that the high effectiveness of the open channel-blocking drugs in the channels with long mean open time is due to their high sensitivity to natural substances, e.g., divalent ions, rather than to their long open time as such.

The fact that acetylcholine- and glutamate-gated channels, which are both permeable to sodium and potassium ions, can be selectively blocked by different compounds implies that the sensitivity of open channels to blocking compounds is not related to their ion selectivities.

7.4. CONCLUSIONS

Selective ganglion-blocking activity in a series of compounds correlates with the rate constant characterizing their binding to the open ionic channels of postsynaptic nicotinic receptors in sympathetic ganglion neurons. This result suggests that selective ganglionic blockade by some compounds is due to their open-channel blocking effect rather than to any other mechanism. This suggestion is supported by theoretical predictions based on the electrical properties of the postsynaptic membrane and showing that the blockade of postsynaptic receptor ionic channels in their open state may result in the blockade of synaptic transmission.

In addition to nicotinic receptors in autonomic ganglia, selective blockade of the open channel has been observed in postsynaptic receptors of many other types: in nerve cells of the central nervous system of vertebrates and in muscle fibers of invertebrates. These results indicate that the ionic channels in many postsynaptic receptors possess selective chemical sensitivity which is not directly related to the species of mean permeant ions or to the type of synaptic transmitter and may be responsible for independent modulation of their normal activity by endogenous compounds, particularly by divalent cations.

Chapter 8

RECOGNITION CENTER OF MUSCARINIC ACETYLCHOLINE RECEPTOR

8.1. STOICHIOMETRY OF INTERACTIONS OF AGONISTS AND ANTAGONISTS WITH THE MUSCARINIC ACETYLCHOLINE RECEPTOR

As was shown in Chapter 2, the stoichiometry of the agonist–receptor interaction can be deduced in electrophysiological experiments from the dependence of the electrical response of a neuron to an agonist on the agonist concentration. For muscarinic acetylcholine receptors, such a relationship was obtained from rabbit superior cervical ganglion neurons (Dun et al., 1976). Muscarinic responses were recorded from individual sympathetic neurons as membrane depolarizations elicited by perfusion of the ganglion with acetylcholine or methacholine in the presence of tubocurarine. An analysis of concentration–response curves revealed that the Hill number indicating the lower limit for the number of agonist molecules interacting with a single acetylcholine receptor [see Eq. (2.1)] was close to 1; the Hill number was not affected by preganglionic denervation of the ganglion (Dun et al., 1976). Thus, at least one agonist molecule is needed to activate the muscarinic acetylcholine receptor in sympathetic ganglion.

This finding is consistent with the data obtained from voltage-clamped amphibian sympathetic ganglion neurons, where, as in mammalian sympathetic ganglion neurons, muscarinic depolarization results from inhibition of an outward potassium current (the M-current: see Chapter 9). The Hill number for the concentration dependence of M-current inhibition by muscarine appears to be 1 (Jones, 1985).

For the concentration dependence of muscarinic hyperpolarization in cardiac muscle cells the Hill number is also about 1 (Glitsch and Pott, 1978; Noma et al., 1978; Osterrieder et al., 1981).

The advantage of the electrophysiological approach for determining the stoichiometry of the agonist–receptor interaction is that it is based on measuring the functionally important result of this interaction, i.e., the membrane conductance change. However, it does not allow one to study the first stage of the agonist–receptor interaction, the binding reaction, or to determine the stoichiometry of the antagonist–receptor interaction, which is not accompanied by electrical response of the cell. The biochemical method of measuring the binding of labeled agonists and antagonists with muscarinic acetylcholine receptors in tissue homogenates is devoid of these disadvantages. This method was used in various tissues including nervous tissue.

The first investigations of this type showed that binding of many classical muscarinic antagonists (propylbenzilylcholine, scopolamine, atropine, etc.) was described by Eq. (2.1) when the Hill number was close to 1 (Birdsall et al., 1976; Birdsall and Hulme, 1976). By contrast, the Hill number for acetylcholine binding was markedly lower, i.e., was about 0.5 (Birdsall et al., 1976; Birdsall and Hulme, 1976; Birdsall et al., 1978). The value of the Hill number equal to 1 unambiguously indicates that one ligand molecule interacts with one receptor molecule, whereas the value of the Hill number lower than 1 implies two alternative explanations. According to one of them, there may be only one binding site for antagonists and two binding sites for agonists in the recognition center of the muscarinic acetylcholine receptor. In this case the Hill number for agonists lower than 1 may result from the interaction between agonist binding sites with negative cooperativity. The second explanation suggests that muscarinic acetylcholine receptors are heterogeneous for agonists whereas they are homogeneous for antagonists. Heterogeneity for agonists implies that there are various muscarinic acetylcholine receptor subtypes, with different K_D values for agonists, thereby shifting their dose–response curves from each other along the abscissa. As a result, the slope of the Hill plot for the overall agonist binding appears to be smaller than that for its binding to any particular subtype of receptors.

As further investigations showed, the second explanation was more likely. Heterogeneity of muscarinic acetylcholine receptors was indeed found either for agonists or for certain antagonists, e.g., pirenzepine (see below). The slope of the dose–response curves for binding of pirenzepine in any particular tissue depends on the number of subtypes of muscarinic acetylcholine receptors present in the tissue. For example, in the majority of autonomic ganglia and in brain, where at least two subtypes of muscarinic receptors were detected, the binding of pirenzepine was characterized by a Hill number close to 1, whereas in cardiac and smooth muscles, as well as in human sympathetic ganglion, where only one receptor subtype was revealed, the Hill number was also equal to 1 (Hammer et al., 1980; Watson et al., 1984). These data favored the above suggestion that the Hill number lower than 1 for binding of agonists with muscarinic acetylcholine receptors was underestimated due to the heterogeneity of receptors. "True" values of the Hill number either for agonists or antagonists characterizing each muscarinic acetylcholine receptor subtype seem not to essentially deviate from 1, which is consistent with electrophysiological data.

The coincidence of the values of the Hill number for agonist binding to a receptor with those for resulting membrane conductance changes indicates that one muscarinic acetylcholine receptor operates one ionic channel. This assumption could not be made *a priori*, since there are indications that muscarinic acetylcholine receptor, unlike nicotinic acetylcholine receptor, is spatially separated from its ionic channel (see Chapter 9).

8.2. FUNCTIONAL ORGANIZATION AND CHEMICAL STRUCTURE OF THE MUSCARINIC ACETYLCHOLINE RECEPTOR RECOGNITION CENTER

8.2.1. Molecular Properties of the Muscarinic Acetylcholine Receptor

The muscarinic acetylcholine receptor isolated from mammalian brain is a protein with molecular mass estimated to be 76,000–80,000 (Birdsall et al., 1979; Rueb and Liefländer, 1979; Avissar et al., 1983; Rinken et al., 1984). In some studies evidence was obtained favoring the possibility of receptor protein heterogeneity. In addition to the main, low-molecular-mass protein, a small fraction of twofold higher-molecular-mass protein (160,000–180,000) was detected (Uchida et al., 1982; Avissar et al., 1983; Rinken et al., 1984). The low-molecular-mass protein, in turn, was able to transform, under certain experimental conditions, into a protein with two times lower molecular mass, about 40,000. From a comparison of molecular masses of different receptor proteins it was concluded that a protein with a molecular mass of 80,000 is dimeric, whereas a protein with molecular mass of 160,000 is tetrameric (Avissar et al., 1983; Sokolovsky et al., 1983).

Recently cDNA clones derived from porcine cerebral mRNA encoding the muscarinic acetylcholine receptor were isolated (Kubo et al., 1986). Nucleotide sequence analysis of the cDNA yielded the complete amino-acid sequence of the receptor. It was found that the muscarinic acetylcholine receptor consisted of 460 amino acid residues with estimated total molecular mass of 51,416. The amino acid sequence of the muscarinic acetylcholine receptor predicted seven transmembrane segments with hydropathicity profile similar to those of rhodopsin and β_2-adrenoceptor. The amino-terminal regions of all these proteins were devoid of the signal sequence and contained two putative N-glycosylation sites. Finally, the above proteins contained short carboxy-terminal tails with serine and threonine residues which may be putative sites for phosphorylation. It is also known that all the proteins considered interact with G proteins, thereby eliciting cellular responses. All of the above data indicate that the muscarinic acetylcholine receptor, rhodopsin, and β_2-adrenoreceptor arose from a common ancestor (Kubo et al., 1986).

Interestingly, in the muscarinic receptor there is a long insert in the cytoplasmic loop connecting transmembrane segments V and VI, which in its sequence is similar to those of the sodium channel, viral transforming proteins, and homoeotic gene products (Kubo et al., 1986; see also Hulme and Birdsall, 1986).

Kubo et al. (1986) also succeeded in expressing cDNA to produce functional muscarinic acetylcholine receptors in *Xenopus* oocytes. It was established that responses to acetylcholine of oocytes injected with the mRNA resemble those of oocytes containing native muscarinic acetylcholine receptors. This finding favored the suggestion that cloned cDNA encoded functional muscarinic receptors. Muscarinic acetylcholine receptors produced in oocytes exhibited high apparent binding affinity for pirenzepine, indicating that they belonged to the M_1 subtype (see below).

8.2.2. General Pharmacological Properties of Neuronal Muscarinic Acetylcholine Receptors

Until recently, the existence of specific (distinct from nicotinic) neuronal muscarinic acetylcholine receptors was debatable. Some of the initial studies carried out on hippocampal neurons provided evidence that neuronal responses elicited by specific muscarinic agonists can be suppressed by nicotinic antagonists, whereas specific nicotinic agonists produced responses similar to those elicited by muscarinic agonists. This led to the conclusion that acetylcholine receptors on hippocampal neurons are of mixed type, with the properties of both nicotinic and muscarinic receptors (Bird and Aghadjanian, 1976; Segal, 1982). However, this conclusion has not been confirmed by the data obtained by other authors either from hippocampal or from other central neurons. For instance, it has been established that excitation produced by acetylcholine in hippocampal neurons is not affected by nicotinic antagonists, while it can be completely abolished by atropine (Biscoe and Straughan, 1966; Cole and Nicoll, 1984b). The nicotinic agonist phenyltrimethylammonium, similarly to muscarinic agonists, produced an excitation in hippocampal neurons, although at concentrations 10–40 times higher than those of muscarinic agonists. Similar results were obtained from neocortex and caudate nucleus neurons (Krnjevič and Phillis, 1963; McLennan and York, 1966). In Renshaw cells two distinct types of acetylcholine receptors, i.e., nicotinic and muscarinic, but not mixed-type receptors, were detected (Curtis and Ryall, 1966). Electrophysiological evidence in favor of the existence of specific muscarinic acetylcholine receptors on central neurons was consistent with the biochemical data that the binding of muscarinic agonists and antagonists in the brain was not affected by nicotinic antagonists (Yamamura and Snyder, 1974).

The existence of mixed-type neuronal acetylcholine receptors has also been suggested based on early results obtained from the study of sympathetic ganglia (Ginsborg, 1965; Hancock and Volle, 1970; Kristic, 1972; Beleslin and Malobabic, 1974). According to these data, the action of nicotinic or muscarinic agonists could not be abolished by nicotinic or muscarinic antagonists taken alone. However, as it appeared later, it was the existence of both nicotinic and muscarinic receptors on sympathetic ganglion neurons which was responsible for the inefficiency of either nicotinic or muscarinic antagonists applied in isolation in blocking the responses to acetylcholine.

A detailed pharmacological analysis of muscarinic receptors in sympathetic ganglia has been performed by Brown et al. (1980a). Based on the effects of antagonists on the responses elicited by muscarinic agonists, it was found that the majority of classical agonists studied (oxotremorine, methylfurmethide, pilocarpine, and AHR-602) are selective in the sense that they do not activate nicotinic acetylcholine receptors, and only acetylcholine, methacholine, and bethanechol are "mixed" agonists, activating both nicotinic and muscarinic acetylcholine receptors.

The potency of muscarinic agonists diminished in the order oxotremorine > methylfurmethide > muscarine > furmethide > pilocarpine > AHR-602. Oxotremorine, pilocarpine, and AHR-602 revealed "partial agonism," which was indicated by lower maximum responses than those produced by other agonists. Muscarinic agonists were almost equipotent when acting on muscarinic acetylcholine receptors in sympathetic ganglion and ileum.

Muscarinic antagonists acted competitively on ganglionic responses produced by muscarine: the dose–response curve for muscarine shifted to the right, without changing its slope, in the presence of antagonists. The potencies of muscarinic antagonists in the ganglion decrease in the order methylscopalamine > lachesin > atropine > hyoscine, and are similar to those in ileum.

Thus, the use of a number of classical muscarinic agonists and antagonists has not revealed any essential difference between muscarinic acetylcholine receptors in sympathetic ganglia and in smooth muscle.

8.2.3. Peculiarities of the Muscarinic Acetylcholine Receptor Recognition Center

The topography of the recognition center of the muscarinic acetylcholine receptor has been thoroughly studied in smooth muscles (see Michelson and Zeimal, 1973; Triggle and Triggle, 1976). On the one hand, an analysis of the chemical structure of traditional muscarinic agonists and antagonists, as well as their analogs, in conjunction with their activities, allowed one to establish the peculiarities of the recognition center of the muscarinic acetylcholine receptor compared with that of the nicotinic receptor. On the other hand, since the potencies of most classical muscarinic agonists and antagonists at sympathetic ganglion and at smooth muscle appear to be similar, it was suggested that the recognition centers of muscarinic acetylcholine receptors in these two preparations are similar.

Like nicotinic agonists and the natural agonist acetylcholine, most muscarinic agonists contain positively charged trimethylammonium groups. The important role of the positive charge of the nitrogen atom in interaction with a muscarinic acetylcholine receptor is ascertained by the fact that the replacement in the acetylcholine molecule of the nitrogen atom by an uncharged carbon atom results in a 3000-fold drop in activity (Burgen, 1965). Such a difference in the activities of acetylcholine and its analog suggests that the ionic bond between the nitrogen atom and some negatively charged group in the receptor (anionic site) is functionally important. At the same time, the role of the positively charged nitrogen atom in muscarinic antagonists is less essential than that in acetylcholine. This is evidenced by the fact that the replacement of the nitrogen atom with a carbon atom results in a much smaller change in the activities of muscarinic antagonists than in the activity of acetylcholine (Barlow and Tubby, 1974).

The methyls in the cationic group are very important for activation of muscarinic receptors. The increase in the size of the cationic head of the acetylcholine molecule as a result of sequential replacement of methyls with ethyls leads to an essential decrease in acetylcholine activity (Holton and Ing, 1949; Triggle and Triggle, 1976). One reason for this could be the increase in the distance between the positive charge of the nitrogen atom in acetylcholine and the negative charge in a receptor. As calculated, the full replacement of methyls by ethyls should increase this distance by 0.8 Å, from an initial distance of 3.3 Å, which is associated with a decrease in the energy of ionic interaction by 2 kcal·M^{-1}, as well as with a 50-fold decrease in the potency of the drug (see Triggle and Triggle, 1976). However, the decrease in muscarinic potency of acetylcholine (1000-fold) resulting from the replacement of $\overset{+}{N}Me_3$ by NEt_3 is much more essential than follows from calculations. Therefore, it seems unlikely that the change in intercharge distance itself underlies the decrease in agonist potency. Probably, the cationic group of acetylcholine interacts with the muscarinic acetylcholine receptor, as with the nicotinic acetylcholine receptor, by acting on a deepened binding site, to which it fits sterically. This explanation has been advanced by Zhorov (1986), who established a correlation between productive conformations of ligands acting on both the acetylcholine receptor and on butyrylcholinesterase, which possess recognition centers similar to each other.

It is assumed that the hydrophobic interaction of methyls with the receptor binding sites surrounding its negative charge provides an anhydrous low dielectric micro-environment where ionic interaction between the positive charge of the nitrogen atom and the negative charge of a receptor by van der Waals forces essentially increases. Probably, the replacement of methyls by ethyls results in a weakened influence of the hydrophobic environment on ionic interaction.

The polarity of the remaining part of the agonist molecule is also of great importance. The decrease in agonist potency resulting from the substitution in its cationic group of methyls by ethyls depends on the structure of the remaining part of the molecule and is greater the less polar the overall agonist molecule. With sequential decrease in polarity of the agonist molecule it exhibits the increasing properties of an antagonist.

Some muscarinic agonists, e.g., pilocarpine (**8.1**), arecoline (**8.2**), and oxotremorine (**8.3**), unlike acetylcholine, possess tertiary, instead of quaternary, nitrogen atoms:

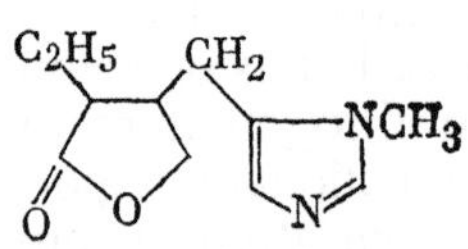

Pilocarpine
8.1

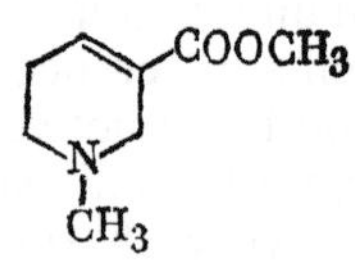

Arecoline
8.2

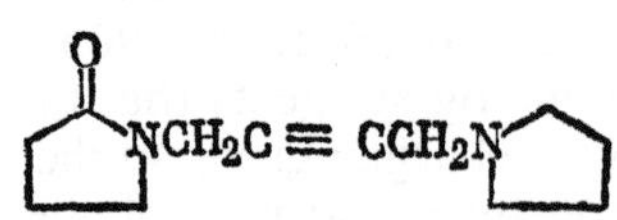

Oxotremorine
8.3

These drugs exhibit a much higher activity than do their analogs containing quaternary nitrogen atoms (Gloge et al., 1966; Hanin et al., 1966). Since at physiological pH values these drugs are cations, they can interact with the anionic site in a muscarinic acetylcholine receptor, whereas their increased activities can also be accounted for by the involvement of hydrogen bonds.

The above-mentioned data show that the requirements for cationic groups of agonists and antagonists for their actions on nicotinic and muscarinic acetylcholine receptors are similar. However, there are essential differences in the requirements for the remaining part of nicotinic and muscarinic agonist molecules. The ester oxygen area in the acetylcholine molecule is much less essential for the interaction with nicotinic versus muscarinic acetylcholinic receptors. For example, aromatic esters of choline are devoid of muscarinic activity, while they possess nicotinic activity, which is higher, the lower the electron density in the molecule area corresponding to the ester oxygen of acetylcholine (Hey, 1952). At the same time, it is known that some muscarinic agonists contain double or triple [e.g., oxotremorine (**8.3**) and McN-A-343 (**8.4**)] bonds between the second and third carbon atoms (relative to the nitrogen atom in the cationic group). The high electron density of such bonds indicates that in the recognition center of a muscarinic acetylcholine receptor these bonds may correspond to a binding site possessing a partial positive charge and located 4 Å apart from the anionic site. Possibly, it is the site with which the ester oxygen of acetylcholine interacts. Indeed, on substitution of the ester oxygen, muscarinic activity of acetylcholine decreases 100-fold.

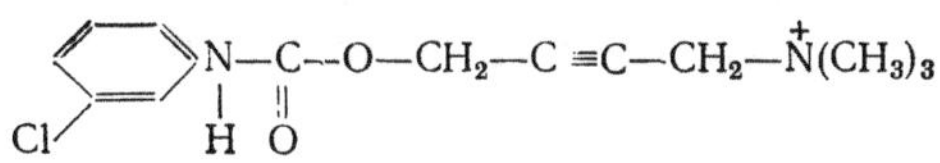

McN-A-343
8.4

It has been known for a long time that the introduction of methyls into the acetylcholine molecule in its middle part produces different effects on its nicotinic and muscarinic activities (Beckett et al., 1961, 1963; Lesser, 1965; Ellenbroeck et al., 1965): the replacement of hydrogen by methyl at the α-carbon abolishes muscarinic activity, whereas a similar replacement at the β-carbon abolishes nicotinic activity. A comparison of the activities of stereoisomers of acetyl-α-methylcholine and acetyl-β-methylcholine indicates that the action of the former on nicotinic acetylcholine receptors is not stereoselective, while the action of the latter on muscarinic acetylcholine receptors is stereoselective. Stereoselectivity is an important feature of muscarinic acetylcholine receptors, differentiating them from nicotinic ones. The activities of series of stereoisomers including those of muscarine (**8.5**) are essentially different. For instance, (+) muscarine is about 110 times as active as (–) muscarine

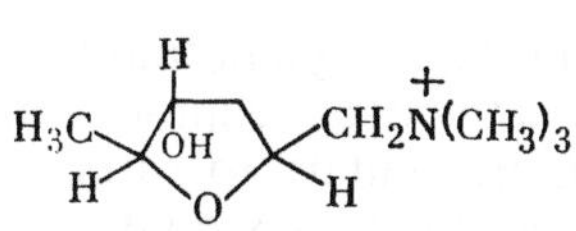

Muscarine
8.5

(see Barlow, 1955). The high stereoselectivity of muscarinic acetylcholine receptors suggests that there are at least three active sites in their recognition center. In addition to the sites mentioned above, of great importance is the site corresponding to 2-methyl in muscarine and in some other muscarinic agonists, including 5-methylfurmethide (**2.10**) and F-2268 (**8.6**), and to methyl in the acid part of the acetylcholine molecule.

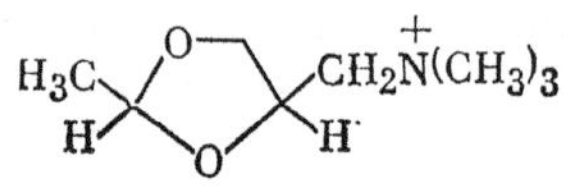

F-2268
8.6

Removal of the 2-methyl from or substitution of one more 2-methyls into the muscarine molecule results in an essential drop in its muscarinic activity (Gyermek and Unna, 1958; Waser, 1961). However, the change in agonist activity following the removal of the terminal methyl group appears to be much higher than one would expect from simple hydrophobic interaction of this group with a receptor. The effect of removal of the terminal methyl group on the agonist activity seems to depend on the structure of the remaining part of its molecule, and it increases when the overall molecule becomes more polar (Pratesi et al., 1968). All these data indicate that the terminal methyl group affects the interactions of other active groups of an agonist with a receptor, causing their strengthening.

Not only is the presence of the terminal methyl group of great importance, but also its position in an agonist molecule. According to the "five-atom rule," the highest muscarinic activity is provided if there are five carbons between the positively charged nitrogen atom and the methyl group (Ing, 1949).

Conformational analysis of acetylcholine and other muscarinic and nicotinic agonists revealed that acetylcholine may interact with muscarinic and nicotinic acetylcholine receptors in two different conformations (Chothia, 1970; Baker et al., 1971; Beers and Reich, 1970). When acting on a nicotinic acetylcholine receptor, acetylcholine seems to exhibit its "carbonyl side," which allows it to interact with a receptor by two active groups: ammonium head and carbonyl oxygen (see Chapter 2). While acting on a muscarinic acetylcholine receptor, acetylcholine probably exhibits its "methyl side," which allows it to interact with a receptor by three active groups: ammonium head, carbonyl oxygen, and terminal methyl group. It should be pointed out that, according to some authors, among the three active groups of acetylcholine is ester oxygen instead of the terminal methyl group (see Michelson and Zeimal, 1973; Haefely, 1980).

The carbonyl oxygen in the acetylcholine molecule, situated 5.9 Å from the nitrogen atom, seems to interact with a receptor through a hydrogen bond. The site in the receptor which is complementary to carbonyl oxygen in the acetylcholine molecule appears to also bind hydroxyls contained in the molecules of various antagonists [e.g., atropine (**8.7**) and scopolamine (**8.8**)].

Interestingly, the stereoselectivity of muscarinic acetylcholine receptors varies in animals at different stages of evolution. In lower-organized animals, unlike in higher-organized animals, muscarinic acetylcholine receptors of the heart and neu-

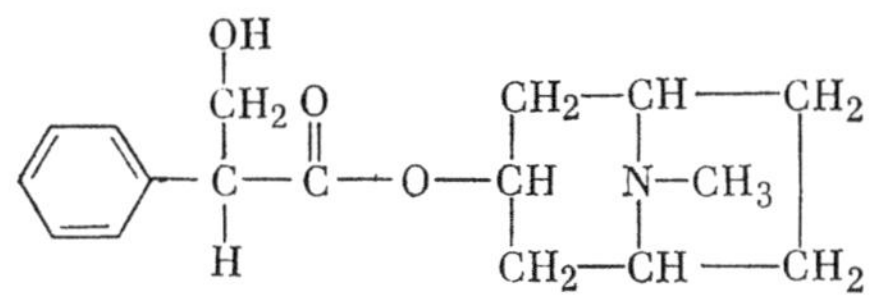

Atropine
8.7

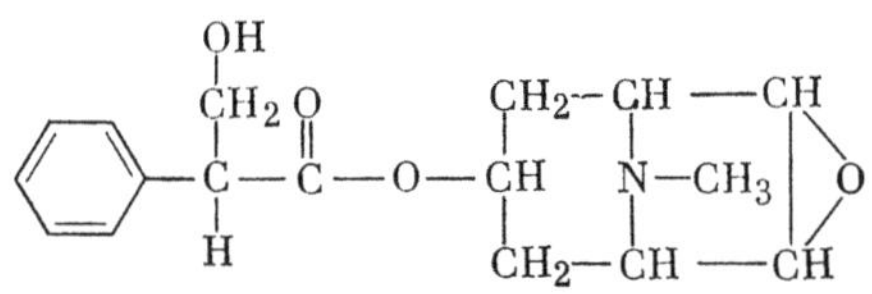

Scopolamine
8.8

rons (in molluscs) are nonstereoselective (Ger et al., 1976; Michelson et al., 1979). Muscarinic receptors of invertebrates also differ from those of vertebrates in that they exhibit low sensitivity to atropine and to agonists containing a triple inter-carbon bond (e.g., oxotremorine).

8.2.4. Heterogeneity of Muscarinic Acetylcholine Receptors

The first indications that muscarinic acetylcholine receptors are heterogeneous were obtained with the use of the muscarinic antagonists diphenylacetoxy-N-methylpiperidine and stercuronium. The former appeared to be stronger when acting on smooth muscle rather than on cardiac muscle, whereas the latter, on the contrary, was more efficient when acting on cardiac muscle (Barlow et al., 1976).

With the use of agonists, it has been found that the receptors not only of different tissues, but also of the same tissue, may be heterogeneous. The binding of labeled agonists in various tissues is characterized by gently sloping curves, with deflections reflecting multiplicity of K_D values for agonists (Birdsall and Hulme, 1976; Birdsall et al., 1976, 1978). By contrast, the binding of traditional muscarinic antagonists is described by steeper curves, fitted by the Langmuir isotherm (Birdsall and Hulme, 1976; Hulme et al., 1978). It seems that some muscarinic antagonists do not differentiate among various subtypes of muscarinic acetylcholine receptors.

Somewhat later a novel muscarinic antagonist, pirenzepine (**8.9**), was obtained, whose binding in the tissue was similar to that of agonists, i.e., was characterized by a more gently sloping curve than the Langmuir isotherm (Hammer et al., 1980).

An analysis of pirenzepine binding revealed that muscarinic acetylcholine receptors in the brain, in glands, and in both smooth and cardiac muscles are characterized by high, medium, and low affinity to pirenzepine, respectively. From an analysis of pirenzepine binding, three major subtypes of muscarinic receptors, with low ($K_D = 1{\cdot}10^{-6}$ M), medium ($K_D = 2{\cdot}10^{-7}$ M), and high ($K_D = 2{\cdot}10^{-8}$ M) af-

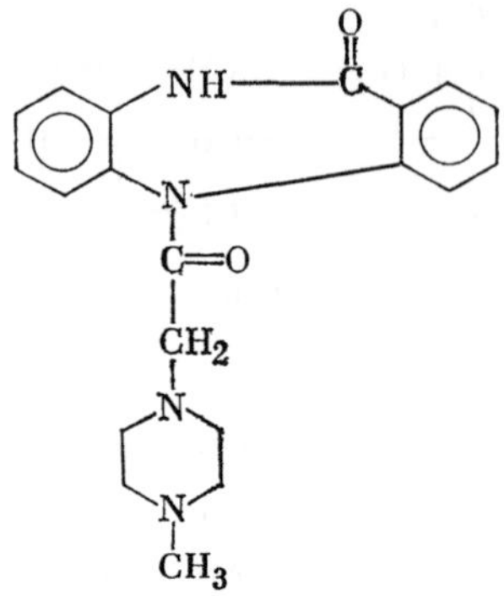

Pirenzepine
8.9

finities, were distinguished. The proportion of muscarinic receptors of different subtypes in various tissues is variable. In brain receptors all three types were detected, whereas in glands either only intermediate-affinity or both intermediate- and high-affinity receptors were revealed. In cardiac and smooth muscles mostly receptors with low affinity for pirenzepine were detected.

The heterogeneity of muscarinic acetylcholine receptors for pirenzepine was used for their classification. Initial attempts to classify muscarinic acetylcholine receptors had been made earlier, based on the results obtained with the gastrointestinal tract (Goyal and Rattan, 1978). It had been suggested that muscarinic receptors of intramural ganglia (termed M_1 receptors) be distinguished from those of smooth muscle (M_2-receptors). It was further established in electrophysiological experiments that ganglionic muscarinic acetylcholine receptors were indeed different in their higher affinity for pirenzepine from ileal muscarinic acetylcholine receptors (Brown et al., 1980b). Now it is generally accepted that muscarinic acetylcholine receptors of M_1 subtype are those with low K_D values for pirenzepine, ranging between 0.01 and 0.2 μM ("high-affinity receptors"), whereas muscarinic receptors of M_2 subtype are those with a high K_D value for pirenzepine, close to 1 μM ("low-affinity receptors").

The distinct sensitivity of ganglionic and smooth muscle muscarinic receptors to pirenzepine allows one to account for the mechanism of the clinical action of pirenzepine, which is widely used for the treatment of ulcers because of its antimuscarinic action (Jaup et al., 1979). Since pirenzepine more effectively blocks ganglionic muscarinic receptors than those of smooth muscle, it is apparently the block of interneuronal transmission in the myenteric plexus that is responsible for its clinical effect.

It was established that not only pirenzepine, but also stercuronium, is much more effective in blocking ganglionic than ileal muscarinic acetylcholine receptors (23 and 100 times, respectively), while another blocker, 4-diphenylacetoxy-N-methylpiperidine methiodide, is more active in smooth muscle than in the ganglion (Brown et al., 1980b). Thus, either M_1- or M_2-receptors can be selectively blocked.

Binding studies revealed that receptors with distinct affinities for pirenzepine are characterized by their different distributions in the brain. Unlike high-affinity receptors, whose density of distribution is almost similar in various parts of the brain, the low-affinity receptors possess a greatly varying density of distribution. In the rat the ratio of the densities of high- and low-affinity receptors varies from 11:1 in the cerebellum to 1:4 in the hippocampus. The ratio of low- and high-affinity receptors in the amphibian brain is close to this, indicating that it has not been changed during evolution (Birdsall et al., 1980). The rate of appearance of muscarinic acetylcholine receptors in postnatal development varies in different parts of the brain (Coyle and Yamamura, 1976; Kuhar et al., 1980). The highest number of muscarinic acetylcholine receptors exists at the time of birth of the animal, and the highest rate of their appearance during postnatal development occurs in the medulla oblongata, compared with those in the diencephalon and the brain cortex. This correlates with the finding that termination of cell cleavage and the beginning of synaptogenesis occur faster in the midbrain, compared with those in the brain cortex (Altman, 1969; Coyle, 1974). Interestingly, there is a delay, lasting 6-7 days, in the appearance of high-affinity receptors, compared with low-affinity receptors, which suggests that the appearance of low-affinity receptors may be associated with the formation of functioning synapses (Kuhar et al., 1980). During postnatal development of chick smooth muscle, a decrease rather than an increase in the number of muscarinic acetylcholine receptors has been detected (Crouch et al., 1982).

The difference in the distribution of low- and high-affinity muscarinic acetylcholine receptors in the brain and their appearance at different stages of postnatal development suggests that these receptors are distinct molecular structures. More evidence favoring this suggestion is that the role of these receptors in cellular metabolism is different: the activity of low-affinity receptors seems to be associated with an increased synthesis of cGMP, whereas inositol turnover is apparently regulated via both low- and high-affinity receptors (Birdsall and Hulme, 1976; Heilbronn and Bartfai, 1978). There are indications that M_1 and M_2 receptors may be encoded by different mRNA species (Kubo et al., 1986; see also Hulme and Birdsall, 1986).

On the other hand, there is evidence in favor of a certain similarity of muscarinic acetylcholine receptors that belong to different subtypes. Among them is the possibility of transformation of the receptors of one subtype into the receptors of another subtype, e.g., transformation of low-affinity receptors into high-affinity receptors. This transformation is assumed to occur following reductive alkylation of membrane sulfhydryl groups (Aronstam et al., 1977, 1978). In addition, the heterogeneity of muscarinic acetylcholine receptors is abolished by certain chemical factors (Berrie et al., 1979; Birdsall et al., 1979).

It was found recently that binding of agonists to certain kinds of muscarinic acetylcholine receptors may be facilitated by divalent cations (Ca^{2+}, Mg^{2+}) and inhibited by guanine nucleotides (see Birdsall et al., 1983). The binding of antagonists under the influence of guanine nucleotides is, on the contrary, facilitated (Ehlert et al., 1981). It has been suggested that the action of guanine nucleotides is related to the involvement of muscarinic acetylcholine receptors in the regulation of adenylate cyclase activity. Indeed, the effect of guanine nucleotides was observed in those cardiac muscarinic acetylcholine receptors which exhibited interaction with the adenylate cyclase system. In the hypothalamus, however, divalent cations and guanine nucleotides almost do not affect muscarinic acetylcholine receptors (Birdsall et al., 1983).

Another interesting question is whether receptors exhibiting equal sensitivity to pirenzepine are heterogeneous. The evidence obtained so far indicates that high-affinity receptors are homogeneous, while low-affinity receptors seem to be heterogeneous. The distinctions are observed between the muscarinic acetylcholine receptors of the heart and blood vessels (Kharkevich et al., 1980) and heart and smooth muscle (Godovnikov et al., 1985), as well as the gastrointestinal tract and urinary bladder (Lambrecht et al., 1985). An analysis of the compounds whose actions on heart and smooth muscle differ indicated that they differ from each other only by the structure of their cationic head. It was therefore suggested that muscarinic acetylcholine receptors of heart and smooth muscle may differ in their anionic binding sites (Godovnikov et al., 1985).

8.2.5. Chemical Nature of the Recognition Center

Experiments with chemical modification of the recognition center of the muscarinic acetylcholine receptor were aimed at localizing the functionally important disulfide and sulfhydryl groups. Investigations of this type were carried out on frog sympathetic neurons (Trinus and Skok, 1979). Modification of the muscarinic acetylcholine receptor was followed by changes in membrane depolarization caused by acetylcholine or 5-methylfurmethide and recorded with intracellular microelectrodes. As is shown in Fig. 8.1, dithiothreitol, a substance that reduces the disulfide (S–S) bonds, inhibits muscarinic depolarization. This action of dithiothreitol was prevented by acetylcholine but not by atropine. An oxidizer, ferricyanide, as well as modifying agent N-maleinimide, also suppressed muscarinic responses, although their actions were not prevented by acetylcholine. Based on the effect of dithiothreitol and the protection from its action by acetylcholine, it has been concluded that there are functionally important disulfide bonds in the recognition center of the muscarinic acetylcholine receptor. On the other hand, the results obtained with N-maleinimide and ferricyanide, as well as the inefficiency of acetylcholine as a protector, suggest that the functionally important sulfhydryl groups in the muscarinic acetylcholine receptor of the sympathetic ganglion neuron are located outside its recognition center.

Earlier studies on molluscan neurons (Sato et al., 1976) and smooth muscle (Stubbins and Hudgins, 1971), in contrast to the above data, did not reveal any effect of dithiothreitol on muscarinic acetylcholine receptors. This difference is probably associated with the peculiarities of muscarinic acetylcholine receptors in these preparations. Indeed, muscarinic acetylcholine receptors in smooth muscle and in molluscan neurons differ from those in sympathetic ganglion neurons in their sensitivities to certain antagonists: The former ones have low sensitivity to pirenzepine, whereas the latter ones are weakly sensitive to atropine (see above).

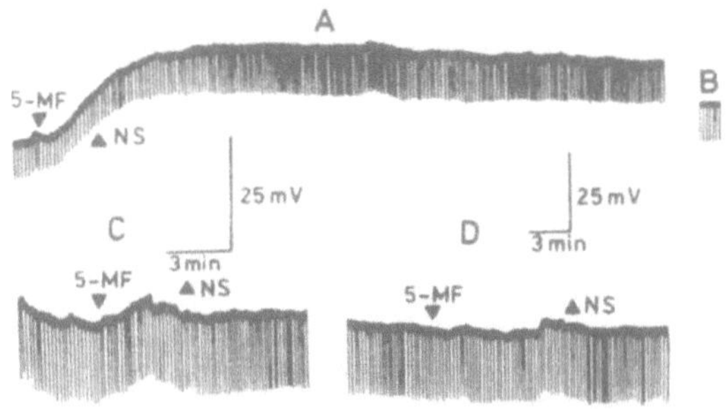

Fig. 8.1. Effects of dithiothreitol and ferricyanide on the responses of frog sympathetic ganglion to 5-methylfurmethide (5-MF, 10^{-4} M). A) Beginning of the 5-MF-induced response recorded in normal Ringer solution (NS). B) Continuation of the response shown in A as recorded 1.5 min later. C) The 5-MF-induced response recorded 20 min after dithiothreitol (10^{-3} M) had been washed out. D) The 5-MF-induced response recorded 20 min after ferricyanide (10^{-4} M) had been washed out. Right and left calibration scales relate to A, B and to C, D, respectively. Vertical bars seen in the records are electrotonic potentials caused by hyperpolarizing current pulses. Triangles indicate the moments when the solutions were changed (Trinus and Skok, 1979).

Sulfhydryl groups have been detected in binding studies in muscarinic receptors in the brain (Aronstam et al., 1978; Hedlund and Bartfai, 1979) and in cardiac muscle (see Turpaev, 1962). However, unlike in sympathetic ganglion, the actions of oxidizing agents on SH groups in cardiac muscle were prevented by acetylcholine. It seems that the majority of functionally important SH groups are located in these receptors in their recognition centers.

Experiments on modification of nonneuronal muscarinic acetylcholine receptors, carried out mostly on smooth muscle using reagents selectively acting on carboxyl and phosphate groups, indicated that these groups may be located in the anionic binding site of the recognition center of the muscarinic acetylcholine receptor (see Vulfius, 1978), as shown above for the nicotinic acetylcholine receptor (see Chapter 2).

8.3. CONCLUSIONS

Interactions of agonists and antagonists with muscarinic acetylcholine receptors are characterized by Hill numbers close to 1, indicating that there is only one binding site for agonists and antagonists in the recognition center of a muscarinic acetylcholine receptor, in contrast to two analogous binding sites in the nicotinic acetylcholine receptor.

The muscarinic acetylcholine receptor isolated from mammalian brain is a protein with molecular mass of about 80,000. There are indications that this protein is dimeric and may also exist in tetrameric form.

Porcine cerebral muscarinic acetylcholine receptor, as revealed by cloning and further sequence analysis of DNA complementary to encoding receptor messenger RNA, consists of 460 amino acid residues, with a calculated molecular mass of 51,416. In its amino acid sequence, as well as in its hydropathicity, the muscarinic acetylcholine receptor resembles β_2-adrenoceptor and rhodopsin, which suggests that the three receptors developed from a common ancestor.

Neuronal muscarinic acetylcholine receptors, as examined in mammalian sympathetic ganglion, do not differ in their sensitivity to the majority of classical muscarinic agonists and antagonists from those of smooth muscle. This implies that the general characteristics of the recognition center of a muscarinic acetylcholine receptor, well studied in smooth muscle, may be similar to those in neurons. The main tentative binding sites in the recognition center of a muscarinic acetylcholine receptor are: i) the anionic site exhibiting ionic interaction with the positively charged nitrogen atom in the acetylcholine molecule, ii) the binding site with a partial positive charge located 4 Å away from the anionic site, which interacts with the ester oxygen in the acetylcholine molecule and, possibly, with the triple intercarbon bond in some muscarinic agonists, and iii) the hydrophobic binding site, interacting with the methyl in the acid part of the acetylcholine molecule and with the 2-methyl in molecules of muscarine and other agonists. The anionic site is less important, whereas the two remaining binding sites are more important for activation of the muscarinic acetylcholine receptor, compared with those in the nicotinic receptor. As in other tissues, in neurons the necessity of simultaneous binding of an agonist to all three active binding sites to activate the muscarinic acetylcholine receptor explains the stereoselectivity found in many muscarinic agonists and antagonists, which contrasts with the lack of agonist stereoselectivity in the case of the nicotinic acetylcholine receptor.

The use of certain muscarinic antagonists, in particular pirenzepine, as well as certain agonists (carbacholine) reveals heterogeneity of muscarinic acetylcholine receptors within the same tissue or among different tissues. Variable sensitivity of the receptors to pirenzepine was used for their classification. Two main subtypes of muscarinic acetylcholine receptors are distinguished, M_2 and M_1 receptors, differing in their K_D values for pirenzepine, which are equal to 1 μM and 0.01–0.02 μM, respectively. In central and autonomic ganglion neurons receptors of both subtypes are present, whereas in gland cells mostly M_2 receptors and, to a lesser extent, M_1 receptors exist. In cardiac and smooth muscle cells only M_2-receptors have been found. It is suggested that M_2-receptors are, in turn, heterogeneous and may differ in the structure of their anionic binding sites.

In addition, the functionally important disulfide and sulfhydryl groups are detectable in neuronal muscarinic acetylcholine receptors.

Chapter 9

MECHANISMS UNDERLYING MUSCARINIC POSTSYNAPTIC RESPONSE

9.1. GENERAL PROPERTIES OF MUSCARINIC RESPONSE

9.1.1. Muscarinic Responses in Various Types of Neurons

Electrical responses of neurons to synaptic activation of muscarinic acetylcholine receptors from turtle (Laporte and Lorente de No, 1950) and rabbit (Eccles, 1952) superior cervical ganglia were first recorded in the 1950s with extracellular electrodes. In response to a single orthodromic stimulus or to a short train of repetitive orthodromic stimuli, a slow biphasic potential wave consisting of early positive and later negative phases was recorded from the surface of the ganglion pretreated with tubocurarine. The muscarinic nature of this potential wave was proved when it was established that it can be abolished by low concentrations of atropine (Eccles and Libet, 1961). Similar slow muscarinic potentials were found in various sympathetic ganglia: in rat, rabbit, and cat superior cervical ganglia (Eccles and Libet, 1961; Dunant and Dolivo, 1967; Libet, 1967), in cat stellate and coeliac ganglia (Libet, 1964, 1967), and in frog sympathetic ganglia (Libet et al., 1968; Nishi and Koketsu, 1968). Slow potentials recorded from rabbit and cat superior cervical ganglia are shown in Fig. 9.1.

In the majority of sympathetic ganglia both slow positive and slow negative potentials were found. Only in guinea-pig superior cervical ganglion under normal conditions were slow negative potentials solely detected (Libet, 1970). However, following acetylcholinesterase inhibition, slow positive as well as slow negative potentials were found even in this ganglion (Dun, 1980).

Among parasympathetic ganglia, attempts to record slow potentials were made only in a few extramural ganglia. It was found that in cat ciliary ganglion

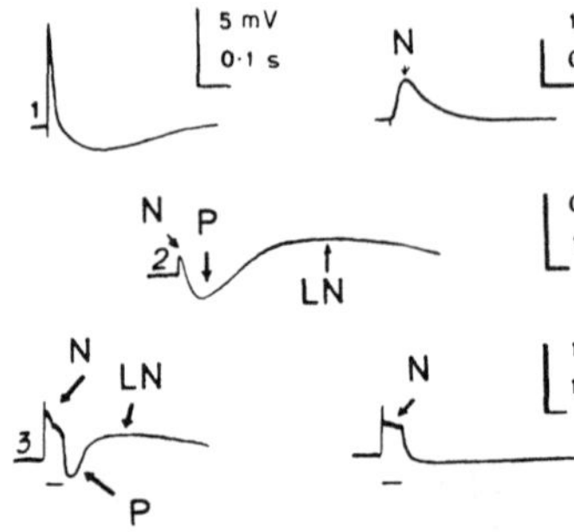

Fig. 9.1. Effects of nicotinic and muscarinic antagonists on the responses to orthodromic stimulation recorded from rabbit (1, 2) and cat (3) superior cervical ganglia by the sucrose gap method. 1) Action potential (left) and negative potential (right) elicited by single orthodromic stimuli applied to cervical sympathetic nerve and recorded in normal solution and in the solution containing tubocurarine ($5 \cdot 10^{-6}$ M), respectively. 2) Response to single orthodromic stimulus recorded in tubocurarine-containing solution ($5 \cdot 10^{-6}$ M) at slower sweep than that in 1. 3) Response to repetitive orthodromic stimulation (30 Hz, 1 s long, as indicated by horizontal bar) recorded in tubocurarine-containing solution ($5 \cdot 10^{-6}$ M) before (left record) and after (right record) atropine (10^{-6} M) was applied to superfusing solution. Records 1, 2, and 3 were obtained from three different ganglion preparations. N) Fast negative potential. P) Positive potential. LN) Late negative potential (Skok, 1980).

slow potentials are lacking (Ivanov and Melnichenko, 1971), although, according to Nishi (1974), slow negative potentials can be recorded here. In cat pelvic ganglion slow positive potentials have been revealed (Griffith et al., 1981).

The origin of slow extracellular potentials has been clarified using the intracellular recording method. In single sympathetic ganglion neurons, repetitive orthodromic stimulation evoked slow hyperpolarizing and depolarizing potentials, which could be abolished by atropine, similarly to slow extracellular potentials (Tosaka and Libet, 1965; Kobayashi and Libet, 1968; Libet et al., 1968; Libet and Tosaka, 1969; Ivanov and Skok, 1980, 1981). Therefore, the hyperpolarizing and depolarizing potentials were called slow inhibitory and slow excitatory postsynaptic potentials (slow i.p.s.p. and slow e.p.s.p.), respectively. These names were given to slow potentials prior to examining their role in neuronal inhibition and excitation (see Chapter 10).

The name "slow" given to muscarinic potentials indicates that they are slower in their time course than fast (nicotinic) e.p.s.p. and shorter than late slow (noncholinergic) e.p.s.p. (Nishi and Koketsu, 1968), which also occur in sympathetic ganglia. On the other hand, some noncholinergic potentials found both in neurons and in other types of cells do not differ from muscarinic potentials in their durations, and they were also called "slow" (see Hartzell, 1981). Because noncholinergic potentials are not specially considered in this book, only muscarinic potentials are referred to here as "slow" ones.

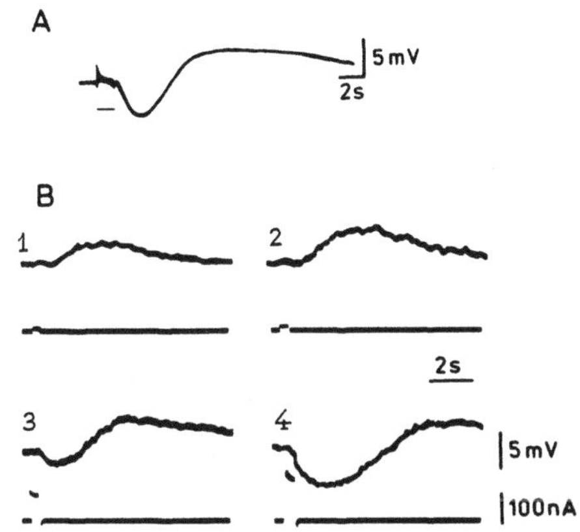

Fig. 9.2. Slow (muscarinic) potentials recorded from rabbit cervical ganglion neurons, as elicited by repetitive orthodromic stimulation (A) and by iontophoretic application of 5-methylfurmethide (B). A) Slow i.p.s.p. and e.p.s.p. elicited by repetitive (10 Hz, 2 s long) stimulation of cervical sympathetic nerve, as recorded in the presence of tubocurarine (50 μM) and neostigmine (1 μM). Resting membrane potential was –60 mV (Ivanov and Skok, 1981). B) One-component depolarizing responses to weak iontophoretic currents (1, 2) and two-component (hyperpolarizing and depolarizing) responses to strong iontophoretic currents (3, 4) passed through the micropipette containing 5-methylfurmethide. Records 1–4 were obtained from the same neuron. Top records indicate membrane potential, bottom records indicate iontophoretic currents. Resting membrane potential was –47 mV (Koval et al., 1982).

Figure 9.2A shows an example of slow i.p.s.p. and slow e.p.s.p. recorded as a biphasic potential change in a single rabbit superior cervical ganglion neuron. Similar biphasic responses can be observed only in some mammalian sympathetic ganglion neurons, while in the rest of the neurons only slow e.p.s.p. appears (Ivanov and Skok, 1981). Slow i.p.s.p. and slow e.p.s.p. can be evoked in the ganglion neurons by local application on their surface of 5-methylfurmethide, a specific muscarinic agonist (Koval et al., 1982). As Fig. 9.2B shows, the application of low doses of 5-methylfurmethide elicits in a neuron only membrane depolarization (records 1 and 2). Biphasic (hyperpolarizing and depolarizing) responses need much higher doses of 5-methylfurmethide (records 3 and 4) to be observed.

In guinea-pig superior cervical ganglion, where under normal conditions only slow negative potentials can be recorded (Libet, 1970), intracellular recording likewise reveals only slow e.p.s.p. (Dun and Karczmar, 1977).

In mammalian sympathetic ganglion, both slow postsynaptic potentials occur either in B- or in C-neurons (Libet and Tosaka, 1969). In amphibian sympathetic ganglion, slow i.p.s.p. and slow e.p.s.p. occur in different neurons. Tosaka et al. (1968) found that slow e.p.s.p. appeared mostly in B-neurons of this ganglion. This observation was confirmed by others (Weight and Padjen, 1973a,b; Weitsen and Weight, 1977; Dodd and Horn, 1983). When nicotine was used (instead of tubocurarine) for blocking nicotinic acetylcholine receptors, slow i.p.s.p. was also observed in B-neurons (Libet and Kobayashi, 1974; Koketsu and Nakamura, 1976).

In parasympathetic ganglion neurons, slow postsynaptic potentials are lacking (in rat submandibular ganglion: Rang, 1981), or they are represented by only one type of response, either by slow i.p.s.p. (in amphibian cardiac ganglion: Hartzell et al., 1977) or by slow e.p.s.p. (in guinea-pig myenteric plexus: North and Tokimasa, 1982). Only slow i.p.s.p., and in a small part of the neurons also slow e.p.s.p., were detected in cat pelvic ganglion (Gallagher et al., 1982).

In central neurons the effects produced by acetylcholine and other muscarinic agonists are mostly excitatory (depolarizing). This has been established initially by using extracellular, and, subsequently, intracellular recording (Krnjevič and Phillis, 1963; Benardo and Prince, 1981; Herrling, 1981; Hounsgaard, 1978; Krnjevič et al., 1971; Egan and North, 1985a). Recently, a synaptic equivalent of muscarinic depolarization, slow e.p.s.p., was recorded from central neurons (Sastry, 1980; Cole and Nicoll, 1983), which in its characteristics (latency, duration) corresponded to slow e.p.s.p. in sympathetic ganglion neurons. Until now muscarinic hyperpolarization has been revealed only in two types of central neurons, in nucleus parabrachialis (Egan and North, 1985b) and in thalamic reticular neurons (McCormick and Prince, 1986).

In molluscan neurons, slow membrane hyperpolarization could be elicited by exogenous acetylcholine or other muscarinic agonists (Kehoe, 1972; Ger et al., 1979; Katchman et al., 1980). This response was insensitive to nicotinic agonists, although its name "muscarinic" is questionable, since it is not abolished by atropine, a classical muscarinic antagonist. Muscarinic depolarization in molluscan neurons has not been revealed as yet.

Neurons of rat superior cervical ganglion (Freschi and Shain, 1980; Freshi, 1983) and hippocampus (Gahwiller and Brown, 1985) grown in culture retain their muscarinic sensitivities and respond to muscarinic agonists with membrane depolarization.

9.1.2. Factors Determining the Time Course of Muscarinic Response

Muscarinic responses both in neurons and in other cell types are characterized by long latencies and long durations, ranging from 30 to 500 ms and from 1 to 60 s, respectively (see Hartzell, 1981). Thus, muscarinic responses are 1–3 orders slower than nicotinic ones. This difference is illustrated in Fig. 1.6. It can be seen that acetylcholine, acting both on nicotinic and muscarinic acetylcholine receptors, elicits in a rabbit superior cervical ganglion neuron a two-component (fast and slow) membrane depolarization (records A1–A3). By contrast, 5-methylfurmethide, selectively activating muscarinic acetylcholine receptors, causes in the same neuron a one-component membrane depolarization (record A4), which in its time course corresponds to the slow component of acetylcholine response. This comparison indicates that the fast component of membrane depolarization caused by

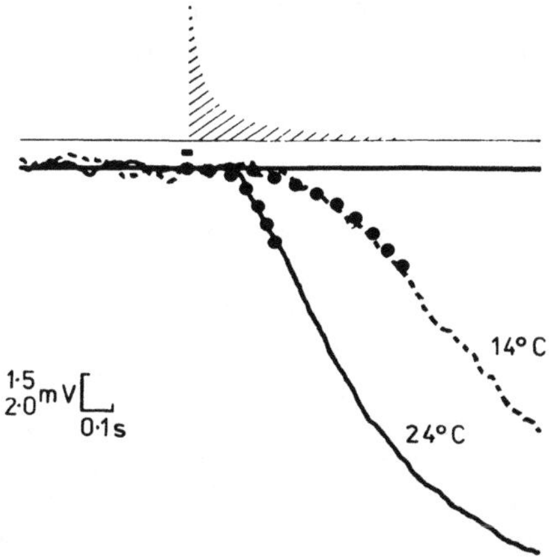

Fig. 9.3. Effect of temperature on the initial part of muscarinic hyperpolarization elicited by iontophoretic application of bethanechol in the neuron of amphibian parasympathetic cardiac ganglion. Two responses obtained at 24 and 14°C are shown (calibration scale is 2 and 1.5 mV, respectively). Each record is the result of averaging of 9 responses. The responses were elicited by the iontophoretic pulses (200 nA, 30 ms long) applied at the moment indicated by the horizontal bar over the records. Dotted lines indicate theoretical curves predicted based on the assumption that the response amplitude depends on the third power of time. Dashed area at the top corresponds to the calculated changes in bethanechol concentration near the cellular membrane as a function of time. For more details see text (Hartzell et al., 1977).

acetylcholine is associated with activation of nicotinic acetylcholine receptors, whereas the slow component is derived from activation of muscarinic acetylcholine receptors. A comparison between fast and slow components of acetylcholine depolarization, therefore, illustrates the difference in the time courses of nicotinic and muscarinic responses.

For a long time the reasons for the slow time course of muscarinic response remained unknown, and only recently were they subjected to detailed analysis. First of all, it was found in various types of neurons that the responses caused by brief applications of muscarinic agonists were of the same long latencies and durations as those in the responses elicited by synaptic activation of muscarinic acetylcholine receptors (see, for instance, reports by Hartzell et al., 1977; North and Tokimasa, 1982; Dodd and Horn, 1983). This result makes it very unlikely that the slow time course of muscarinic response is derived from the long acetylcholine release from the presynaptic nerve ending. Other factors, potentially able to limit the rate of development of the muscarinic response, are (see Hartzell, 1981): (i) slow diffusion of an agonist to a receptor; (ii) low rate of agonist binding to, or its dissociation from, a receptor; (iii) slow transfer of a signal from activated receptor to ionic channel; and (iv) slow kinetics of opening and closing of a channel operated by a receptor.

The role of agonist diffusion to the membrane of a neuron in the development of muscarinic response has been clarified in amphibian cardiac parasympathetic neurons (Hartzell et al., 1977). For this purpose, the experimentally obtained curves of the initial phase of muscarinic hyperpolarization were compared with the calculated curves of changes in the agonist concentration in the vicinity of the neuronal membrane (Fig. 9.3). This comparison showed that the voltage and concentration curves essentially differ from each other: When the concentration of the agonist at the membrane reaches its maximum, membrane hyperpolarization does not yet appear; it appears when the agonist concentration drops already to 20% of its maximal value. This study found that the initial phase of muscarinic hyperpolarization can be described by the equation $V = k \cdot t^3$, where V is the amplitude of the response at the time t after the beginning of the agonist application, and k is the apparent rate constant. The temperature dependence of the initial phase of muscarinic hyperpolarization is very high: the Q_{10} value for k is 11, i.e., it is much higher than the Q_{10} for diffusion. The difference between the time courses of muscarinic response and agonist concentration change at the membrane, as well as the high temperature dependence of the initial phase of the muscarinic response, makes it unlikely that diffusion of the agonist determines the latency and amplitude of the initial phase of muscarinic hyperpolarization. On the other hand, the third-order dependence of the initial part of muscarinic hyperpolarization on time suggests that there are at least three steps leading to the conductance change underlying muscarinic hyperpolarization. This makes it unlikely that the latency or the rise time of muscarinic response is determined by any one particular factor in isolation, e.g., by binding of the agonist to the muscarinic receptor.

In isolated molluscan neurons, an equivalent of muscarine hyperpolarization, recorded as a slow outward current, also reveals the third-order dependence of its rising phase on time (Ger et al., 1979). The temperature dependence for k here is even higher than in amphibian neurons ($Q_{10} = 15$).

Similarly, in cardiac muscle cells the initial phase of muscarinic hyperpolarization demonstrates both the third-order dependence on time and the high temperature dependence (Hartzell et al., 1977; Pott and Pusch, 1979). This has allowed the authors to conclude that the delay in the onset of muscarinic response is associated with some steps following the binding of the agonist to the muscarinic receptor. Kinetic analysis of the activity of potassium channels activated by acetylcholine in cardiac muscle cells indicates that the kinetics are slow enough to explain the latency of muscarinic hyperpolarization as long as 30 ms (Noma and Trautwein, 1978; Osterrieder et al., 1980, 1981). In neurons the latency of muscarinic hyperpolarization is very close to that in cardiac muscle cells, and it therefore may likewise be associated with the kinetics of potassium channels activated by acetylcholine.

In cardiac muscle cells the nature of the most durable part of muscarinic hyperpolarization, its decay, has been thoroughly studied. According to Hartzell, the duration of muscarinic hyperpolarization decay depends on the nature of the agonist that causes a response (Hartzell, 1981). For instance, with the use of oxotremorine the decay of the hyperpolarizing response is several times slower than with the use of carbacholine. Since oxotremorine exhibits an affinity for the muscarinic acetylcholine receptor several times higher than that of carbacholine (Birdsall et al., 1978; Hartzell, 1980), it has been suggested that the decay of muscarinic hyperpolarization is determined by the affinity of the agonist to a receptor, and it is slower the higher its affinity. This suggestion has been confirmed by the results of

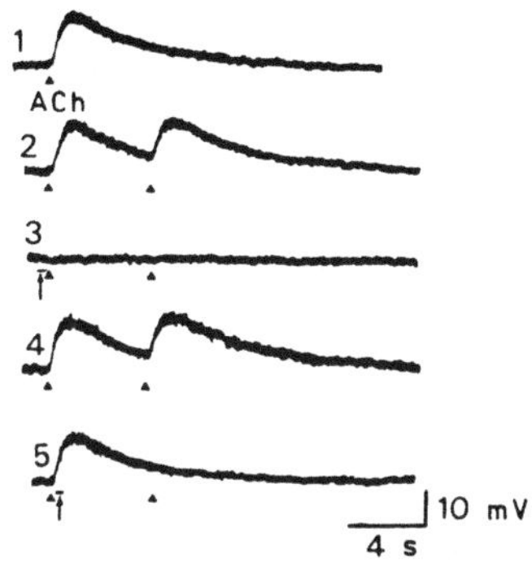

Fig. 9.4. Effect of hyoscine on the time course of muscarinic depolarization recorded from the neuron of the guinea-pig myenteric plexus. Nicotinic response was eliminated by hexamethonium (200 μM). The responses were elicited by single (1) or paired (2) iontophoretic applications of acetylcholine (ACh) using current pulses 10 nA, 150 ms (indicated by triangles). 3, 5) The same as 2, but with hyoscine applied by pressure (500 ms) to neuronal surface immediately prior to the first ACh pulse (3) or immediately after the first ACh pulse, as indicated by arrows (5). 4) 10 min after hyoscine was applied both ACh-induced responses recovered (North and Tokimasa, 1984).

investigation of the actions of fast-dissociating antagonists on the decay of muscarinic hyperpolarization (Pott, 1983). Acceleration of the decay observed under the influence of these antagonists can be attributed to the antagonist-induced increase in the apparent rate constant of dissociation of an agonist from a receptor.

Factors determining the time course of muscarinic depolarization have been studied in guinea-pig myenteric neurons (North and Tokimasa, 1984). Muscarinic depolarization in these neurons results from a decrease in calcium-dependent membrane potassium conductance (Gräfe et al., 1980; Johnson et al., 1980; Morita et al., 1982). This conductance can be turned off by external application of Ba^{2+} ions, which obviously affect the potassium channels either directly or through changing their sensitivity to intracellular Ca^{2+} ions. Interestingly, membrane depolarizations caused by brief applications of muscarinic agonists and Ba^{2+} ions are similar in their latencies and durations (North and Tokimasa, 1984). These results indicate that the steps preceding the transfer of a signal from activated muscarinic receptors to potassium channels (diffusion and binding of an agonist to receptors) are not rate limiting. The time constant of relaxation of persistent Ca-dependent potassium current (2.1 s at 60 mV and 35–37°C: North and Tokimasa, 1987) is comparable with the rise time of muscarinic response (1–5 s: Morita et al., 1982). This fact provides evidence that it is the closure of potassium channels which is the rate-limiting step in eliciting muscarinic response in myenteric neurons.

The temperature dependences for rising phase and decay are high ($Q_{10} = 3$) both for muscarinic response and for the response elicited by brief application of Ba^{2+} ions. More evidence favoring the suggestion that the time course of muscarinic depolarization in myenteric neurons is determined by the steps following activation of muscarinic acetylcholine receptors rather than by their activation itself was obtained with the use of the muscarinic antagonist hyoscine, which was briefly applied during various phases of muscarinic response (North and Tokimasa, 1984). The results of such an experiment are shown in Fig. 9.4. It can be seen that hyoscine is ineffective if it is applied just after the onset of muscarinic depolarization. This finding differs from that mentioned above as obtained with brief applications of fast-dissociating antagonists during muscarinic hyperpolarization in cardiac muscle cells (Pott, 1983). Thus, the mechanisms determining the durations of muscarinic hyperpolarization and depolarization seem to be different. The duration of muscarinic hyperpolarization is determined by the rate of dissociation of an agonist from a receptor or (in cardiac muscle cells) by its diffusion coupled with a low density of receptors. On the contrary, the duration of muscarinic depolarization is probably determined by the kinetics of potassium channels operated via muscarinic acetylcholine receptors.

9.2. IONIC MECHANISMS UNDERLYING MUSCARINIC DEPOLARIZATION

9.2.1. General Characteristics

The dependence of slow e.p.s.p. on membrane potential was first studied in amphibian and mammalian sympathetic ganglia (Kobayashi and Libet, 1968). In frog sympathetic ganglion neurons the response increased with membrane depolarization and decreased with its hyperpolarization. In rabbit superior cervical ganglion neurons a somewhat different dependence of slow e.p.s.p. on membrane potential was obtained: The response decreased with membrane depolarization and increased with its slight hyperpolarization, although it disappeared with stronger membrane hyperpolarization. Both in frog and in rabbit sympathetic ganglion neurons, membrane conductance during slow e.p.s.p. was either unchanged or decreased.

The decrease of slow e.p.s.p. with membrane hyperpolarization and the decrease in membrane conductance during the response in frog sympathetic ganglion neurons have been confirmed by other workers (Nishi et al., 1969; Weight and Votava, 1970; Kuba and Koketsu, 1976a). Some authors also reported a reversal of the initial part of slow e.p.s.p. (Kobayashi and Libet, 1970) or the overall slow e.p.s.p. (Weight and Votava, 1970) with membrane hyperpolarization. In the latter case the reversal potential for the response coincided with the reversal potential for post-spike afterhyperpolarization. This result led to the conclusion that slow e.p.s.p. is marked by decreased membrane conductance for K^+ ions (Weight and Votava, 1970). However, Libet (1971) suggested that the reversed response was actually a slow i.p.s.p. augmented with membrane hyperpolarization. Indeed, according to Kobayashi and Libet (1974), the reversal of the initial part of slow e.p.s.p. occurred at a membrane potential level less negative than V_K. By contrast, other workers (Nishi et al., 1969; Kuba and Koketsu, 1976a) failed to detect any reversal of slow e.p.s.p.

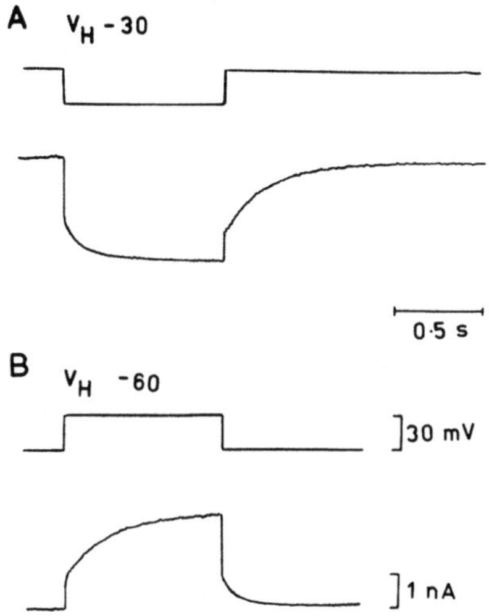

Fig. 9.5. M-current recorded from frog sympathetic ganglion neuron. Membrane currents were evoked by hyperpolarizing 30 mV voltage step from –30 mV holding (V_H) potential level (A) and by depolarizing 30 mV voltage step from –60 mV holding potential level (B). Top records indicate membrane potential, bottom records indicate membrane current. Slow relaxations of inward and outward currents are due to deactivation and activation of M-current, respectively (Adams et al., 1982a).

Further investigations carried out on amphibian sympathetic ganglion neurons (Nishi et al., 1969; Kuba and Koketsu, 1974, 1976a) indicated that the decrease of slow e.p.s.p. with membrane hyperpolarization and the decrease in membrane conductance during slow e.p.s.p. could be obtained only in some sympathetic ganglion neurons. In the rest of the neurons, slow e.p.s.p. was either increased or unaffected by membrane hyperpolarization. Membrane conductance during such a response was increased or unchanged, respectively.

Since the reversal potential for slow e.p.s.p. was not determined at this stage of investigation, the effect of modified external ionic medium was studied only on the amplitude, not on the reversal potential, for slow e.p.s.p. or its analogs elicited by muscarinic agonists. In accordance with previous data indicating that a decrease in membrane potassium conductance may be responsible for the generation of muscarinic depolarization, it was found that elevation of the external concentration of K^+ ions strongly decreases muscarinic depolarization (Kuba and Koketsu, 1976a). Diminishing the concentration of Na^+ ions in the external solution (containing an increased amount of K^+ ions for suppressing the response marked by a decrease in membrane conductance for these ions) decreases muscarinic depolarization (Kuba and Koketsu, 1976a). A complete removal of Cl^- ions from the external solution did not affect slow e.p.s.p. or muscarinic depolarization in rabbit and frog sympathetic ganglion neurons (Kobayashi and Libet, 1968; Weight and Votava, 1970;

Kuba and Koketsu, 1976a). A decrease in the concentration of Ca^{2+} ions in the external solution containing an increased amount of K^+ ions suppressed muscarinic depolarization, while in isotonic calcium solution this depolarization increased (Kuba and Koketsu, 1976a).

The above changes in membrane conductance during slow e.p.s.p., the dependence of the response on membrane potential, and the effects of modification of the external ionic medium on muscarinic depolarization indicate that there are three distinct types of slow e.p.s.p. in amphibian sympathetic ganglion neurons (see Kuba and Koketsu, 1978): i) slow e.p.s.p. resulting from decreased membrane potassium conductance and therefore enhanced by membrane hyperpolarization, ii) slow e.p.s.p. arising from increased membrane conductance for certain cations (Na^+ and Ca^{2+}), which increases with membrane hyperpolarization and decreases with its depolarization, and iii) slow e.p.s.p. of mixed type, marked by decrease in membrane conductance for K^+ ions and by simultaneous increase in membrane conductance for other cations. The amplitude of such a compound response is weakly dependent on the membrane potential.

A similar heterogeneity of slow e.p.s.p. was found in rabbit sympathetic ganglion neurons (Ivanov and Skok, 1981). In some neurons (about 28%) the amplitude of slow e.p.s.p. increased with membrane hyperpolarization and decreased with its depolarization, while being accompanied by membrane conductance increase. Similar properties were established earlier for slow negative potentials recorded from the ganglion surface (Ivanov and Skok, 1972). Some of the remaining ganglion neurons (about 55%) revealed an increase in slow e.p.s.p. with membrane depolarization and its disappearance with membrane hyperpolarization (at about –79 mV). In these neurons, no changes in membrane conductance were observed during a response. And, finally, in 17% of the ganglion neurons, slow e.p.s.p. increased with membrane depolarization and was unaffected by membrane hyperpolarization.

An apparent drop in potassium membrane conductance during muscarinic depolarization is consistent with the finding that muscarine suppresses $^{86}Rb^+$-efflux from rat isolated superior cervical ganglion (Brown et al., 1980a).

In other types of neurons, muscarinic depolarization is more homogeneous in its ionic mechanisms than in sympathetic ganglion neurons. For example, in myenteric plexus (Gräfe et al., 1980; Johnson et al., 1980; Morita et al., 1982), in cortex and hippocampus (Krnjevič et al., 1971b; Dodd et al., 1981; Cole and Nicoll, 1983), as well as in spinal cord (Nowak and MacDonald, 1983a), the muscarinic depolarization decreases at membrane hyperpolarization and is accompanied by a drop in membrane conductance, mainly for K^+ ions. Only in one type of central neurons – locus coeruleus neurons – does muscarinic depolarization seem to be followed by an increase in membrane conductance for Na^+ ions (see North, 1986).

Thus, initial investigations indicated that a decrease in membrane potassium conductance is the most common mechanism of muscarinic depolarization in neurons. In some neurons (sympathetic or locus coeruleus) muscarinic depolarization may also be associated with an increased cationic membrane conductance. A more detailed analysis of each of these mechanisms, carried out under voltage-clamp conditions, is given below.

9.2.2. Decrease in Membrane Potassium Conductance

A decrease in membrane conductance for K^+ ions elicited by muscarinic agonists implies that in the absence of agonists a steady-state potassium current, nonactivated in time and operated via muscarinic receptors, should continuously flow through the membrane.

Two main types of such a current were revealed in various types of neurons: voltage dependent and voltage independent.

It should be pointed out that, in addition to depression of the steady-state potassium current, muscarinic agonists also diminish the inactivating potassium currents contributing to post-spike afterhyperpolarization (see Chapter 10). However, since these currents are constantly inactivated at resting membrane potential level (except current underlying afterhyperpolarization in myenteric neurons, which will be discussed below), their inhibition via muscarinic acetylcholine receptors cannot result in membrane depolarization.

A steady-state voltage-dependent potassium current was first revealed by Brown and Adams (1980) in amphibian sympathetic ganglion neurons. Since this current was inhibited by muscarinic agonists, it was termed the "M-current." However, it was revealed later that the M-current was sensitive both to muscarinic agonists and to several other transmitters, including neuropeptides (see Chapter 11). Moreover, in one type of neuron (C-cells of amphibian sympathetic ganglion neurons) M-current appeared to be insensitive to muscarine and, on the contrary, sensitive to neuropeptides (Jones, 1984). Despite this confusion, steady-state voltage-sensitive potassium current present in sympathetic and in some other neurons is still called the M-current, and we will use this term throughout the text. The characteristics of M-current in B-neurons of frog sympathetic ganglion are given below (see Adams et al., 1982a).

First of all, it is worthwhile to emphasize the distinctions of the M-current compared with three other major potassium currents which are also present in ganglion neurons. These currents are: i) delayed rectifying current (I_K) activated at membrane potentials less negative than –25 mV; ii) calcium-dependent current (I_c), activated at membrane potentials close to –25 mV, which reaches its maximal value at membrane potentials ranging between 20 and 60 mV; iii) "transient" current (I_A), activated by membrane repolarization to a level less negative than –60 mV, preceded by membrane hyperpolarization; this current is rapidly and completely inactivating at all membrane potentials at which it is activated. M-current differs from other potassium currents in its more negative activation potential and in lacking inactivation in time. Because of these two peculiarities, the M-current can be recorded in isolation from other potassium currents, when the membrane potential is stepped between –60 and –30 mV. M-currents, observed under this condition, are shown

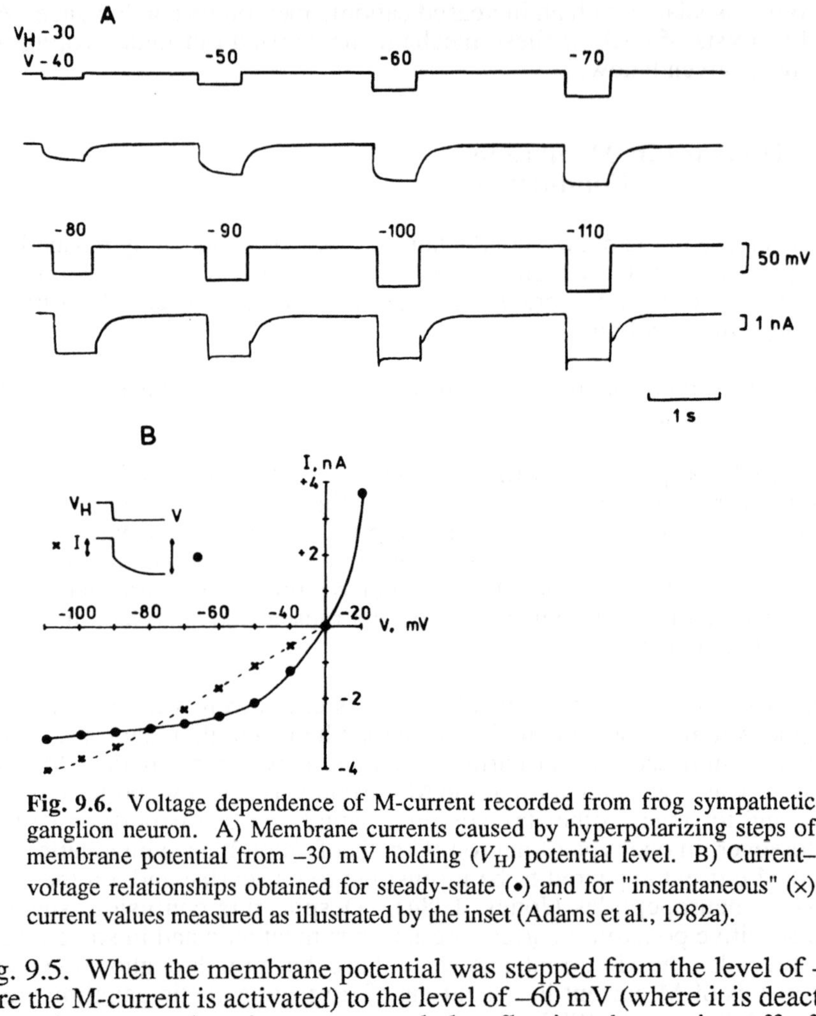

Fig. 9.6. Voltage dependence of M-current recorded from frog sympathetic ganglion neuron. A) Membrane currents caused by hyperpolarizing steps of membrane potential from –30 mV holding (V_H) potential level. B) Current–voltage relationships obtained for steady-state (•) and for "instantaneous" (×) current values measured as illustrated by the inset (Adams et al., 1982a).

in Fig. 9.5. When the membrane potential was stepped from the level of –30 mV (where the M-current is activated) to the level of –60 mV (where it is deactivated), an inward current relaxation was recorded, reflecting the turning-off of the M-channels. With membrane repolarization, an outward current relaxation was recorded, reflecting the turning-on of a part of the M-channels (Fig. 9.5A). Similar current relaxations, although opposite in sign, were recorded with a membrane potential step from the holding level of –30 mV (Fig. 9.5B). One can see that the amplitudes of the corresponding current relaxations at holding membrane potential levels of –30 and –60 mV are equal to each other, which indicates the lack of M-current inactivation at a membrane potential of –30 mV.

Figure 9.6A illustrates the dependence of the M-current on membrane potential obtained by stepping the membrane potential to various levels from a holding level of –30 mV. The reversal of the M-current is indicated by a change in the direction of relaxation. The reversal potential can be estimated from a crossover of "instantaneous" and steady-state current–voltage relationships (Fig. 9.6B). The reversal potential for the M-current (V_r) thus obtained is about –80 mV, i.e., it is

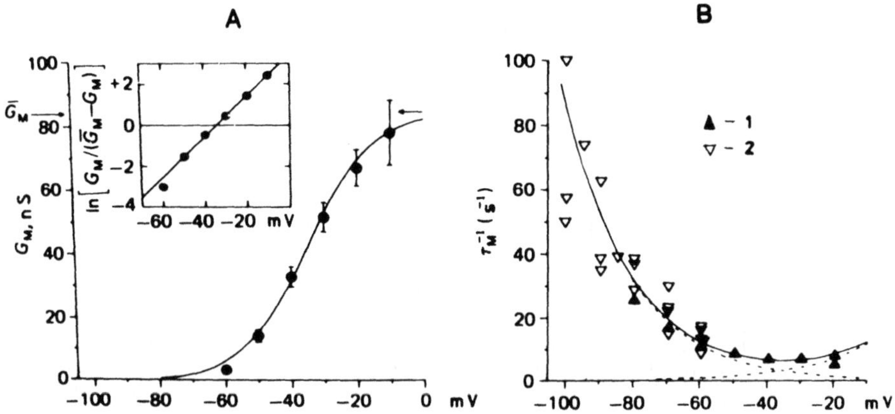

Fig. 9.7. Voltage dependences of M-conductance (A) and relaxation time constant of M-current (B) in frog sympathetic ganglion neurons. Maximum M-conductance value is indicated by arrow. Vertical bars correspond to S.E. of the mean conductance values. The inset in A is a semilogarithmic plot of M-conductance vs. membrane potential level. Different symbols in B correspond to τ_M^{-1} measurements in solutions with different external concentration of K^+ ions, 2.5 mM (1) and 8 or 25 mM (2). For more details see text (Adams et al., 1982a).

close to the potassium equilibrium potential. The V_r for the M-current depends only on one ion species, i.e., on potassium ions, changing by about 50 mV with a tenfold change in their external concentration. This indicates that the M-current is carried by K^+ ions.

The amplitude of the M-current relaxation recorded with membrane repolarization from testing to holding potential level, divided by the e.m.f., that is, by the difference between the testing potential level and V_r, provides the value of the change in membrane conductance for the M-current (ΔG_M) at the testing membrane potential level. The full value of the membrane conductance for the M-current (M-conductance, G_M) at the testing membrane potential level is determined as the difference between the value obtained with the membrane potential step to the level of –70 mV (where M-current is completely deactivated) from the level of –30 mV, and the value of G_M obtained with the membrane potential step from testing to holding level. Figure 9.7A shows the dependence of G_M on membrane potential. It is seen that this dependence is S-shaped, with a maximum value of M-conductance ($\bar{G}_M$) of 84 nS, a value of $1/2\bar{G}_M$ corresponding to a membrane potential level of –35 mV, and a slope corresponding to an e-fold increase in G_M with membrane depolarization by 10 mV.

The kinetics of M-current relaxations are determined by the testing level of the membrane potential and is independent of the preceding level of membrane potential, the direction of the M-current, or the concentration of external K^+ ions. At the same time, the kinetics of M-current relaxations vary with the change in testing membrane potential level. Figure 9.7B shows the dependence of the rate constant (which is inverse to the time constant τ_M) of M-current relaxation on membrane potential. It is evident that M-current relaxation slows down with membrane depolarization and accelerates with its hyperpolarization.

The authors of this investigation (Adams et al., 1982a) described the voltage-dependence of G_M according to the Hodgkin–Huxley model (Hodgkin and Huxley, 1952), based on an assumption that M-channel opening and closing are determined by a single voltage-sensitive particle. It has been suggested that M-conductance is the sum of identical conductances of single channels, each of which is in one or two states, open or closed. Then the M-conductance is determined as

$$G_M = \bar{G}_M \cdot Y_M(\infty), \tag{9.1}$$

where $Y_M(\infty)$ is the fraction of channels remaining in the open state, or the probability of a channel remaining in open state, whereas $\bar{G}_M$ is the maximum M-conductance at $Y_M(\infty) = 1$. It is also suggested that the probability of a channel remaining in the open state obeys Boltzmann's distribution:

$$Y_M(\infty) = \left\{1 + \exp\frac{ze}{kT}(V_o - V)\right\}^{-1}, \tag{9.2}$$

where z is the effective valence of a voltage-sensitive particle, V is the membrane potential, and V_o is the membrane potential at which $Y_M(\infty) = 0.5$.

As follows from Eqs. (9.1) and (9.2),

$$G_M = \bar{G}_M\left\{1 + \exp\frac{ze}{kT}(V_o - V)\right\}^{-1}. \tag{9.3}$$

As Fig. 9.7A shows, an experimentally obtained $G_M(V)$ relationship is satisfactorily described by Eq. (9.3) at $\bar{G}_M = 84$ nS, $z = 2.5$, and $V_o = -35$ mV.

The $\tau_M(V)$ relationship is determined by the expression

$$\tau_M^{-1} = \alpha_M + \beta_M, \tag{9.4}$$

where α_M and β_M are the voltage-dependent rate constants of M-channels openings and closings, which are equal to

$$\alpha_M = \alpha_M(0)\exp\frac{+ze}{2kT}(V - V_o), \tag{9.5a}$$

and

$$\beta_M = \beta_M(0)\exp\frac{-ze}{2kT}(V - V_o), \tag{9.5b}$$

at $\alpha_M(0) = \beta_M(0) = 3.3\ s^{-1}$ and $V = V_o$.

As follows from Fig. 9.7B, the experimentally determined $\tau_M^{-1}(V)$ relationship can be satisfactorily fitted with the theoretical one, calculated according to Eqs. (9.4) and (9.5). This finding indicates that the above assumptions are correct.

The M-current was studied in the most detail in frog sympathetic ganglion neurons, where it was recorded using the two-electrode voltage-clamp method. At the same time, the M-current was also revealed in other types of neurons, where it was studied using the one-electrode voltage-clamp method: in mammalian sympathetic (Constanti and Brown, 1981; Brown and Selyanko, 1985a, b), hippocampal (Halliwell and Adams, 1982), cortical (Constanti and Galvan, 1983), and spinal neurons (Nowak and MacDonald, 1983b). The characteristics of the M-current in all these neurons appeared to be qualitatively similar to those in amphibian sympathetic ganglion neurons, with only a few minor quantitative differences. First, in rat sympathetic ganglion neurons the activation potential for the M-current is 5–20 mV more negative than in frog sympathetic neurons. This correlates with the more negative resting membrane potential level in rat neurons. Second, in rat sympathetic ganglion neurons the maximum value of M-conductance ($\bar{G}_M$) is half as large as that in frog sympathetic ganglion neurons (Constanti and Brown, 1981). However, taking into account the difference in somata diameters between these groups of neurons, the densities of M-channels seem to be similar, provided the conductances of single M-channels in both types of neurons are identical.

Third, at room temperature the kinetics of M-current relaxations both in amphibian and mammalian sympathetic ganglion neurons, as well as in hippocampal neurons, is about the same (Constanti and Brown, 1981; Adams and Brown, 1982). However, taking into account the difference in body temperatures in poikilotherms and homoiotherms and the temperature dependence of τ_M ($Q_{10} = 5$: Halliwell and Adams, 1982), it may be concluded that at body temperatures normal for these animals, the relaxations of M-current are much faster in homoiotherms than in poikilotherms.

The M-current differs from other potassium currents both in its voltage dependence, kinetics, and pharmacological properties. The M-current, when compared with other potassium currents, is much less sensitive to the potassium channel blocker tetraethylammonium (TEA) and much more sensitive to Ba^{2+} ions (Brown and Adams, 1980; Adams et al., 1982b). However, the most distinctive peculiarity of the M-current is its sensitivity to muscarinic agonists. Figure 9.8A shows the effect of muscarine on the M-current recorded from an amphibian sympathetic ganglion neuron at the characteristic range of its membrane potentials (–30 to –60 mV). The membrane potential was stepped in succession between –30 and –60 mV. The advantage of this potential range was that it allowed one to compare the actions of muscarine under conditions when the majority of M-channels were activated (at –30 mV) and when all of them were deactivated (at –60 mV). It was obvious that muscarine caused an inward current at –30 mV, while no current appeared at –60 mV. The following observations indicated that the observed steady-state inward current resulted from inhibition of the M-current (see Fig. 9.8A): (i) muscarine depressed relaxations of the M-current produced by the hyper-

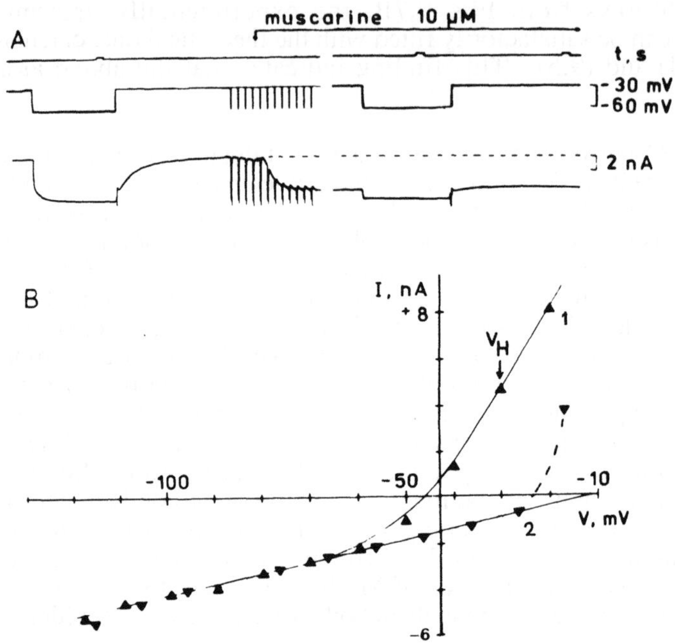

Fig. 9.8. Effect of muscarine on frog sympathetic ganglion neuron. A) Effect of muscarine (10 μM) on membrane current (bottom record) measured at two membrane potential levels, –30 mV and –60 mV (top record). B) Current–voltage relationship obtained from the same neuron (A) in normal solution (1) and in muscarine-containing solution (2) (Adams et al., 1982b).

polarizing voltage step to the level of –60 mV; this was apparently due to the fact that the M-channels, already closed by muscarine at –30 mV, could not be additionally closed by membrane hyperpolarization, and (ii) in muscarine, ohmic current jumps accompanying voltage steps from –30 to –60 mV and from –60 to –30 mV were equal to each other; this indicated that under the influence of muscarine membrane conductance became equal to –30 and –60 mV.

The link between muscarine-induced M-current inhibition and the appearance of a steady-state inward current is illustrated in more detail in Fig. 9.8B. Here the current–voltage curves obtained before and during the action of muscarine are shown. It is evident that muscarine does not induce current at the membrane potential range where the M-current is deactivated (more negative than –60 mV), and it induces an inward current within the membrane potential range where the M-current is activated (less negative than –60 mV).

In the neuron shown in Fig. 9.8, the depression of the M-current by muscarine (10 μM) is virtually complete. On the average, at this or higher concentrations, muscarine inhibits 80–90% of the M-current; half-inhibition of the M-current was obtained for 0.7–3 μM of muscarine (Adams et al., 1982b; Jones, 1985). With partial inhibition of the M-current, the degree of M-current inhibition is the

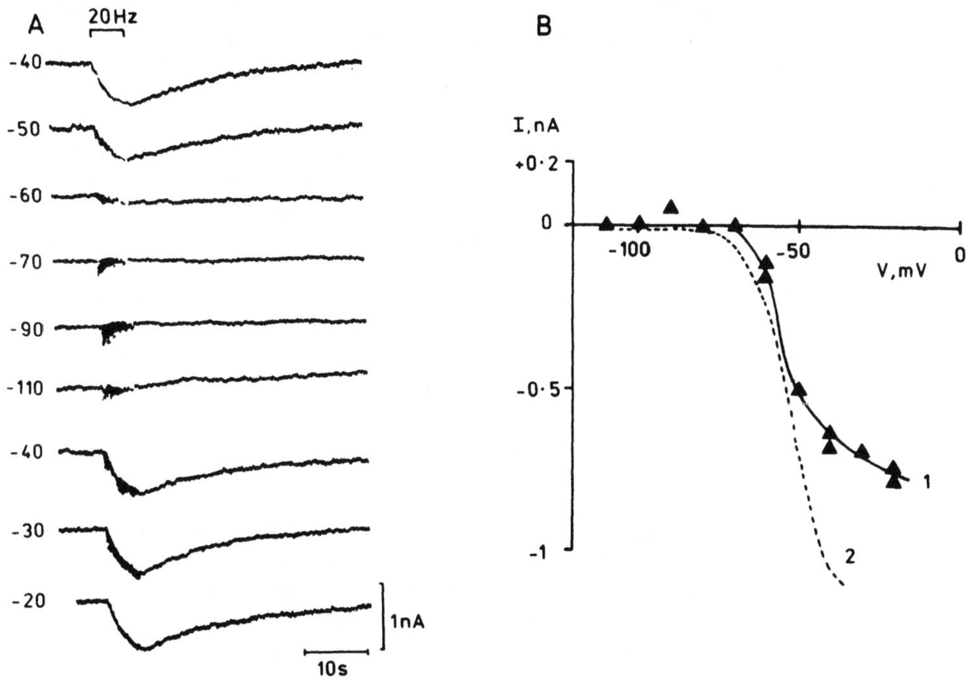

Fig. 9.9. Voltage dependences of slow e.p.s.c. and of muscarine-induced response recorded from rat superior cervical ganglion neuron. A) Slow e.p.s.p.'s elicited by repetitive (20 Hz, 5 s) stimulations of cervical sympathetic nerve, as recorded from the same neuron at different membrane potential levels (indicated in mV near each record). Nicotinic acetylcholine receptors were blocked by tubocurarine (0.4 mM). B) Voltage dependences of the slow e.p.s.c. (1) and of the current induced by muscarine (10 μM) (2) obtained from the same neuron (A) (Brown and Selyanko, 1985b).

same at all potentials at which the M-current is normally activated (less negative than –60 mV). This indicates that the fraction of M-channels inhibited by muscarine does not depend on the membrane potential. On the other hand, the amount of current by which the M-current is diminished by muscarine (determined from the decrease in M-current relaxation caused by the repolarizing voltage step to 60 mV) exactly corresponds to the steady-state current induced by muscarine at this membrane potential. This is additional evidence favoring the suggestion that steady-state muscarine-induced inward current results, at least in the range of membrane potentials where the M-current is activated, solely from M-current inhibition. With partial inhibition of the M-current, the kinetics of M-current relaxations and its voltage dependence remain unchanged.

The voltage-independent inhibition of the M-current resulting from activation of muscarinic acetylcholine receptors indicates that the voltage dependences of muscarinic response and of the M-current itself should be similar. Figure 9.9 shows the dependence of slow excitatory postsynaptic current (slow e.p.s.c.) on membrane potential in a rat superior cervical ganglion neuron. In this particular neuron both slow e.p.s.c. and muscarine-induced current result solely from M-current inhibition. Figure 9.9B shows that both muscarinic responses exhibit S-shaped I(V) curves inherent to that of the M-current. Like the M-current, both muscarinic responses disappear at –70 mV (the activation potential for the M-current in neurons of this type: Constanti and Brown, 1981) and do not reverse with further membrane hyperpolarization. The lack of reversal of muscarinic depolarization, accompanied by a decrease in membrane conductance, with membrane hyperpolarization was reported earlier (see above), but it could not be explained until the M-current was revealed. Now it is possible to explain this observation by deactivation of the M-current at membrane potentials more negative than –70 mV.

That it is deactivation of the M-current which is responsible for the lack of reversal in muscarinic response is proved by the fact that muscarinic response does reverse when V_K is shifted to membrane potential levels less negative than –60 mV by elevating the external concentration of K^+ ions (Adams et al., 1982a, b).

The finding that Ba^{2+} ions suppressed the M-current (Constanti et al., 1981; Adams et al., 1982a) allowed one to explain the well-known similarity in action of these ions and of muscarinic agonists on sympathetic (Ambache, 1949; Takeshige and Volle, 1964a; Tashiro and Nishi, 1972; McLachlan, 1977; McAfee and Yarowsky, 1979), cortical (Krnjevič et al., 1971a), and hippocampal neurons (Hotson et al., 1979; Hotson and Prince, 1980). This similarity is manifested by membrane depolarization, by an increase in input resistance, and by discharges resulting from the actions of both Ba^{2+} ions and muscarinic agonists. However, one difference has been found between the actions of Ba^{2+} and muscarine on the M-current: The action of muscarine is abolished by atropine, whereas the action of Ba^{2+} ions is not. The most likely explanation for this difference is that Ba^{2+} ions suppress the M-current by acting directly on M-channels, while muscarine causes the same effect by acting indirectly, through activation of muscarinic acetylcholine receptors (Constanti et al., 1981). The inhibitory action of Ba^{2+} ions on the M-current persists in the presence of Cd^{2+} ions, the calcium channel blocker, preventing the entry of Ba^{2+} ions into the cell through calcium channels; thus, M-channels are blocked by Ba^{2+} ions from the outer, rather than the inner, side of the membrane (Constanti et al., 1981).

Another conclusion that follows from the difference in the action of atropine on M-current inhibition caused by muscarine and by Ba^{2+} ions is that atropine interacts with the recognition center of the muscarinic acetylcholine receptor and does not affect the M-channels operated by this receptor. Otherwise, if acting on M-channels, atropine would mimic the excitatory action of muscarine, as Ba^{2+} ions do, rather than prevent it.

In addition to occurring in amphibian sympathetic ganglion neurons, inhibition of the M-current resulting from activation of muscarinic acetylcholine receptors has also been revealed in other types of neurons: in mammalian sympathetic (Constanti and Brown, 1981; Hashiguchi et al., 1982; Brown and Selyanko, 1985a, b), hippocampal (Halliwell and Adams, 1982), cortical neurons (Constanti and Galvan, 1983), and in cultured spinal (Nowak and MacDonald, 1983a) and cultured mammalian sympathetic neurons (Freschi, 1983). In all these neurons,

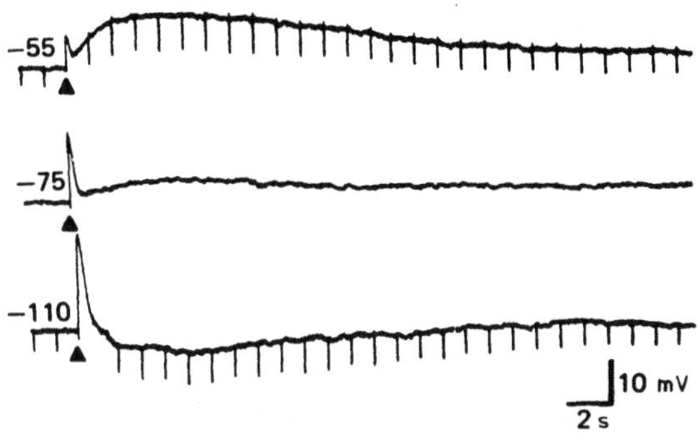

Fig. 9.10. Effect of membrane potential on muscarinic depolarization recorded from guinea-pig myenteric neuron. At resting membrane potential level (–55 mV), iontophoretically applied acetylcholine (the moment of application is indicated by a triangle) caused fast, or nicotinic, and slow, or muscarinic, depolarization. Hyperpolarization of the membrane caused a decrease (at –75 mV) or a reversal (at –110 mV) of muscarinic depolarization, while nicotinic depolarization increased. Vertical bars seen in the top and bottom records are electrotonic potentials caused by testing hyperpolarizing current pulses (North and Tokimasa, 1982).

except mammalian sympathetic ganglion neurons, M-current inhibition is the only mechanism underlying muscarinic depolarization.

In guinea-pig myenteric neurons (Gräfe et al., 1980; Johnson et al., 1980; Morita et al., 1982) and in amphibian as well as mammalian sensory neurons (Morita and Katayama, 1984), muscarinic depolarization likewise results from a decrease in membrane potassium conductance. However, there are indications that this potassium conductance differs from the M-conductance described above. This follows from the characteristics of muscarinic depolarization in these neurons: i) it is linearly dependent on membrane potential, ii) it reverses at a hyperpolarized membrane potential that corresponds to V_K (Fig. 9.10), and iii) it depends on the presence of Ca^{2+} ions in external solution. Recently, a persistent calcium-sensitive potassium current, whose inhibition by muscarinic agonists might contribute to muscarinic depolarization, was identified in AH myenteric neurons (North and Tokimasa, 1987). This current indeed appeared to be distinct in several respects from the M-current: i) it persisted only in the presence of Ca^{2+} ions outside the cell, ii) it was itself low voltage-dependent, and its actual voltage dependence seemed to be derived from the voltage dependence of the entry of calcium ions into the cell, and iii) its relaxations accompanying voltage steps were slowed down, rather than accelerated, with membrane hyperpolarization.

However, there were indications that the above calcium-dependent potassium conductance could only partially be responsible for muscarinic depolarization in myenteric neurons (North and Tokimasa, 1987). This was evidenced by the fact that muscarinic agonists decrease membrane potassium conductance at membrane potentials even more negative than V_K, where persistent calcium-dependent potassium current is turned off due to deactivation of persistent calcium current. Fur-

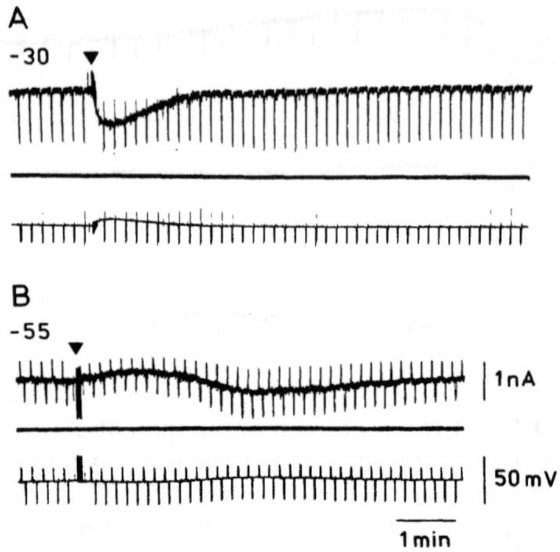

Fig. 9.11. Biphasic synaptic current recorded from rat superior cervical ganglion neuron. Records were obtained at membrane potential levels –30 mV and –55 mV with testing hyperpolarizing (top record) and both depolarizing and hyperpolarizing (bottom record) 1 s long current pulses. Synaptic current was evoked by repetitive (20 Hz, 5 s) orthodromic stimulation (indicated by a triangle). The solution contained 0.4 mM tubocurarine. For further details see text (Brown and Selyanko, 1985b).

thermore, the contribution of persistent calcium-dependent potassium current to resting membrane conductance is very low (about 15%), whereas the drop in resting membrane conductance induced by muscarinic agonists is about two- or threefold. North and Tokimasa (1987) have concluded that in myenteric neurons there may be more than one steady-state potassium current sensitive to muscarinic agonists.

9.2.3. Increase in Membrane Conductance

It was revealed in early experiments carried out on amphibian and mammalian sympathetic ganglion neurons that in some neurons muscarinic depolarization is accompanied by an increase, rather than a decrease, in membrane conductance, and the amplitude of such depolarization is augmented with membrane hyperpolarization. This result has been confirmed in experiments performed on ganglion neurons with the use of the voltage-clamp method. Adams and co-workers have revealed that in a small part of amphibian sympathetic ganglion neurons, an inward current caused by muscarine is not abolished by membrane hyperpolarization strong enough to deactivate the M-current (Adams et al., 1982b). This response is accompanied by an increase, rather than a decrease, in membrane conductance.

Akasu and co-workers have found that in 14% of amphibian sympathetic ganglion neurons slow e.p.s.c., if examined at all levels of membrane potential, is accompanied by an increase in membrane conductance (Akasu et al., 1984a). Since the reversal potential for this response was about 5 mV, it was concluded that the response resulted from a membrane conductance increase for cations.

In a number of investigations carried out on amphibian (Kuffler and Sejnowski, 1983; Jones, 1985) and mammalian sympathetic ganglion neurons (Brown and Selyanko, 1985a, b), an inward current accompanied by an increased membrane conductance was found as constituting an additional component of muscarinic response (besides that derived from M-current inhibition). This inward current never occurred in isolation as, for instance, did a component associated with the M-current inhibition. Another peculiarity of this component was that it was much slower than a component resulting from M-current inhibition and therefore appeared as a delayed component of muscarinic response. An example of the delayed component of muscarinic response is shown in Fig. 9.11. It was suggested that a delayed component of muscarinic response in mammalian sympathetic ganglion neurons might result from an increased membrane conductance for Cl^- ions (Brown and Selyanko, 1985a, b). This suggestion was consistent with the fact that the reversal potential for a delayed component was close to V_{Cl}, which is near 0 mV, due to diffusion of Cl^- ions into the cell from a microelectrode. An increased chloride membrane conductance may be a secondary effect, resulting from the accumulation of Cl^- ions in the cell due to inhibition of steady-state chloride membrane conductance by an acetylcholine-transmitter or by muscarine (Brown and Selyanko, 1985a, b).

Thus, due to some controversy, the ionic basis for an additional component of muscarinic depolarization in sympathetic ganglion neurons remains to be clarified.

As has been mentioned above, in one type of central neuron – locus coeruleus – muscarinic depolarization results from an increase in membrane conductance, presumably for Na^+ ions (see North, 1986).

9.2.4. Depolarization without Changes in Membrane Conductance

Muscarinic depolarization, not dependent on membrane potential and not accompanied by any changes in membrane conductance, was detected in early studies carried out on amphibian and mammalian sympathetic ganglion neurons (see above). Later, Akasu et al. (1984a) confirmed these data by finding that in about 21% of the amphibian sympathetic ganglion neurons slow e.p.s.c. was weakly dependent on membrane potential and was not accompanied by any changes in membrane conductance at the resting membrane potential. However, these authors also revealed that at membrane potentials less negative than resting potential level, membrane conductance during slow e.p.s.c. decreased, whereas at more negative potentials it increased. Based on these findings, Akasu and co-workers concluded that slow e.p.s.c. in a given group of neurons consisted of two components exhibiting reciprocal dependences on membrane potential and accompanied by reciprocal changes in membrane conductance. One of these components, associated with M-current inhibition, increased with membrane depolarization and decreased

TABLE 9.1. Ionic Mechanisms Underlying Muscarinic Depolarization (D) and Hyperpolarization (H) in Various Types of Neurons

Preparation	Animals	Response	Conductance change: increase (↑) or decrease (↓)	References
Central nervous system	Mammalia	D	↓ K	Krnjevič et al., 1971
				Dingledine et al., 1977
				Halliwell and Adams, 1982
				Constanti and Galvan, 1983
				Cole and Nicoll, 1983
				Nowak and MacDonald, 1983a
			↑ Na	Egan and North, 1985a
		H	↓ K	Egan and North, 1985b
Sympathetic ganglion	Mammalia	D	↓ K, ↑ cations	Ivanov and Skok, 1981
			↓ K	Constanti and Brown, 1981
			↓ K, ↑ CI?	Brown and Selyanko, 1985a, b
			↓ K, without changes	Hashiguchi et al., 1982
		H	↑ K	Ivanov and Skok, 1980
			↑ K	Cole and Shinnick-Gallagher, 1984
			↓ CI	Brown and Selyanko, 1985a, b
	Amphibia	D (B-cells)	↓ K	Weight and Votava, 1970
			↓ K	Brown and Adams, 1980
			↓ K, ↑Na, Ca	Kuba and Koketsu, 1976a
				Akasu et al., 1984a

		H	↓ Na	Weight and Padjen, 1973a, b
		(C-cells)	↓ Na, ↑K	Weight and Smith, 1981
			↑ K	Dodd and Horn, 1983
			↑ K, electrogenic Na^+-pump	Nishi and Koketsu, 1967, 1968 Akasu and Koketsu, 1976, 1983 Akasu et al., 1978
Myenteric plexus	Mammalia	D	↓ K	Gräfe et al., 1980 Johnson et al., 1980 Morita et al., 1982
Parasympathetic ganglion	Mammalia	H	↑ K	Gallagher et al., 1982
	Amphibia	H	↑ K	Hartzell et al., 1977
Sensory ganglion	Mammalia	D	↓ K	Morita and Katayama, 1984
Ganglion	Mollusc	H	↑ K	Kehoe, 1972 Ger et al., 1979a,b

TABLE 9.2. Evidence in Favor (A) and Against (B) Disynaptic Mediation of Muscarinic Hyperpolarization in Mammalian Sympathetic Ganglion Involving Dopamine as a Second Transmitter

A	B
1. Presumably interneurons (SIF-cells) contain dopamine and enzymes synthesizing it (Bjorklund et al., 1970; Libet and Owman, 1974).	1. There are no anatomical signs that SIF-cells are interneurons (Dun, 1980).
2. Dopamine content in SIF-cells as well as muscarinic hyperpolarization decrease following orthodromic stimulation or action of muscarinic agonists. After the treatment of the ganglion with dopamine its content in SIF-cells and muscarinic hyperpolarization are restored (Libet, 1970; Libet and Tosaka, 1970; Libet and Owman, 1974).	2. ^{3}H-dopamine is not released following orthodromic stimulation (Noon, 1975). Slow i.p.s.p. inhibition can be associated with the desensitization of muscarinic receptors, since it is accompanied by depression of slow e.p.s.p.; slow i.p.s.p. restoration following dopamine treatment does not occur (Dun, 1980).
3. Exogenous dopamine, as well as inhibitory transmitter, cause membrane hyperpolarization in ganglion neurons which is not accompanied by membrane conductance change (Libet, 1970; Kobayashi and Libet, 1970).	3. Slow i.p.s.p. is accompanied by membrane conductance increase and reverses at $V \cong E_K$. Similar properties exhibit membrane hyperpolarization elicited by exogenous acetylcholine (Nishi, 1979; Dun and Karczmar, 1980; Cole and Shinnick-Gallagher, 1984).
4. Factors inhibiting transmitter release (low $[Ca^{2+}]_o$ to $[Mg^{2+}]_o$ ratio, TTX) selectivity abolish acetylcholine-induced membrane hyperpolarization without affecting muscarinic depolarization (Libet, 1970; Nishi, 1979; Dun and Karczmer, 1978).	4. Membrane hyperpolarization induced by exogenous acetylcholine remains unchanged in the absence of Ca^{2+} ions in external solution and is depressed only following prolonged perfusion of the ganglion with Ca^{2+}-free solution; TTX does not affect acetylcholine-induced membrane hyperpolarization (Cole and Shinnick-Gallagher, 1984).
5. Adrenergic antagonists blocked either slow i.p.s.p. or membrane hyperpolarization produced by muscarinic agonists (Libet, 1970; Libet and Tosaka, 1970; Ashe and Libet, 1982).	5. Adrenergic antagonists do not affect slow i.p.s.p. (Cole and Shinnick-Gallagher, 1981, 1984; Dun, 1980; Dun and Karczmar, 1980).

with its hyperpolarization, while another component, associated with an increased membrane conductance, increased with membrane hyperpolarization and decreased with membrane depolarization. The two components of the response, when superimposed, provided the characteristics of slow e.p.s.c. described above. Similar data were obtained in frog sympathetic ganglion neurons by other workers (Kuffler and Sejnowski, 1983).

According to Hashiguchi et al. (1982), in mammalian sympathetic neurons investigated at hyperpolarized membrane potentials where the M-current is deactivated, muscarine induces membrane depolarization that is not dependent on membrane potential and is not accompanied by changes in membrane conductance. The appearance of this component of muscarinic depolarization is indicated by a parallel shift in the steady-state $I(V)$ relationship, under the influence of muscarine, at membrane potentials more negative than –60 mV. The ionic mechanism for a component of muscarinic depolarization not accompanied by changes in membrane conductance is not yet clear. It is known only that this component can be mimicked by the effect of cGMP derivatives, which implies that cGMP may serve as a link in eliciting muscarinic depolarization without changes in membrane conductance. Interestingly, in contrast to mammalian sympathetic ganglion neurons, this component of muscarinic depolarization is not detectable in amphibian sympathetic ganglion neurons (Hashiguchi et al., 1982).

The results of an investigation of ionic mechanisms underlying muscarinic depolarization in various types of neurons are summarized in Table 9.1.

9.2.5. Ionic Mechanisms of Muscarinic Depolarization in Other Types of Cells

Membrane depolarization resulting from activation of muscarinic acetylcholine receptors was also revealed in some smooth muscle (Bolton, 1979; Benham et al., 1985) and gland cells (Nishiyama and Petersen, 1984; Maruyama and Petersen, 1982a, b; Petersen and Maruyama, 1984). Muscarinic responses in these cells reverse at nearly 0 mV and result from an increase in membrane conductance for cations, mostly for Na^+ ions. In this respect such a depolarization is similar to one of the components of muscarinic depolarization in amphibian sympathetic neurons and to that in locus coeruleus neurons.

Evidence has been obtained that at least in one type of smooth muscle cells – amphibian stomach muscle fibers – muscarinic depolarization results from M-current inhibition (Sims et al., 1985).

Muscarinic response in oocytes is characterized by a rather positive reversal potential (–19 mV) and is either depolarizing (in amphibia: Kusano et al., 1982) or hyperpolarizing (in man: Eusebi et al., 1984), depending on the resting membrane potential level. In amphibian oocytes, muscarinic response is associated with increased membrane conductance for Cl^- ions (Kusano et al., 1982), whereas in man it is due to increased membrane conductance for unidentified ions (Eusebi et al., 1984).

9.3. MECHANISMS UNDERLYING MUSCARINIC HYPERPOLARIZATION

9.3.1. Cellular Mechanisms Underlying Muscarinic Hyperpolarization in Mammalian Sympathetic Ganglion

Muscarinic depolarization in the majority of neurons is a monosynaptic response, i.e., it is derived from the direct action of acetylcholine or other muscarinic agonists on neurons. By contrast, as concerns muscarinic hyperpolarization in mammalian sympathetic ganglion neurons, the monosynaptic mechanism of its appearance had for a long time been questioned.

The proposal that slow i.p.s.p. in mammalian sympathetic ganglion neurons was disynaptic was originally made by Eccles and Libet (1961). They suggested that acetylcholine released from preganglionic nerve terminals excited adrenergic interneurons in the ganglion by activation of their muscarinic acetylcholine receptors. This, in turn, resulted in the release of catecholamines from interneurons, which, acting on α-adrenoreceptors, induced slow i.p.s.p. in ganglion neurons. The disynaptic hypothesis of the origin of slow i.p.s.p. was initially based on the observations that slow i.p.s.p. was abolished either by atropine or by the α-adrenoreceptor antagonist dibenamine, and that catecholamines were re-released in the ganglion following its orthodromic stimulation (Bülbring, 1944). Later, Libet and co-workers concretized this hypothesis. They suggested, in particular, that the role of interneurons might be played by small intensely fluorescing cells (SIF-cells), whereas the role of the second transmitter might be played by dopamine released by SIF-cells. An updated set of evidence favoring the disynaptic origin of muscarinic hyperpolarization in mammalian sympathetic ganglion is given in Table 9.2. This evidence takes into account dopamine synthesis and storage in SIF-cells, its release following orthodromic stimulation, and imitation of muscarinic hyperpolarization by exogenous dopamine. At the same time, in the chain of evidence in favor of the disynaptic mechanism one important link is lacking: there are no indications that SIF-cells possess anatomical signs of interneurons. On the other hand, most of Libet and co-workers' data are inconsistent with the results obtained by other investigators (see Table 9.2). According to the latter, slow i.p.s.p. in its characteristics, e.g., in its ionic mechanism and sensitivity to certain blockers, is distinct from dopamine-induced hyperpolarization and, at the same time, is similar to acetylcholine-induced membrane hyperpolarization. Thus, at present the disynaptic mechanism of muscarinic hyperpolarization in mammalian sympathetic ganglion is supported by data originating mostly from only one group of investigators, i.e., from Libet and co-workers. By contrast, the data obtained by other authors indicate that muscarinic hyperpolarization in these neurons is monosynaptic.

There is evidence that in amphibian sympathetic ganglion slow i.p.s.p. is purely cholinergic and monosynaptic in its origin (see Smith and Weight, 1986).

9.3.2. General Characteristics of Ionic Mechanisms

As was mentioned above, when analyzing muscarinic depolarization, the preliminary conclusion about the ionic mechanisms underlying postsynaptic response

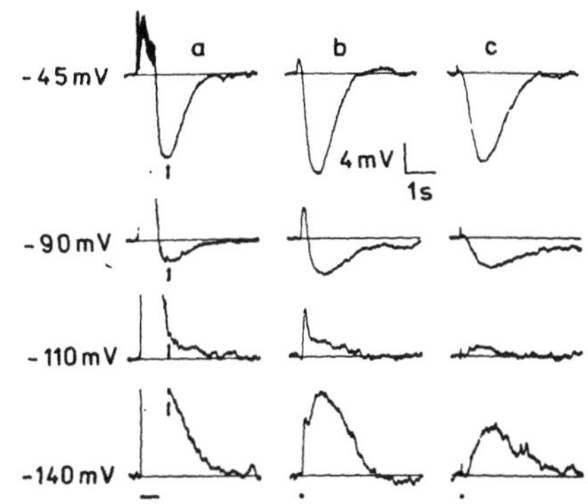

Fig. 9.12. Effect of membrane potential on muscarinic hyperpolarization recorded from amphibian cardiac parasympathetic ganglion neuron. Slow i.p.s.p.'s (a) evoked by a repetitive (30 Hz, 0.7 s) orthodromic stimulation and hyperpolarizing responses to iontophoretic application of acetylcholine (20 nA, 10 ms, b) and bethanechol (170 nA, 10 ms, c) at different membrane potential levels. The moments of orthodromic stimulation and of iontophoretic application are indicated by horizontal bars and by dots beneath the bottom records. Nicotinic responses were partially blocked by dihydro-β-erythroidine ($5 \cdot 10^{-8}$ M) (Hartzell et al., 1977).

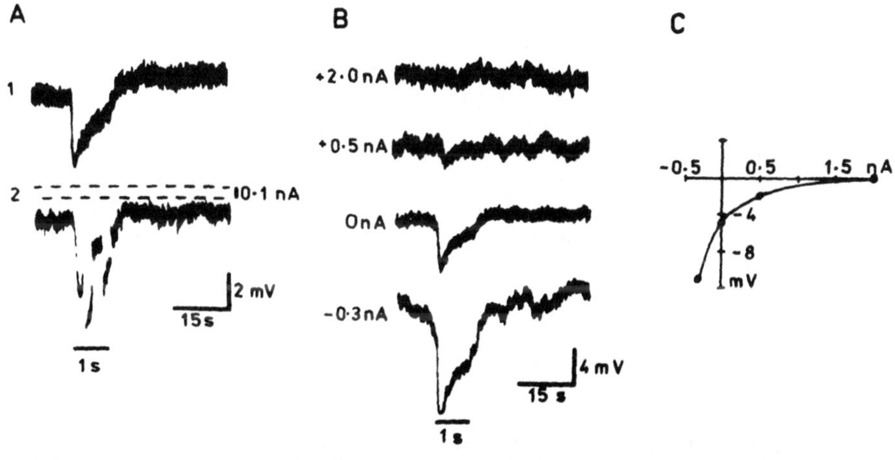

Fig. 9.13. Slow i.p.s.p. recorded from C-neuron of frog sympathetic ganglion. A: 1) Slow i.p.s.p. elicited by a repetitive (50 Hz, 1 s) stimulation of preganglionic C-fibers in the 9th connective ramus; nicotinic acetylcholine receptors are blocked by 30 μM nicotine; 2) Top record shows testing hyperpolarizing current pulses, bottom record is slow i.p.s.p. with the testing potentials superimposed. Bridge circuit was preliminarily balanced to eliminate testing potentials at rest (but not during a response). B: Effects of depolarizing (+) and hyperpolarizing (–) currents on slow i.p.s.p. C: Relationship between the amplitude of slow i.p.s.p. and polarizing current (Weight and Padjen, 1973a).

can be deduced from membrane conductance changes accompanying a response and from its dependence on membrane potential.

Slow i.p.s.p. was first recorded in mammalian sympathetic ganglion, and no changes in membrane conductance were detected during this response (Tosaka et al., 1968; Libet and Kobayashi, 1969; Kobayashi and Libet, 1970). In subsequent, more detailed studies carried out on these neurons, it was established that slow i.p.s.p. and membrane hyperpolarization caused by acetylcholine were accompanied, at least in some ganglion neurons, by increased membrane conductance (Ivanov and Skok, 1980; Cole and Shinnick-Gallagher, 1984). Since this response diminished with membrane hyperpolarization and reversed at V_K, one might assume that it resulted from increased membrane conductance for K^+ ions. Similar properties were found in muscarinic hyperpolarization in amphibian parasympathetic cardiac ganglion neurons (Hartzell et al., 1977), amphibian sympathetic ganglion C-neurons (Dodd and Horn, 1983), mammalian central neurons (Egan and North, 1985b, 1986; McCormick and Prince, 1986), and molluscan neurons (Kehoe, 1972; Ger et al., 1979a; as was mentioned above, this hyperpolarization in molluscan neurons differed from that in other types of neurons by its weak sensitivity to atropine). According to other data, muscarinic hyperpolarization in the C-neurons of amphibian sympathetic ganglion was accompanied by decreased, rather than increased, membrane conductance, and was enhanced, rather than diminished, with membrane hyperpolarization (Weight and Padjen, 1973a, b).

In some mammalian sympathetic ganglion neurons, slow i.p.s.p. also increases with membrane hyperpolarization, although it is not accompanied by changes in membrane conductance (Ivanov and Skok, 1980). A compound slow positive potential, recorded from the whole mammalian sympathetic ganglion by a sucrose gap technique, revealed a complicated dependence on membrane potential: It was enhanced at weak membrane hyperpolarization and diminished at stronger membrane hyperpolarization (Ivanov and Skok, 1971). Such a nonmonotonic dependence of a compound ganglion response on membrane potential is apparently associated, as mentioned above, with the increase or decrease in slow i.p.s.p., following membrane hyperpolarization, in different groups of ganglion neurons.

In amphibian sympathetic ganglion pretreated with nicotine to block nicotinic acetylcholine receptors, slow positive potential was enhanced with membrane hyperpolarization (Nishi and Koketsu, 1968; Koketsu, 1969; Kobayashi and Libet, 1974). Using tubocurarine instead of nicotine, this relationship, according to some authors, was not observed (Akasu and Koketsu, 1976; Akasu et al., 1978), in contrast to what was reported by others (Weight and Smith, 1981).

The variability of voltage dependences of muscarinic hyperpolarization, as well as in the accompanying membrane conductance changes in various types of neurons, implies that it may be marked by various ionic mechanisms. The evidence obtained so far indicates that muscarinic hyperpolarization may be marked by at least four distinct ionic mechanisms: i) increased membrane potassium conductance, ii) decreased membrane sodium conductance, iii) decreased membrane chloride conductance, and iv) activation of an electrogenic Na^+ pump.

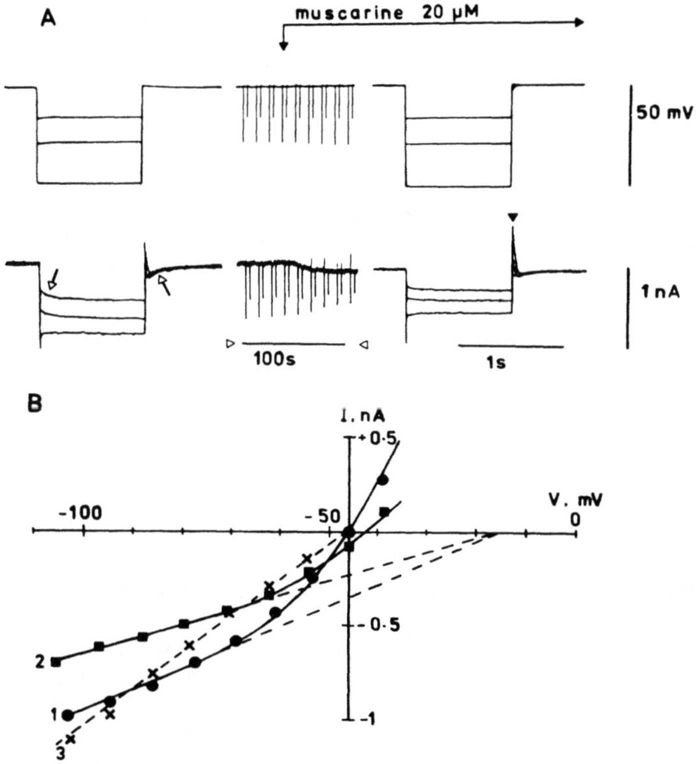

Fig. 9.14. Effect of muscarine on rat superior cervical ganglion neuron. A) Membrane currents (bottom record) elicitied by muscarine (20 μM) at four different levels of membrane potential (top record): -46 (holding potential), -62, -78, and 96 mV (command potentials). Arrows denote M-current relaxations caused by voltage steps. Perfusion solution contained 0.2 mM $CdCl_2$. B) Control steady-state (1) and instantaneous (3) current–voltage relationships and steady-state $I(V)$ relationship obtained in the presence of muscarine (2). The intersection point of curves 1 and 3 gives the reversal potential for the M-current (-89 mV) (Brown and Selyanko, 1985a).

9.3.3. Increase in Membrane Potassium Conductance

This is the most common mechanism for eliciting muscarinic hyperpolarization. It was initially revealed in molluscan neurons (Kehoe, 1972; Ger et al., 1979a,b) and subsequently identified in some mammalian sympathetic ganglion neurons (Ivanov and Skok, 1980; Cole and Shinnick–Gallagher, 1984), in C-neurons of amphibian sympathetic ganglia (Dodd and Horn, 1983), in nucleus parabrachialis neurons (Egan and North, 1985b, 1986), in thalamic reticular neurons (McCormick and Prince, 1986), in amphibian parasympathetic cardiac ganglion neurons (Hartzell et al., 1977), and in mammalian parasympathetic ganglion neurons (Gallagher et al., 1982).

Several observations support the suggestion that increased potassium conductance underlies muscarinic hyperpolarization in certain types of neurons. First, muscarinic hyperpolarization is accompanied by an increase in membrane conductance. Second, the reversal potential for muscarinic hyperpolarization is close to the reversal potential for post-spike afterhyperpolarization and to V_K, calculated from intracellular and extracellular K^+ ion concentrations. Examples of muscarinic hyperpolarizing responses elicited in the same cardiac parasympathetic neuron by orthodromic stimulation, bethanechol, and acetylcholine are shown in Fig. 9.12. It is obvious that all three responses reversed at the same level of membrane potential, close to V_K. Third, the reversal potential for muscarinic hyperpolarization reveals a dependence (approximately Nernstian) on the external concentration of K^+ ions and a lack of dependence on the presence of other ions in external solution.

In molluscan and amphibian parasympathetic neurons, the increased potassium conductance is the only mechanism responsible for muscarinic hyperpolarization, whereas in amphibian and mammalian sympathetic neurons some additional mechanisms described below are involved.

9.3.4. Decrease in Membrane Sodium Conductance

At present, only one type of neuron is known in which muscarinic hyperpolarization is associated with this mechanism: C-neurons of amphibian sympathetic ganglion neurons (Weight and Padjen, 1973a, b). The decrease in membrane sodium conductance during muscarinic response is apparent from the following observations. Muscarinic hyperpolarization is accompanied by an essential drop in membrane conductance (Fig. 9.13A). Furthermore, slow i.p.s.p. increases with membrane hyperpolarization and decreases, or even disappears, with membrane depolarization (Fig. 9.13B, C). The reversal potential for slow i.p.s.p. of this type has not been determined, although it is clear that the dependence of slow i.p.s.p. on membrane potential is similar to that for the spike. Finally, muscarinic hyperpolarization diminishes and disappears following the removal of Na^+ ions from external solution and does not depend on the presence of other ions.

These data are apparently inconsistent with the potassium conductance increase hypothesis if applied to C-neurons of amphibian sympathetic ganglion (Dodd and Horn, 1983). However, some observations suggest that actually both mechanisms, sodium and potassium, may be involved in the generation of muscarinic hyperpolarization in these neurons, as follows from an analysis of the slow positive potential in amphibian sympathetic ganglion (Weight and Smith, 1981). The slow positive potential is characterized by the reversal at membrane potentials more negative than the presumable potassium equilibrium potential estimated from the reversal of post-spike afterhyperpolarization. Moreover, its reversal potential shifts toward 0 mV with increased external potassium concentration. An assumed simultaneous increase in membrane potassium conductance and decrease in sodium conductance is consistent with the early observation that slow i.p.s.p. in sympathetic ganglion neurons may appear at resting membrane potential level without detectable changes in membrane conductance (Kobayashi and Libet, 1968).

9.3.5. Activation of Electrogenic Na^+ Pump

The hypothesis of the existence of this mechanism was originally proposed based on the finding that the slow positive potential recorded from amphibian sympathetic ganglion pretreated with nicotine can be abolished by ouabain, a cardiac glycoside selectively inhibiting the Na^+ pump (Nishi and Koketsu, 1967, 1968). Further investigations revealed that in the ganglion pretreated with tubocurarine, the slow positive potential is smaller and is inhibited by ouabain to a lesser extent than in the ganglion pretreated with nicotine (Akasu and Koketsu, 1976; Akasu et al., 1978; Akasu and Koketsu, 1983).

Another difference was that in the presence of nicotine slow i.p.s.p. decreased, but did not disappear, with membrane hyperpolarization, while in the presence of tubocurarine it reversed with membrane hyperpolarization. In the ganglion pretreated with tubocurarine only, the slow positive potential in its amplitude and voltage dependence was similar to that in the ganglion pretreated with both nicotine and ouabain. The difference in the action of ouabain on preparations pretreated with nicotine or tubocurarine indicated that the slow positive potential recorded from the ganglion pretreated with nicotine consisted of two distinct components: ouabain-sensitive, resulting from the activation of the electrogenic Na^+ pump, and ouabain-insensitive, probably resulting from increased membrane potassium conductance. At the same time, the slow positive potential recorded from the ganglion pretreated with tubocurarine apparently possessed only the ouabain-insensitive component, associated with increased membrane potassium conductance. The results of an investigation of the slow positive potential were subsequently confirmed by the results of an analysis of slow i.p.s.p. and of its analog, recorded as a slow outward current, under voltage-clamp conditions (Akasu and Koketsu, 1983). As has been mentioned above, in the same preparation (amphibian sympathetic ganglion) Weight and Padjen (1973a, b) suggested another mechanism of slow i.p.s.p. – a decreased membrane sodium conductance. Thus, the results obtained by these two groups of investigators appeared to be contradictory. Smith and Weight (1977) later verified the electrogenic Na^+ pump hypothesis and obtained data that were at variance with it.

These authors established that in amphibian sympathetic ganglion, as in other preparations, the electrogenic Na^+ pump could be inhibited by both ouabain and a potassium-free external solution. However, these two factors affected slow i.p.s.p. in different ways. The action of ouabain was similar to that described by Koketsu and co-workers: It slightly depressed slow i.p.s.p. in the ganglion pretreated with nicotine, while not affecting the response in the ganglion pretreated with tubocurarine. On the contrary, in potassium-free solution, slow i.p.s.p. increased, both in the absence and in the presence of ouabain. These data indicated that slow i.p.s.p. could not be explained by the activation of the electrogenic Na^+ pump and therefore the action of ouabain on slow i.p.s.p. might not be associated with its effect on the electrogenic Na^+ pump. Thus, the role of the electrogenic Na^+ pump in the generation of muscarinic hyperpolarization is not yet clear.

It should be pointed out that in molluscan neurons Na^+ ions, though not involved in the generation of muscarinic-like hyperpolarization directly, nevertheless are able to affect its amplitude. This was observed in *Aplysia* neurons, where an increase in membrane potassium conductance caused by carbacholine was suppressed by an increase in intracellular or a decrease in extracellular Na^+ ion concentrations (Chesnoy-Marchais and Ascher, 1983). This response was also inhibited by elevations of extracellular or intracellular concentrations of Ca^{2+} ions.

9.3.6. Decrease in Membrane Chloride Conductance

So far this mechanism has been identified only in mammalian sympathetic ganglion neurons (Brown and Selyanko, 1985a, b). The decrease in membrane chloride conductance is most easily revealed when the M-current is deactivated by membrane hyperpolarization. Figure 9.14A illustrates the action of muscarine on a neuron of rat superior cervical ganglion. One can see that at a resting membrane potential level (–46 mV) muscarine suppresses the M-current and induces a steady-state *inward* current. At the same time, at more negative membrane potentials, at which the M-current is normally deactivated (≤–60 to –70 mV), muscarine elicits a steady-state *outward* current. The reversal of the muscarinic response is indicated by a crossover of the $I(V)$ relationship obtained before and during the action of muscarine (Fig. 9.14B). Interestingly, in amphibian sympathetic ganglion neurons, where the action of muscarine is almost entirely associated with the M-current inhibition, such a crossover of $I(V)$ curves, as a rule, is not observed (see Fig. 9.8B).

Figure 9.15 demonstrates the drop in membrane conductance, under the influence of muscarine, not only in the range of potentials where the M-current is activated, but also where it is deactivated. Thus, an outward current elicited by muscarine at hyperpolarized membrane potentials is derived from the inhibition of another, separate steady-state inward current flowing through the membrane at potential levels more negative than the resting potential level.

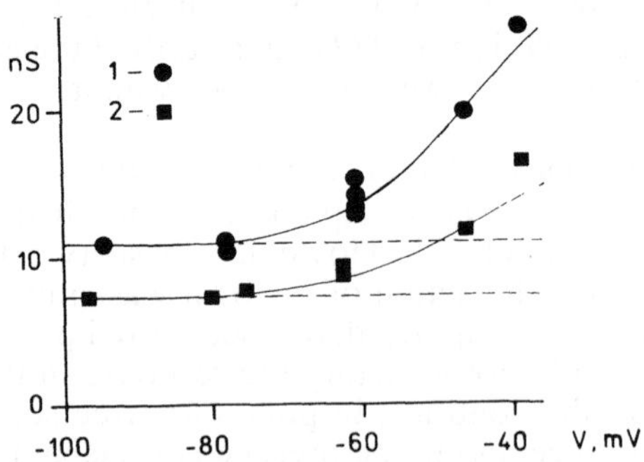

Fig. 9.15. Effect of muscarine on steady-state membrane conductance in rat superior cervical ganglion neuron. Steady-state membrane conductance as measured before (1) and during (2) the action of muscarine (20 μM) is plotted against membrane potential. To suppress time- and voltage-dependent chloride current (Selyanko, 1984), both control and testing solutions contained 0.2 mM $CdCl_2$ (Brown and Selyanko, 1985a).

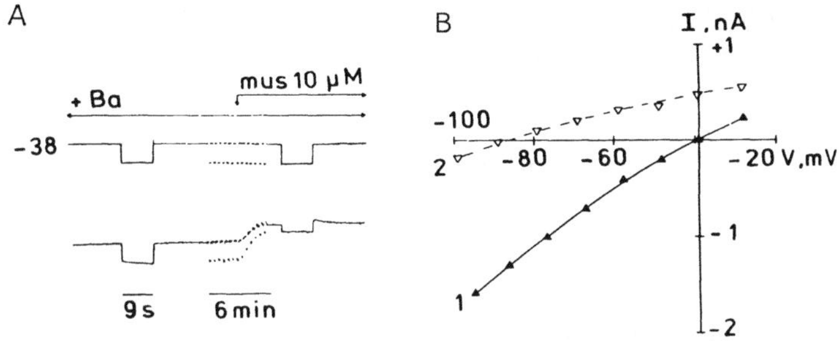

Fig. 9.16. Effect of muscarine on rat superior cervical ganglion neuron as examined after suppression of the M-current by Ba^{2+}. A) Membrane currents elicted at holding membrane potential –38 mV and command potential –58 mV. B) Current–voltage relationships obtained before (1) and during (2) the action of muscarine (20 μM). Control and testing solutions contained both 0.2 mM $CdCl_2$ and 5 mM $BaCl_2$ (Brown and Selyanko, 1985a).

With the use of M-current blockers, i.e., Ba^{2+} ions, acting from the outer side of the membrane, or Cs^+ ions, acting from the inner side of the membrane, it is possible to eliminate the M-current and to establish that an outward current, caused by activation of muscarinic acetylcholine receptors, persists at the resting membrane level as well. This result is demonstrated in Fig. 9.16.

In order to clarify the ionic basis for steady-state inward current observed as an outward (hyperpolarizing) muscarinic current, provisionally termed the "X-current," the effect of modification of external medium content on the reversal potential for an outward current was studied. The reversal potential for the X-current was estimated as a crossover potential for $I(V)$ relationships obtained before and during inhibition of the X-current by muscarine or by orthodromic stimulation, under conditions when the M-current was preliminarily blocked by Cs^+ ions acting from the inner side of the membrane. It was established that the reversal potential for the X-current did not depend on the presence of Na^+ or Ca^{2+} ions in the external medium and, thus, the X-current could not be carried by these ions.

At the same time, data were obtained indicating that Cl^- ions were involved in the generation of the X-current. This was evidenced by the dependence of the reversal potential for an outward muscarinic current and for the accompanying conductance change on the anion species present in the intracellular microelectrode. With the use of KCl or CsCl, as opposed to K-acetate or Cs-acetate, for filling microelectrodes, the reversal potential appeared more positive, and the accompanying conductance change was larger. The reversal potential for an outward muscarinic current coincided with that for GABA-induced response. As is well known, GABA selectivity activates membrane chloride conductance, thereby producing a response with a reversal potential close to V_{Cl} (Adams and Brown, 1975). Reversal potentials for muscarine-induced outward current and for GABA-induced current exhibited similar dependences on the presence of Cl^- ions in the intracellular microelectrode. Then the reversal potential for an outward muscarinic current depended on the presence of Cl^- ions in external solution, becoming more positive on removal of Cl^- ions from the external solution.

Unlike the M-current, the X-current could not be recorded directly. This was manifested by the lack of any current relaxations associated with the X-current as recorded following voltage steps. This peculiarity of the X-current could be related either to too slow voltage-clamping or to the lack of voltage dependence of the X-current. The latter explanation is consistent with the fact that the membrane current elicited by X-current suppression is weakly dependent on membrane potential. The lack of voltage dependence in the X-current and the related lack of current relaxations impedes its isolation from the rest of the leak current, which is voltage independent, too. The only way to isolate the X-current is to inhibit it by muscarine. However, the fraction of X-current inhibited by muscarine remains unknown.

Further investigation of the X-current was impeded by the fact that various substances known to block membrane chloride conductance, i.e., penicillin G, furosemide, the disulfonic stilbene derivative 4,4'-diisothiocyanostilbene-2,2'-disulfonate (DIDS), and Zn^{2+} ions, failed to inhibit the X-current (Brown and Selyanko, 1985a).

As the above data show, at least two resting membrane conductances in mammalian sympathetic ganglion neurons – the potassium (M) and chloride (X) conductances – are operated via muscarinic acetylcholine receptors. In addition, another kind of membrane chloride conductance was identified in those neurons, which differed from the X-conductance in its time and voltage dependence as well as in its insensitivity to muscarinic agonists (Selyanko, 1984). This conductance was activated at the resting membrane potential level and, therefore, it contributed to maintenance of the resting potential. Thus, the action of muscarinic agonists on M- and X-conductances is rather selective and does not affect other components of the resting membrane conductance.

9.3.7. Unusual Membrane Hyperpolarization Caused by Acetylcholine

In amphibian sympathetic ganglion neurons an unusual membrane hyperpolarization was revealed, which could be elicited by both exogenous acetylcholine and orthodromic stimulation (Koketsu and Yamamoto, 1975). It was termed "unusual" since it could be observed in the presence of both nicotine and atropine, i.e., following a complete blockade of nicotinic and muscarinic acetylcholine receptors. This action of acetylcholine apparently involved a special mechanism, since it was not reproduced by either carbacholine or by bethanechol. Membrane hyperpolarization caused by acetylcholine was not affected by a reduction in the $[Ca^{2+}]_o$-to-$[Mg^{2+}]_o$ ratio and thus it was not associated with the involvement of di- or polysynaptic mechanisms. It was inhibited by ouabain and slightly enhanced with membrane hyperpolarization. It was therefore similar to one of the components of slow i.p.s.p. in amphibian sympathetic ganglion neurons, presumably marked by activation of the electrogenic Na^+ pump (Nishi and Koketsu, 1967, 1968). Since an unusual membrane hyperpolarization could be elicited by orthodromic stimulation, it apparently contributed to slow i.p.s.p. in amphibian sympathetic ganglion neurons.

The ionic mechanisms underlying muscarinic hyperpolarization in various types of neurons are summarized in Table 9.1.

9.3.8. Ionic Mechanisms Underlying Muscarinic Hyperpolarization in Other Types of Cells

Muscarinic hyperpolarization was revealed in cardiac muscle cells (Purves, 1976; Hartzell et al., 1977; Glitsch and Pott, 1978; Pott, 1979), in some gland cells (Petersen, 1976; Maruyama et al., 1983; Pearson et al., 1984; Pearson and Petersen, 1984), and in amphibian oocytes (Dascal et al., 1984). It resulted here, as in most types of neurons, from an increased membrane potassium conductance.

In cardiac muscle and gland cells, the characteristics of single potassium channels activated by acetylcholine were established. A single potassium channel revealed strong rectifying properties: Its conductance was several times higher for inward than for outward current. Single-channel conductance for inward current recorded at resting membrane potential (about –60 mV) and at $[K^+]_o = 70$ mM and $[K^+]_i = 140$ mM was about 35 pS (Sakmann et al., 1983b). Single-channel activity was represented mostly by bursts of channel openings. At 30–32°C each burst was, on the average, 11 ms long and consisted of 5.3 current pulses separated by gaps of 1.3 ms long each. Assuming each burst results from the single occupation of a receptor by acetylcholine, and based on Eqs. (3.16) and (3.27)–(3.29), the authors obtained the following values of rate constants characterizing the acetylcholine–receptor interaction: $k_{-1} = 150\ s^{-1}$; $\beta = 460\ s^{-1}$, and $\alpha = 510\ s^{-1}$ (Sakmann et al., 1983b). These data indicate that the kinetics of interaction of acetylcholine with the muscarinic receptor is much slower than the kinetics of its interaction with the nicotinic receptor (see Chapter 3).

An interesting feature of acetylcholine-sensitive potassium channels in cardiac muscle cells is that they open even in the absence of acetylcholine, although the rate of their openings in the absence of acetylcholine is much lower than in its presence. The action of acetylcholine results in a higher probability of the channel being open, due to a shortening of its stay in the closed state (Soejima and Noma, 1984).

In acinar cells of guinea-pig pancreas, single potassium channels activated by acetylcholine have a very large conductance (about 200 pS); an integral membrane potassium conductance, activated by acetylcholine, reveals here a dependence both on membrane potential and on intracellular Ca^{2+} ions (Petersen and Maruyama, 1984).

9.4. IONIC MECHANISMS UNDERLYING ACTIVITIES OF MUSCARINIC ACETYLCHOLINE RECEPTORS OF VARIOUS SUBTYPES

Muscarinic responses in neurons are greatly variable in their underlying mechanisms (see Table 9.1). The question arises whether this variability is associated with the heterogeneity of muscarinic acetylcholine receptors themselves (see Chapter 8) or with the heterogeneity of their ionic channels. Based on their distinct sensitivities to certain antagonists, two main muscarinic acetylcholine receptor subtypes, M_1 and M_2, were distinguished. One of the most selective antagonists, pirenzepine, exhibited a much higher affinity to M_1-receptors ($K_D \cong 0.02$ μM) than to M_2-receptors ($K_D \cong 1$ μM) (Hammer et al., 1980). Because of this property, pirenzepine was used in recent studies for the identification of receptors mediating various muscarinic responses.

In mammalian sympathetic ganglion, muscarinic depolarization is inhibited by low concentrations of pirenzepine (Brown et al., 1980b). The dissociation equilibrium constant (K_D) for this inhibition (0.004 μM) indicates that receptors responsible for muscarinic depolarization here are of M_1 subtype. Since muscarinic depolarization in mammalian sympathetic ganglion neurons mostly results from the M-current inhibition (Constanti and Brown, 1981; Brown and Selyanko, 1985a, b), one may conclude that it is the M_1-receptors that mediate suppression of the M-current. Interestingly, in smooth muscle cells from toad stomach (Sims et al., 1985) and in the neurons of guinea-pig olfactory cortex (Constanti and Sim, 1987), where M-current inhibition seems to be the only mechanism underlying muscarinic depolarization, the responses exhibit very low sensitivities for pirenzepine, indicating that they may be mediated by M_2-receptors. Moreover, in hippocampal neurons, where muscarinic excitation also results solely from M-current inhibition (Halliwell and Adams, 1982), muscarinic excitation seems to be mediated by muscarinic receptors of both subtypes (Muller and Misgeld, 1986).

In myenteric neurons, muscarinic depolarization is also derived from potassium current inhibition, although this current differs from M-current in its voltage independence and high sensitivity to extracellular Ca^{2+} ions (Morita et al., 1982; North and Tokimasa, 1983). Mucarinic depolarization in myenteric neurons is antagonized by low concentrations of pirenzepine ($K_D = 0.005$–0.01 μM), which suggests that it is mediated via M_1-receptors (North et al., 1985).

Activation of M_1-receptors is also probably involved in the decrease of membrane chloride conductance. This is evident from the fact that in mammalian sympathetic ganglion neurons, where muscarinic depolarization results from simultaneous inhibition of potassium M-current and chloride X-current, pirenzepine antagonizes both these components of muscarinic response to the same extent (Brown and Selyanko, 1985b).

Other types of muscarinic responses seem to be evoked through activation of M_2-receptors.

Muscarinic depolarization in locus coeruleus marked by an increase in membrane conductance (Presumably for Na^+ ions) is resisted for pirenzepine ($K_D \cong 0.23$ μM: Egan and North, 1985a), indicating that M_2-receptors are involved. In this respect, it is interesting that muscarinic depolarizations in smooth muscle cells (Bolton, 1979; Benham et al., 1985) and in gland cells (Maruyama and Petersen, 1982a, b; Nishiyama and Petersen, 1974; Petersen and Maruyama, 1984) result from increased membrane conductance for cations too, and the resultant muscarinic effects in these tissues (smooth muscle contraction and secretion from gland cells, respectively) are mediated by M_2-receptors (see North, 1986).

There is evidence indicating that M_2-receptors mediate muscarinic hyperpolarization coupled with increased membrane potassium conductance. Muscarinic hyperpolarization in sympathetic ganglion neurons (Ashe and Yarosh, 1984; Newberry et al., 1985), cortical neurons (McCormick and Prince, 1985), thalamic neurons (McCormick and Prince, 1986), and the nucleus parabrachialis (Egan and North, 1986) is resistant to high concentrations of pirenzepine. The K_D value for the effect of pirenzepine, estimated in nucleus parabrachialis neurons, is 0.6 μM (Egan and North, 1986), which is inherent in M_2-receptors. Furthermore, muscarinic hyperpolarization (but not depolarization) in rabbit superior cervical ganglion neurons is abolished by gallamine, preferentially acting on M_2-receptors (Ashe and Yarosh, 1984; Newberry et al., 1985). Finally, in sympathetic ganglion (Newberry et al., 1985) and in cortical neurons (McCormick and Prince, 1985), carbacholine, exhibiting high affinity to M_2-receptors, is more effective in eliciting muscarinic hyperpolarization, as opposed to muscarinic depolarization.

Interestingly, muscarinic acetylcholine receptors that open potassium channels in cardiac cells (Sakmann et al., 1983) are also of M_2 subtype (Chassaing et al., 1984).

Muscarinic acetylcholine receptors located on presynaptic nerve endings and inhibiting transmitter release exhibit low sensitivity to pirenzepine and, therefore, are of M_2 subtype (Fuder et al., 1982; Kilbinger et al., 1984; North et al., 1985). The mechanism of inhibition of transmitter release mediated by muscarinic acetylcholine receptors is not yet clear. One possibility is that it may be associated with decreased entry of Ca^{2+} ions into the nerve terminal. This assumption is supported by the fact that in many types of neurons muscarinic agonists suppress the calcium component of the action potential (see Chapter 10). Moreover, as was revealed in cultured mammalian sympathetic ganglion neurons, inhibition of the Ca-component of the action potential is antagonized by low (10 μM) concentrations of AF-DX 116 and is resistant to high (up to 100 μM) concentrations of pirenzepine. This result suggests that inhibition of the Ca-component of the action potential is mediated by M_2-receptors (Mochida and Kobayashi, 1986).

The data presented above do not support the idea that the heterogeneity of muscarinic acetylcholine receptors may underlie the variability of the resulting ionic processes. This is evident, on the one hand, from the fact that the same ionic mechanism, e.g., inhibition of membrane potassium conductance, may be coupled with either M_1- or M_2-receptors or with both. On the other hand, muscarinic receptors of the same subtype, M_2, may mediate a great variety of ionic processes, including increased sodium and potassium conductance and decreased calcium conductance. However, it should be emphasized that the division of muscarinic acetylcholine receptors into two main subtypes, M_1 and M_2, is somewhat conditional, and some results obtained suggest that receptors belonging to the M_2-subtype are, in turn, heterogeneous (see Chapter 8). Therefore, one may expect that with the new muscarinic acetylcholine receptor subtypes identified, the link between receptor heterogeneity and the variability of associated ionic processes will become more clear.

9.5. COUPLING OF MUSCARINIC ACETYLCHOLINE RECEPTORS WITH IONIC CHANNELS INVOLVED IN THE GENERATION OF POSTSYNAPTIC RESPONSE

Unlike the nicotinic acetylcholine receptor, which is coupled with its ionic channel in the same protein molecule, the muscarinic acetylcholine receptor seems to be located apart from its channel. This is indirectly evidenced by the following properties of muscarinic response. First, the high temperature dependence of the initial part of muscarinic hyperpolarization (Hartzell et al., 1977; Hartzell, 1981; Pott and Pusch, 1979; Ger et al., 1979a, b) and depolarization (North and Tokimasa, 1984) indicates that binding of an agonist to the muscarinic receptor and opening or closing of the ionic channels are separated by several intermediate steps. Second, muscarinic depolarization (see Brown, 1983) and hyperpolarization (Ascher and Chesnoy-Marchais, 1982) may interact with the responses elicited by other transmitters via activation of their own receptors, distinct from muscarinic ones. It is very unlikely that an ionic channel and several distinct associated receptors may be located in the same protein macromolecule.

Indirect coupling between a receptor and an associated ionic channel may occur either with or without intracellular (intramembrane) messengers. The former suggestion, though purely hypothetical, implies that a receptor and its channel may independently exist and move in the membrane (Kehoe and Marty, 1980). It is assumed that the channel may be activated by its interaction with a receptor, provided that the latter has already been bound with the agonist. This hypothesis may explain the long delay in the appearance of muscarinic response and the existence of several steps on the way from binding of an agonist to a receptor to the membrane conductance change. The following criteria for messenger involvement and for its identification are commonly accepted (see Hartzell, 1981): i) an assumed messenger should accumulate in the cell during postsynaptic response; ii) intracellular injection of an assumed messenger should elicit a response, analogous to the postsynaptic one, iii) the postsynaptic response is expected to be affected by changes in the intracellular metabolism of an assumed messenger.

According to the first criterion, possible messengers in the generation of muscarinic response are as follows. Biochemical studies indicate that activation of muscarinic acetylcholine receptors in nervous tissues results in activation of guanylate cyclase, followed by elevation of the intracellular cyclic GMP level, as revealed in both cortex (see Krnjevič, 1978) and sympathetic ganglia following either synaptic activation of muscarinic receptors (Weight et al., 1974; Volle et al., 1981) or the action of muscarinic agonists (Kebabian et al., 1975). However, Brown et al. (1980a) have not detected any changes in cyclic GMP level in rat superior cervical ganglion following muscarine action, even under conditions where phosphodiesterase has been inhibited.

Besides cyclic GMP accumulation, activation of muscarinic receptors leads in sympathetic ganglion to accelerated phosphatidylinositol turnover (Larrabee and Leicht, 1965; Lapetina et al., 1976; Pickard et al., 1977) and in other tissues to the accumulation of intracellular Ca^2 ions (Putney, 1978; Petersen et al., 1981) and transmethylation of phosphatidylethanolamine to phosphatidylcholine by S-adenosylmethionine (Axelrod and Hirata, 1981).

It should be kept in mind, when considering the role of biochemical changes evoked in the nervous system by activation of muscarinic receptors, that they may not necessarily be linked to generation of tissue electrical response. First of all, biochemical events may occur in nonneuronal cells. This possibility is supported by the finding that in sympathetic ganglion cyclic nucleotides are located mostly in glial cells (Briggs et al., 1982). Furthermore, biochemical and electrical events may be triggered via activation of different subtypes of muscarinic acetylcholine receptors. The evidence favoring this assumption is that in sympathetic ganglion acceleration of phosphatidylinositol turnover and accumulation of cyclic GMP can be evoked by both synaptic stimulation (Larrabee and Leicht, 1965) and muscarinic agonists (Lapetina et al., 1976; Pickard et al., 1977). At the same time, DMPP and 4-aminopyridine, acting on muscarinic acetylcholine receptors, cause only the former effect (Volle et al., 1981; Patterson and Volle, 1984).

Despite the ambiguity of the biochemical data, the possible involvement in muscarinic response of various substances accumulated in the cells following activation of their muscarinic acetylcholine receptors is considered below, since it provides valuable information on the nature of muscarinic response.

9.5.1. Cyclic GMP

Attempts were made first on autonomic ganglia neurons and subsequently on central neurons to verify whether cyclic GMP is able to mimic muscarinic excitation. Since the most common mechanism of muscarinic excitation is a decrease in resting membrane potassium conductance (see Table 9.1), one could expect that cyclic GMP, when applied intracellularly, should cause membrane depolarization accompanied by a decrease in membrane conductance.

In sympathetic ganglion, applications of cyclic GMP or its membrane-penetrating analogs produced controversial results: In rabbit superior cervical ganglion dibutyryl cyclic GMP caused a negative change in extracellular potential (McAfee and Greengard, 1972; Hashiguchi et al., 1978), while in rat superior cervical ganglion, 8-bromo-cyclic GMP elicited no response (Brown et al., 1980a).

Intracellular investigation of the effects produced in rabbit and rat superior cervical ganglion neurons with intracellular injection of cyclic GMP or extracellular application of its analogs revealed membrane depolarization accompanied by an increase in membrane conductance (Dun et al., 1977, 1978; Gallagher and Shinnick-Gallagher, 1977; Kobayashi et al., 1977; Hashiguchi et al., 1978). Since muscarinic depolarization in these neurons was mostly followed by a decrease, rather than an increase, in membrane conductance, this result was considered to be inconsistent with the possible involvement of cyclic GMP in the generation of muscarinic depolarization in sympathetic neurons.

However, as was found later, dibutyryl cyclic GMP caused membrane depolarization with an increase in membrane conductance only when used at high ($\geq 2.5 \cdot 10^{-4}$ M) concentrations; at lower concentrations dibutyryl cyclic GMP elicited membrane depolarization which, like one of the components of muscarinic depolarization, was not accompanied by any changes in membrane conductance (Kobayashi et al., 1977; Hashiguchi et al., 1978). Hence, at least one of the components of muscarinic depolarization in mammalian sympathetic ganglion may be mediated by cyclic GMP.

In amphibian sympathetic ganglion neurons, cyclic GMP and its analogs failed to elicit any responses when applied either extracellularly or intracellularly (Suria, 1976; Busis et al., 1978; Weight et al., 1978; Adams et al., 1982b). This result is consistent with the finding that in these neurons, unlike in mammalian sympathetic ganglion neurons, a component of muscarinic depolarization not accompanied by changes in membrane conductance is lacking (Hashiguchi et al., 1982).

In spinal motoneurons, intracellular injection of cyclic GMP caused membrane depolarization with increased conductance (Krnjevic and Van Meter, 1976; Krnjevič et al., 1976). This result contrasts with the drop in membrane conductance accompanying muscarinic depolarization in motoneurons (Krnjevič et al., 1971).

According to initial observations made on hippocampal neurons, cyclic GMP analogs, when applied extracellularly, appeared to be inefficient in producing any change in resting membrane potential (Benardo and Prince, 1982b; Segal, 1982). By contrast, later observations by Cole and Nicoll (1984b) indicated that 8-bromo-cyclic GMP, applied extracellularly, reproduced all acetylcholine effects: membrane depolarization, decrease in membrane conductance, and inhibition of post-spike afterhyperpolarization. 8-Bromo-cyclic GMP, when preliminarily added to the solution filling the microelectrode, also abolished post-spike afterhyperpolarization within several minutes following impalement of the cell.

It should be noted in conclusion that cyclic GMP may be involved in the generation of muscarinic depolarization resulting from decreased membrane conductance in only one type of neuron – the hippocampal neuron. In mammalian sympathetic ganglion neurons, cyclic GMP may be responsible for the generation of an additional component of muscarinic depolarization, appearing without any changes in membrane conductance. However, not all criteria for the role of cyclic GMP in muscarinic suppression of K-channel activity given above have presently been verified.

In cardiac muscle cells, where cyclic GMP is accumulated following the activation of muscarinic receptors (George et al., 1973), neither resting membrane potential nor muscarinic hyperpolarization were affected by intracellular injection of cyclic GMP (Trautwein et al., 1982). Moreover, electrophysiological evidence was obtained here against the involvement of any secondary messengers that might diffuse through cytoplasm to activate potassium channels (Soejima and Noma, 1984): Acetylcholine was efficient in activating single potassium channels only when applied to the same membrane areas where the channels were disposed, and it was not efficient when applied to neighboring membrane areas.

9.5.2. Ca^{2+} Ions

Initial observations of amphibian sympathetic ganglion neurons signaled against the possible involvement of Ca^{2+} ions in the generation of muscarinic depolarization (Adams et al., 1982b). Intracellular application of Ca^{2+} ions and their removal from extracellular solution, as well as extracellular application of calcium channel blocker Cd^{2+} ions, did not affect the M-current and, thus, failed to reproduce the main effect of muscarine. Consistent with these observations, in mammalian sympathetic ganglion neurons both the M-current and its inhibition by muscarine persisted in Ca^{2+}-free or Cd^{2+}-containing solutions (Brown and Selyanko, 1985a).

Recently, Tokimasa (1985a) observed that Ca^{2+} ions entering amphibian sympathetic ganglion neuron during calcium action potential suppressed the M-current, thereby producing a steady-state inward current. Thus, in this particular study, elevation of the intracellular concentration of Ca^{2+} ions seemed to reproduce muscarinic response.

Another interesting observation regarding the role of divalent cations in the generation of muscarinic response was made in cultured amphibian sympathetic ganglion cells (Galvan et al., 1984). The addition of Mg^{2+} ions to an intracellular medium resulted in almost complete suppression of the M-current. Interestingly, with artificial intracellular medium, the M-current lost its sensitivity to muscarine, while retaining its sensitivity to extracellularly applied Ba^{2+} ions. It seems that in amphibian sympathetic ganglion neurons intracellular divalent cations are important for coupling of muscarinic acetylcholine receptors with the M-channels.

In accordance with the above data we found (unpublished observations) that the M-current persisted in intracellularly perfused rat superior cervical ganglion neurons when no divalent cations were added inside the cell. However, unlike in cultured amphibian sympathetic ganglion neurons, in this preparation acetylcholine, when applied iontophoretically onto the cell, was able to strongly inhibit the M-current, thus producing the steady-state inward current. On the other hand, when the M-current was preliminarily abolished by replacing intracellular K^+ ions with Na^+ ions, acetylcholine produced a steady-state outward current accompanied by a decrease in membrane conductance. This indicates that in mammalian sympathetic ganglion neurons intracellular divalent cations are not very important for coupling muscarinic acetylcholine receptors with both M-channels and X-channels.

In cardiac muscle cells, intracellular injection of Ca^{2+} ions failed to mimic muscarinic hyperpolarization (Trautwein et al., 1982).

9.5.3. Other Substances

In amphibian sympathetic ganglion neurons, intracellular injections of GTP-γ-S (nonhydrolyzable GTP analog) or S-adenosylmethionine failed to affect the M-current and, thus, to imitate muscarinic depolarization (Adams et al., 1982b).

It was also found that the M-current in neuroblastoma cells could be markedly suppressed by phorbol esters, which thereby produced the steady-state inward current (D. A. Brown, personal communication). Therefore, it seems likely that turning off the M-current as the necessary step for eliciting muscarinic depolarization in various types of neurons may occur via activation of protein kinase C.

According to other recent findings, in cardiac muscle cells muscarinic acetylcholine receptors are coupled with potassium channels without cyclic nucleotides. Instead, activation of potassium channels here by muscarinic agonists appears to require intracellular GTP (Pfaffinger et al., 1985; Breitwieser and Szabo, 1985; Kurachi et al., 1986). All three of the above criteria for the role of GTP as a secondary messenger in the appearance of muscarinic hyperpolarization were proved to be fulfilled.

9.6. CONCLUSIONS

Slow (muscarinic) postsynaptic potentials, i.e., slow e.p.s.p. and slow i.p.s.p., are detected in various types of central and autonomic neurons. In most types of neurons possessing muscarinic acetylcholine receptors, only one kind of response, either slow e.p.s.p. or slow i.p.s.p., can be observed, and only in a small fraction of sympathetic ganglion neurons are both kinds of response detectable.

Slow i.p.s.p. and e.p.s.p. are characterized by long latency and duration (30–500 ms and 1–60 s, respectively). The initial parts of muscarinic hyperpolarization and depolarization have high temperature dependences (with Q_{10} equal to 11–15 and 3, respectively) and a high (third order for hyperpolarizing response) dependence on time. It is therefore suggested that the process resulting in the appearance of a muscarinic response, either hyperpolarizing or depolarizing, is a multistep process and the unique features of muscarinic response are not determined by the events preceding activation of the muscarinic acetylcholine receptors (acetylcholine release and diffusion to a receptor). The long latency of muscarinic hyperpolarization apparently is derived from the slow kinetics of potassium channels activated via muscarinic receptors, while its slowest part, the decay, is determined by an apparent rate of dissociation of the agonist–receptor complex. By contrast, the duration of muscarinic depolarization is determined by the steps following agonist binding to a receptor.

Ionic mechanisms underlying muscarinic responses in various types of neurons are greatly variable. Muscarinic depolarization is associated with at least three distinct mechanisms: i) a decreased membrane potassium conductance; ii) an increased membrane conductance for sodium or (and) other cations and, probably, for chloride ions, and iii) a process not accompanied by any changes in membrane conductance. Steady-state membrane potassium conductance, whose inhibition results in membrane depolarization, is represented by two kinds of conductances: voltage-dependent (M-conductance) and voltage-independent calcium-dependent conductance. Inhibition of M-conductance is the most common mechanism of muscarinic depolarization which is detected in sympathetic, hippocampal, and cortical neurons. Suppression of voltage-independent membrane potassium conductance underlies muscarinic depolarization in myenteric and sensory neurons. Muscarinic activation of membrane conductance for cations, and presumably for anions, occurs in sympathetic ganglion neurons, where these mechanisms are accessories to M-conductance inhibition and are especially exaggerated with membrane hyperpolarization. Activation of membrane conductance for cations, as the only mechanism of muscarinic depolarization, is detectable in locus coeruleus neurons, while membrane depolarization not accompanied by changes in membrane conductance occurs as an accessory mechanism of muscarinic depolarization only in mammalian sympathetic ganglion neurons, where it is most prominent at hyperpolarized levels of the membrane potential.

Muscarinic hyperpolarization is associated with four distinct mechanisms: (i) increased membrane potassium conductance, (ii) decreased membrane sodium and (iii) chloride conductances, and (iv) activation of an electrogenic sodium pump. Among these mechanisms the most common one is the increase in membrane potassium conductance. This mechanism has been revealed in all types of neurons responding to muscarinic agonists with membrane hyperpolarization: in sympathetic and parasympathetic ganglia neurons, in certain brain neurons, and in molluscan neurons. Other mechanisms are detectable only in sympathetic ganglion neurons, where they are, as a rule, accessories to increased membrane potassium conductance.

It has been established that a decrease in membrane chloride conductance is mediated via muscarinic acetylcholine receptors of M_1 subtype, whereas a decrease in membrane potassium conductance is mediated through either M_1 receptors (in sympathetic ganglion and myenteric neurons) or both M_1 and M_2 receptors (in certain brain neurons). At the same time, an increase in membrane steady-state sodium and potassium conductances, as well as an increase in membrane voltage-dependent calcium conductance, are mediated through M_2 receptors.

Indirect evidence, i.e., the high temperature dependence of muscarinic response and its interaction with the responses to various other transmitters, suggests that the muscarinic acetylcholine receptor, unlike the nicotinic acetylcholine receptor, exists in the cell membrane apart from its ionic channel. However, for most kinds of muscarinic responses the secondary messengers have not yet been identified. There are indications that muscarinic suppression of membrane conductance may be mediated by intracellular Ca^{2+} ions (in sympathetic ganglion neurons) or by cyclic GMP (in hippocampal neurons). Cyclic GMP is supposed to be involved in the generation of muscarinic depolarization in sympathetic ganglion not accompanied by changes in membrane conductance. In cardiac muscle cells, it has been proved that GTP mediates the potassium channel opening which underlies muscarinic hyperpolarization.

Chapter 10

MECHANISMS OF MUSCARINIC REGULATION OF NEURONAL ELECTRICAL ACTIVITY

The role of nicotinic acetylcholine receptors in neuronal electrical activity is to initiate action potentials by suprathreshold membrane depolarization or to prevent, through hyperpolarization of the membrane (in some invertebrate neurons), the initiation of action potentials elicited by other sources. In addition, nicotinic receptor activation may result in a decrease in the amplitude of the action potential due to shunting of the cellular membrane.

By contrast, the mechanisms of regulation of electrical activity mediated via muscarinic receptors are much more complicated, and they may include changes in membrane potential, membrane conductance, action potential characteristics, and neuronal electrical excitability.

10.1. REGULATION OF ACTION POTENTIAL CHARACTERISTICS

The modification of action potential characteristics by activation of muscarinic acetylcholine receptors was most thoroughly studied in B-cells of amphibian sympathetic ganglion (Kuba and Koketsu, 1975, 1976a, 1978; Tokimasa, 1984, 1985b).

Figure 10.1 shows the effect produced by acetylcholine ($5 \cdot 10^{-5}$ g/ml) in the presence of tubocurarine ($1 \cdot 10^{-4}$ g/ml) on the antidromic action potential in a ganglion neuron. One can see that acetylcholine diminishes both the spike and the afterhyperpolarization of the action potential.

Such an effect of acetylcholine, as well as other muscarinic agonists, is apparently mediated by the activation of muscarinic acetylcholine receptors. This is evident from the fact that the actions of muscarinic agonists on the action potential can be abolished by muscarinic antagonists and are characterized by desensitization resulting in a decrease in the effect during continuous action of the agonist (Akasu and Koketsu, 1980).

Evidence was obtained that the changes in the characteristics of the action potential caused by muscarinic agonists are not secondary to the changes in the membrane potential or membrane shunting during the postsynaptic response: i) the depression of spike or of afterhyperpolarization persisted when the membrane potential was retained at its initial level by passing the appropriate hyperpolarizing current through the microelectrode (Fig. 10.1); ii) membrane depolarization elicited by muscarinic agonists was accompanied by a decrease rather than an increase in membrane conductance, and therefore the decrease in the action potential could not be explained by membrane shunting; iii) acetylcholine, used at very low concentrations, could suppress the post-spike afterhyperpolarization without eliciting postsynaptic response (Tokimasa, 1984).

The decrease in the spike could be observed either in normal solution, where it should mostly be associated with a decrease in active sodium conductance, or in calcium-rich solution, where it was apparently due to a decrease in active calcium conductance (Kuba and Koketsu, 1976b, 1978). Based on these observations, it was concluded that the depression of the spike is marked by a decrease in both sodium and calcium active membrane conductance, whereas the depression of afterhyperpolarization is probably due to a decrease in membrane potassium conductance.

In order to verify whether the acetylcholine-induced depression of the action potential and the appearance of postsynaptic response to acetylcholine are associated with the same mechanism, Kuba and Koketsu studied the relationship between both these effects. They found no correlation between the depression of the spike and the amplitude of postsynaptic response (Kuba and Koketsu, 1976b, 1978). At the same time, there was a weak, although significant, correlation between the depression of post-spike afterhyperpolarization and postsynaptic response. These results indicate that the depression of sodium and calcium membrane conductances underlying the spike occurs independently of postsynaptic response, whereas the depression of one or several components of potassium conductance underlying afterhyperpolarization may share some mechanisms with postsynaptic response. As Kuba and Koketsu suggested, the common mechanism underlying the depression of post-spike afterhyperpolarization and the postsynaptic response could be the turning-off of the same muscarine-sensitive potassium channels. Potassium channels of this type could be M-channels, which are sensitive to muscarinic agonists and are activated at membrane potentials less negative than –70 mV and not inactivated in time (Brown and Adams, 1980; Adams et al., 1982a, b). One could thus suggest that the short-lasting membrane depolarization which occurred during the spike might activate the M-current, resulting in the appearance of afterhyperpolarization. However, the time constant of the M-current activation (about 150 ms at resting membrane potential level) is much longer than the spike duration. Therefore, the M-current cannot be significantly activated during the spike, and one should not expect an essential contribution of the M-current to the afterhyperpolarization. The low contribution of the M-current to the afterhyperpolarization can explain the weak

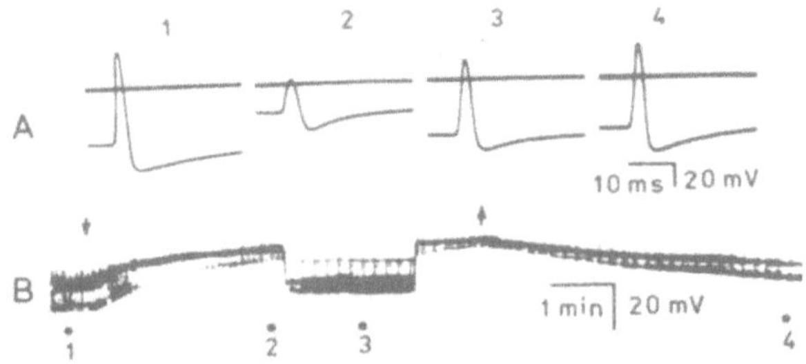

Fig. 10.1. Effect of acetylcholine on action potential recorded from frog sympathetic ganglion neuron. A) Action potentials evoked by antidromic stimulus before acetylcholine ($5 \cdot 10^{-5}$ g/ml) was applied (1) and during development of acetylcholine effect (2–4). Tubocurarine ($1 \cdot 10^{-4}$ g/ml) was used to block nicotinic acetylcholine receptors. B) Muscarinic depolarization during which antidromic action potentials were evoked at a 0.5 Hz rate (some of them are shown in A as recorded at the moments indicated by dots). In the course of the development of muscarinic depolarization membrane potential was transiently shifted toward its resting level using hyperpolarizing current (Kuba and Koketsu, 1975).

correlation between the depression of afterhyperpolarization and the amplitude of postsynaptic response.

Unlike the spike, another depolarizing signal, fast e.p.s.p., has a duration more comparable than that of the spike, with the time constant of the M-current activation. One could therefore assume that fast e.p.s.p. may noticeably activate the M-current, and thus it should be accompanied by its own afterhyperpolarization.

Indeed, it was observed a long time ago that fast e.p.s.p. was followed by membrane hyperpolarization whose nature remained unknown until recently (Blackman et al., 1963; Skok and Savich, 1967; Dennis et al., 1971). This hyperpolarization did not result from the action of inhibitory transmitter, which was indicated by the following observations: i) it was produced not only by fast e.p.s.p. but also by catelectrotonic potential and ii) under the voltage-clamp conditions no outward current equivalent to afterhyperpolarization was observed following fast e.p.s.c. (Selyanko et al., 1979a, b; Kuba and Nishi, 1979). Tosaka and co-workers showed recently that membrane hyperpolarizations which followed fast e.p.s.p. and catelectrotonic potential in amphibian sympathetic ganglion neurons were abolished by muscarine but not by TEA (Tosaka et al., 1983). It is known that muscarine and TEA affect distinct components of the potassium current: Muscarine selectively suppresses the M-current, whereas TEA selectively diminishes calcium-dependent and delayed rectifying potassium currents (Adams et al., 1982a, b, c). Thus, the action of muscarine (but not TEA) on the fast e.p.s.p.'s afterhyperpolarization, as well as the similarity in voltage dependences of this hyperpolarization and of the M-current (Tosaka et al., 1983), allow one to conclude that the fast e.p.s.p.'s afterhyperpolarization is associated with the M-current activation during the fast e.p.s.p. The amplitude of afterhyperpolarization is markedly lower than that of fast e.p.s.p., although its duration, due to the slow M-current deactivation at resting membrane potential level, essentially exceeds the duration of fast e.p.s.p.

It was suggested that the fast e.p.s.p.'s afterhyperpolarization may normally exert an inhibitory effect on ganglionic transmission, the effect being abolished by the activation of muscarinic acetylcholine receptors (Tosaka et al., 1983).

Further investigation of the effects of muscarinic agonists on membrane conductances underlying the action potential was carried out in amphibian sympathetic ganglion neurons under voltage-clamp conditions. This method allows one, first, to exclude the effect of the membrane potential change during postsynaptic response on membrane conductances and, second, to study this effect in isolation on the conductances forming spike and afterhyperpolarization. Figure 10.2 illustrates the effect of acetylcholine (10 μM), in the presence of 100 μM tubocurarine, on an inward calcium current, observed under conditions when the inward sodium current is preliminarily abolished by TTX (1.4 μM). One can see that acetylcholine suppresses calcium current without affecting its voltage dependence and diminishes the outward potassium current. Acetylcholine also depresses the fast (sodium) inward current recorded in the absence of TTX (Akasu and Koketsu, 1982).

It is known that an outward potassium current elicited in amphibian sympathetic ganglion neurons by a depolarizing voltage step contains at least three main components: Ca-dependent current, delayed rectifying current, and M-current (Adams et al., 1982a). Therefore, one may assume that all these current components may contribute to the action potential's afterhyperpolarization. One more component of potassium current, a transient outward current, can be activated by a depolarizing voltage step from a membrane potential level more negative than resting potential level to the levels equal or less negative if compared with resting potential level (Adams et al., 1982a). Among the above components of potassium current, only the M-current, as has been consistently reported by many investigators, is suppressed by muscarinic agonists in B-neurons (in C-neurons the M-current is insensitive to muscarine; see Jones, 1984). As far as other potassium current components are concerned, the data on their sensitivity to muscarinic agonists obtained by different authors are contradictory. According to Adams et al. (1982b), muscarine does not affect potassium current components other than the M-current. In agreement with their voltage-clamp data, Adams and co-workers have not found any effect of muscarine on the configuration of the action potential in amphibian sympathetic neurons. By contrast, according to other investigators, acetylcholine, through the activation of muscarinic receptors, causes, in addition to the M-current, inhibition of the delayed rectifying current, as well as the caffeine-induced calcium-dependent potassium current (Akasu and Koketsu, 1982; Koketsu et al., 1982b). Tokimasa (1984, 1985b) has found that acetylcholine suppresses the slow, calcium-dependent component of the post-spike afterhyperpolarization.

The contradiction between the data obtained by different authors on the effect of muscarinic agonists on calcium-dependent potassium current underlying post-spike afterhyperpolarization has been resolved by the finding that this current in amphibian sympathetic ganglion neurons includes two distinct components – fast and slow (Pennefather et al., 1985; Tanaka et al., 1986). The fast component, larger in amplitude, corresponds to that previously studied in detail (Adams et al., 1982a, b, c), which is insensitive to muscarine and sensitive to TEA. Another, slower component has smaller amplitude and can be inhibited by muscarine (although to a lesser extent than the M-current), apamine, and tubocurarine, but not TEA. The

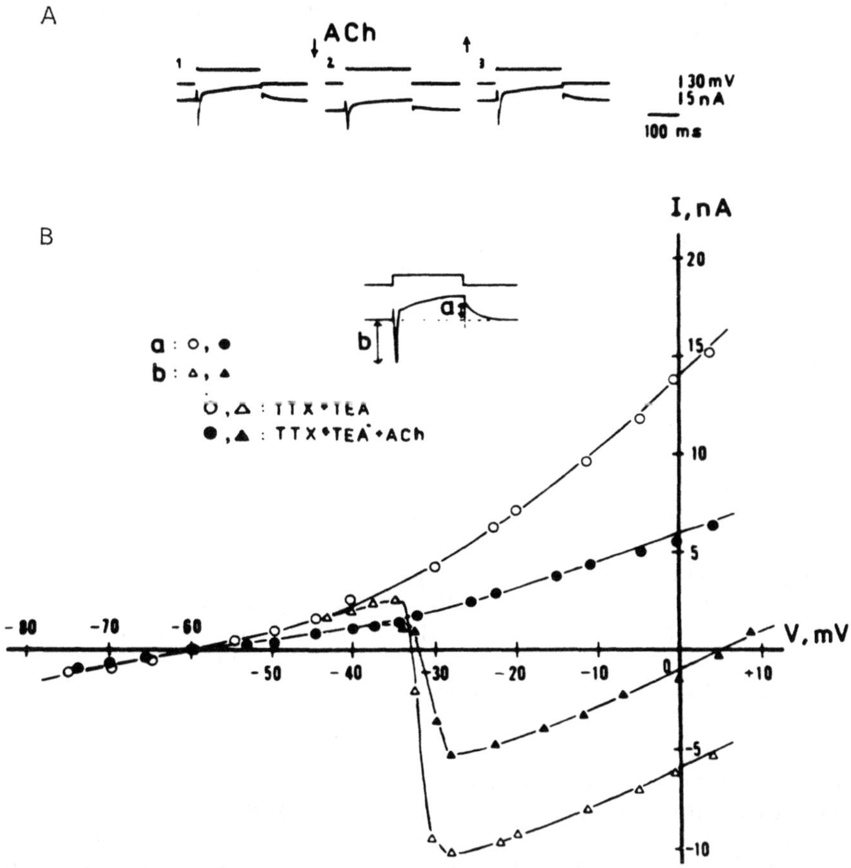

Fig. 10.2. Effect of acetylcholine on membrane currents underlying calcium action potential in frog sympathetic ganglion neuron. A) Membrane currents (bottom records) caused by depolarizing voltage steps (top records) from holding membrane potential level –60 mV, as recorded before (1), during (2), and after (3) the action of acetylcholine (10 μM). Tubocurarine (100 μM), TEA (20 mM), and TTX (1.4 μM) were continuously present in the solution throughout the experiment. B) Effect of acetylcholine on the current–voltage relationship for outward (a) and inward (b) currents as examined in the presence of tubocurarine, TEA, and TTX. Inset illustrates how the outward and inward currents were measured. Holding membrane potential was –60 mV (Akasu and Koketsu, 1982).

two components of calcium-dependent potassium current form two corresponding components, fast and slow, in post-spike afterhyperpolarization, and it was the slow component inhibition by acetylcholine which was initially observed by Tokimasa.

In mammalian sympathetic ganglion neurons, in accordance with earlier results obtained from amphibian sympathetic ganglion neurons, no changes in action potential configuration were detected as being due to the action of muscarine

(Brown and Constanti, 1980). Later, a small reduction by muscarinic agonists of the voltage-dependent calcium and potassium currents recorded under voltage-clamp conditions was reported (Belluzzi et al., 1985), although without any changes in post-spike afterhyperpolarization (Brown et al. , 1986). By contrast, in cultured mammalian sympathetic neurons, acetylcholine acting on muscarinic acetylcholine receptors diminished calcium-dependent components of the spike and afterhyperpolarization (Mochida and Kobayashi, 1986).

The effect of muscarinic agonists on the action potential was also studied in other types of neurons, although less thoroughly than in amphibian sympathetic ganglion.

In guinea-pig myenteric ganglion neurons, muscarinic agonists diminished post-spike afterhyperpolarization due to suppression of the calcium-dependent potassium current (Grafe et al., 1980; North and Tokimasa, 1983). This effect was observed at concentrations of acetylcholine as low as 0.1 μM, which was ineffective in eliciting postsynaptic response (Morita et al., 1982). It was therefore concluded that muscarine-sensitive potassium conductances contributing to post-spike afterhyperpolarization and to resting membrane potential were of two distinct types. However, in recent voltage-clamp experiments performed on AH myenteric neurons, evidence was obtained that calcium-dependent potassium current contributing to the post-spike afterhyperpolarization results from an increase in the same persistent calcium-dependent potassium current which contributes to resting membrane potential (North and Tokimasa, 1987).

According to the initial explanation, acetylcholine suppressed the post-spike afterhyperpolarization in myenteric neurons indirectly, by diminishing the voltage-dependent calcium current (Grafe et al., 1980). However, subsequent studies by other investigators did not confirm this suggestion. It was found that iontophoretic application of acetylcholine to ganglion neurons or their synaptic stimulation effectively suppressed afterhyperpolarizations caused by trains of action potentials even if acetylcholine was applied after the action potentials, i.e., after the entry of Ca^{2+} ions into the cell (North and Tolimasa, 1983; Morita and North, 1985). Acetylcholine seems, therefore, to act either directly on potassium channels or on some steps of their activation by intracellular Ca^{2+} ions. The latter suggestion is consistent with the evidence that Ca^{2+} ions activate potassium channels indirectly, via binding to an intracellular component and a change in enzyme activity (North and Tokimasa, 1987): post-spike afterhyperpolarization has a delay as long as tens of milliseconds (North and Tokimasa, 1983; Hirst et al., 1985a); it has rather high temperature dependence (North and Tokimasa, 1983) and is characterized by a nonlinear dependence on the entry of Ca^{2+} ions into the cell (Hirst et al., 1985a, b).

Muscarinic inhibition of post-spike afterhyperpolarization, but not of the spike, has been found in mammalian central neurons (Krnjevič et al., 1971b; Dodd et al., 1981; Brown, 1983; Cole and Nicoll, 1983, 1984a, b; Constanti and Sim, 1985). This hyperpolarization is mostly associated with membrane calcium-dependent potassium conductance (Alger and Nicoll, 1980; Hotson and Prince, 1980; Schwartzkroin and Stafstrom, 1980; Gustafsson and Wigstrom, 1981). Cole and

Nicoll (1984a) have found that in hippocampal neurons muscarinic agonists almost completely abolish post-spike afterhyperpolarization at a negligible decrease in spike amplitude. It was therefore concluded that msucarinic agonists affected potassium conductance at the step following the entry of Ca^{2+} ions into the cell. Suppression of calcium-dependent afterhyperpolarization in central neurons was also observed by other workers (Benardo and Prince, 1982). However, the outward calcium-dependent potassium current, equivalent to this hyperpolarization, reveals only a small decrease in the presence of muscarine (Brown and Griffith, 1983; however, see Constanti and Sim, 1985).

Summing up the data presented above, one can conclude that in most types of neurons the action of muscarinic acetylcholine receptors results in changes in the action potential characteristics. The decrease in active sodium conductance underlying the spike has been found only in amphibian sympathetic ganglion neurons. The decrease in voltage-dependent calcium conductance underlying the calcium component of the action potential has been found in both amphibian and mammalian sympathetic ganglion neurons. Inhibition of calcium-dependent potassium conductance or of some of its components underlying post-spike afterhyperpolarization occurs in sympathetic, parasympathetic, and brain neurons. There are also indications that delayed rectifying potassium conductance contributing to the post-spike afterhyperpolarization in amphibian sympathetic ganglion neurons can be suppressed by muscarinic agonists.

A priori one could assume that neuronal electrical activity should be inhibited with a decrease in the spike and enhanced with a decrease in post-spike afterhyperpolarization. However, the decrease of the spike by muscarinic agonists was relatively small and could not prevent the spike generation. It seems likely that the decrease of the spike is not related to regulation of neuronal spike activity. The decrease of the spike, especially of its calcium component, may underlie inhibition of transmitter release mediated via presynaptic muscarinic acetylcholine receptors (Koketsu and Yamada, 1982; Morita et al., 1982) or the interaction between cytoplasmic and membrane processes (see Kostyuk, 1984).

As far as the role of post-spike afterhyperpolarization in normal restriction of the rate of neuronal discharges is concerned, its inhibition may serve as an excitatory modulatory factor. This has been directly observed in guinea-pig myenteric neurons (Morita and North, 1985). Post-spike afterhyperpolarization here consists of two distinct components, fast and slow, of which only the slow one is inhibited via activation of the muscarinic acetylcholine receptors. As shown in Fig. 10.3, the rate of discharge in a spontaneously discharging neuron is markedly increased by synaptic stimulation. With the use of weak stimulation, the facilitation of discharges occurs due solely to a decrease in the slow component of afterhyperpolarization. With stronger stimulation, the rate of discharge is additionally increased due to the slow depolarizing postsynaptic response. Since the depression of afterhyperpolarization needs weaker synaptic stimulation than the initiation of postsynaptic response, Morita and North (1985) suggested that the former event may be a more sensitive indicator of muscarinic acetylcholine receptor activation than the latter one. Similarly sensitive to muscarinic agonists is afterhyperpolarization in amphibian sympathetic ganglion neurons, which can be markedly inhibited even by a few quanta of acetylcholine (Tokimasa, 1985b).

Fig. 10.3. Effect of shortening of post-spike afterhyperpolarization, caused by field stimulation, on the rate of spontaneous discharges in guinea-pig myenteric neuron. The shortening of post-spike afterhyperpolarization was elicited by repetitive (10 Hz, 3 s) field stimulation (arrow). Postsynaptic depolarization was lacking at weak (6 V) field stimulation, but appeared at stronger (10 or 12 V) field stimulation (not shown). The apparent resting membrane potential was –55 mV (Morita and North, 1985).

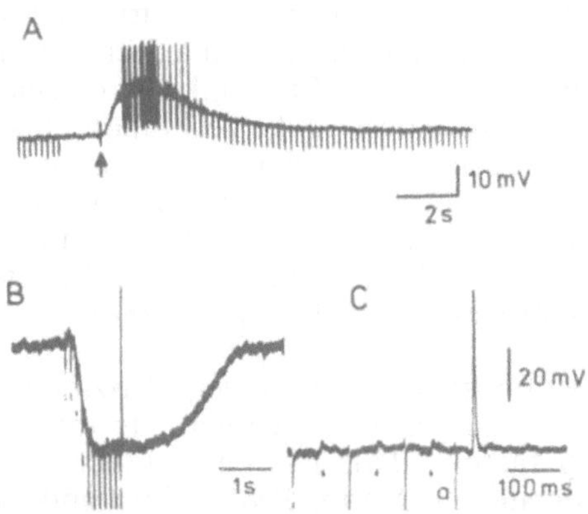

Fig. 10.4. Discharges in guinea-pig myenteric neuron triggered by slow e.p.s.p. (A) and elicited in C-neuron of frog sympathetic ganglion neuron by antidromic stimulation during slow i.p.s.p. (B). A) Slow e.p.s.p. preceded by much smaller fast e.p.s.p., elicited by single field stimulus (denoted by arrow). Vertical lines are electrotonic potentials produced by testing hyperpolarizing current pulses (North and Tokimasa, 1982). B) Slow i.p.s.p. caused by 10 orthodromic stimuli and action potential evoked by single antidromic stimulus during the maximum of a response. C) Same record as in B, shown in detail, with three small fast e.p.s.p.'s (denoted by triangles), evoked by single orthodromic stimuli, and with antidromic action potential (a). Fast e.p.s.p. were partially inhibited by tubocurarine (100 μM) (Dodd and Horn, 1983).

In rat hippocampal neurons, synaptic stimulation decreases the post-spike afterhyperpolarization and facilitates the discharges caused by direct cell stimulation (Cole and Nicoll, 1984a). In some neurons, facilitation of the discharges caused by increased electrical excitability of a neuron was observed in the absence of any postsynaptic depolarization, which was presumably due exclusively to inhibition of afterhyperpolarization.

The examples presented above indicate that the muscarinic effect on characteristics of the action potential provides an effective way of modulating neuronal electrical activity.

It should be pointed out that the effect of acetylcholine on the action potential in neurons can be mediated not only through muscarinic acetylcholine receptors, but also through receptors for other transmitters. The characteristics of the action potential are affected by serotonin (Wood and Mayer, 1979), noradrenaline (Horn and McAfee, 1980; Dunlap and Fischbach, 1978, 1981; Galvan and Adams, 1982; Madison and Nicoll, 1982; Haas and Konnerth, 1983), histamine (Haas and Konnerth, 1983), γ-aminobutyric acid (Kato and Kuba, 1980; Dunlap and Fischbach, 1978, 1981; Cherubini and North, 1984), and neuropeptides (Akasu et al., 1983d). The transmitters may control the action potentials both in neurons and in other types of cells (see Siegelbaum and Tsien, 1983).

The mechanism of the linkage between muscarinic acetylcholine receptors and voltage-sensitive channels in neurons is still poorly studied. Because the same voltage-sensitive channels are operated via various transmitter receptors, one may assume that these channels are separated from their receptors and are connected to them through intramembrane or intracellular messengers.

The involvement of a messenger in the inhibition of voltage-sensitive channel activity by a transmitter initially was shown only for one kind of channel, the potassium channels, and only in *Aplysia* sensory neurons (Siegelbaum et al., 1982). Here serotonin affects the activity of one kind of potassium channels via cAMP. Serotonin also inhibits calcium current in snail neurons, although the data on the role of cAMP in this effect are contradictory (cf. Kononenko and Scherbatko, 1985 and Akopyan et al., 1984, 1985).

In chick dorsal root ganglion neurons, G-protein-mediated inhibition by neurotransmitters (GABA, noradrenaline) of voltage-dependent calcium current has been observed (Holtz et al., 1986). In mammalian (Hescheler et al., 1986) and frog (Fischmeister and Hartzell, 1986) ventricular cardiac myocytes, acetylcholine suppresses calcium current by inhibiting cAMP production.

Activation of muscarinic acetylcholine receptors (by synaptic stimulation or by muscarinic agonists) in sympathetic ganglion neurons results in elevation of the intracellular level of several presumable messengers, including cAMP and cGMP (McAfee et al., 1971; Greengard and Kebabian, 1974; Weight et al., 1974; Kebabian et al., 1975; Volle et al., 1981). However, the role of these probable messengers in muscarinic regulation of the action potential is not yet clear.

10.2. EFFECT OF MUSCARINIC POSTSYNAPTIC RESPONSE ON NEURONAL ELECTRICAL ACTIVITY

Early studies of slow (muscarinic) e.p.s.p. and i.p.s.p. in amphibian and mammalian sympathetic ganglia indicated that these potentials differed in at least two aspects from fast postsynaptic potentials. First, they were very small in amplitude (slow e.p.s.p. was usually a subthreshold for spike initiation) and, second, long repetitive orthodromic stimulation was needed for their appearance (see Libet, 1970; Skok, 1973; Nishi, 1974). The small amplitude of slow e.p.s.p. and the fact that the conditions when it appeared were far from physiological ones led to the conclusion that in sympathetic ganglia slow e.p.s.p. can at best only modulate synaptic transmission, while not being able to elicit neuronal discharges (Skok, 1973, 1981). As was found recently, an excitation produced by repetitive orthodromic stimulation could be transmitted through sympathetic ganglion via muscarinic and noncholinergic transmissions. Nonnicotinic transmission was also observed to be caused by reflectory stimulation of sympathetic ganglia, although the evidence that this transmission is muscarinic was not obtained (Jänig et al., 1983).

It should be pointed out that in sympathetic neurons the physiological role of muscarinic transmission, as well as of other nonnicotinic transmissions, may not necessarily be associated with modulation of nicotinic transmission, but can rather be related to humoral or synaptic regulation of cellular metabolism and development (see Skok, 1981).

An entirely different situation has been revealed in enteric ganglia (North and Tokimasa, 1982; Morita and North, 1985). In the majority of myenteric neurons (about 50% of S-neurons and 20% of AH-neurons) slow e.p.s.p. can be elicited by a single orthodromic stimulus, and a train of several stimuli often elicits discharges. In some ganglion neurons, even slow e.p.s.p. elicited by a single stimulus can cause discharges (see Fig. 10.4A). The amplitude of slow e.p.s.p. can be increased by increasing either the number of stimuli applied or the rate of stimulation, when the number of stimuli is fixed. The characteristics of slow e.p.s.p. in S- and AH-neurons are different: in S-neurons the amplitude of slow e.p.s.p. is larger and the duration is shorter than in AH-neurons.

It was found recently that, similarly to myenteric neurons, in the neurons of mammalian and amphibian sympathetic ganglia and in brain neurons slow e.p.s.p. could be obtained with a single orthodromic stimulus. However, even being elicited with the repetitive stimulation, slow e.p.s.p. in sympathetic neurons remained subthreshold for spike initiation (Adams and Brown, 1982; Cole and Nicoll, 1984a; Brown and Selyanko, 1985b; Brown et al., 1986). Since in all these types of neurons the main mechanism underlying slow e.p.s.p. is the M-current inhibition, one could assume that the inability of this response to reach the threshold for spike generation is related to insufficiently strong M-current inhibition.

Indeed, as calculated by Adams and Brown (1982), synaptic stimulation of amphibian sympathetic ganglion neurons produces, under normal conditions (with noninhibited acetylcholinesterase), only a partial inhibition of the M-current (5–10% with single orthodromic stimulation and less than 50% with repetitive orthodromic

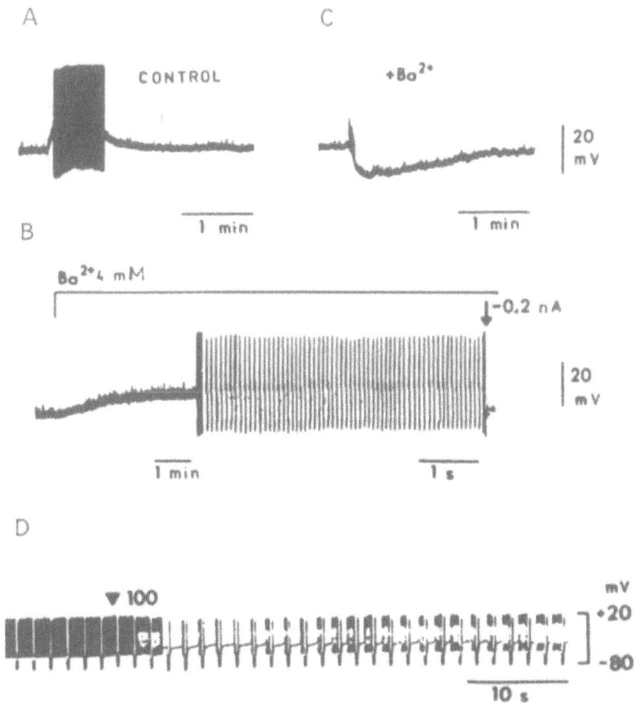

Fig. 10.5. Electrical activity evoked in rat superior cervical ganglion neuron by repetitive orthodromic stimulation. A) Slow e.p.s.p. and discharges evoked by repetitive (20 Hz, 5 s) orthodromic stimulation in the presence of tubocurarine (0.4 mM) and neostigmine (1 μM). B) Membrane depolarizaton and discharges elicited by adding the Ba^{2+} (4 mM) into perfusion solution. At the moment denoted by the arrow, membrane potential was adjusted near to the initial resting membrane potential level (–57 mV) by constant current injection. C) Membrane hyperpolarization caused by repetitive orthodromic stimulation similar to that in A, at membrane potential level of –57 mV after adding the Ba^{2+}. D) Effect of repetitive orthodromic stimulation (100 shocks at 20 Hz at the moment denoted by arrow) on discharges evoked by Ba^{2+} ions. Regular firing of the neuron was maintained by intermittent –0.4-nA current injections. All records were obtained from the same neuron (Brown and Selyanko, 1985b).

stimulation), which corresponds to subthreshold membrane depolarization by 5 mV. More evidence supporting the suggestion that the lack of discharges during slow e.p.s.p. is associated with incomplete inhibition of the M-current is the fact that discharges appear in mammalian sympathetic ganglion neurons when slow e.p.s.p. is elicited after inhibition of acetylcholinesterase or when the M-current is almost completely inhibited by Ba^{2+} ions, as shown in Fig. 10.5 (Brown and Selyanko, 1985b).

The role of the M-current in preventing continuous discharges was further ascertained by recent observations that the tonic-type neurons in guinea-pig inferior mesenteric ganglion lack the M-current, while the phasic-type neurons exhibit the M-current (Cassel et al., 1986).

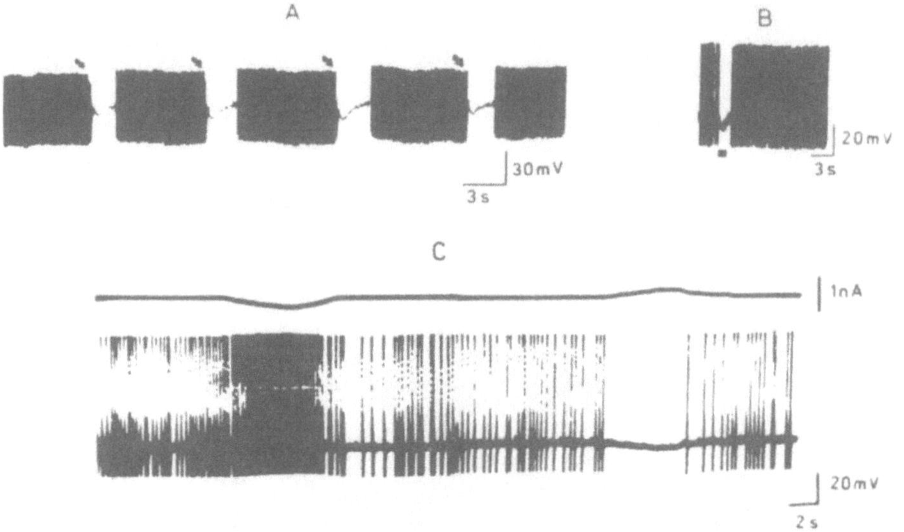

Fig. 10.6. Effect of slow i.p.s.p. on spontaneous discharges in C-neuron of frog sympathetic ganglion (A) and in cat pelvic ganglion neurons (B, C). A) Spontaneous discharges accompanying late slow (peptidergic) e.p.s.p. evoked by long orthodromic stimulation (1 Hz, 4 min). Continuous discharges are interrupted by slow i.p.s.p.'s elicited by short trains (5 Hz, 1 s) of orthodromic stimuli (Dodd and Horn, 1983). B) Arresting spontaneous discharges by slow i.p.s.p. evoked by repetitive (40 Hz, 1 s) orthodromic stimulation (denoted by a horizontal bar) (Gallagher et al., 1982). C) Effects of membrane depolarization or hyperpolarization on the rate of spontaneous discharges in another neuron of cat pelvic ganglion. Top record: polarizing current; bottom record: membrane potential (Griffith and Gallagher, 1980).

In central neurons, besides the incomplete suppression of the M-current, there is another factor restricting the amplitude of slow e.p.s.p. and thus the probability of neuronal discharge. This is the more negative (than in sympathetic ganglion neurons) resting membrane potential level (about –70 mV), which is close to the level for M-current activation (see Brown, 1983).

In mammalian sympathetic ganglion neurons, the amplitude of slow e.p.s.p. is restricted by a special mechanism, found so far only in these neurons. This mechanism is indicated by the appearance of an outward (hyperpolarizing) current concurrent with an inward (depolarizing) current causing slow e.p.s.p. (Brown and Selyanko, 1985a, b), which can be revealed when the M-current is suppressed by Ba^{2+} ions (Fig. 10.5A, B). Since this hyperpolarizing response is associated with a decrease in membrane conductance rather than with an increase, it was not clear *a priori* whether the resulting effect of this response should be excitatory or inhibitory. Suppression of discharges, caused by Ba^{2+} ions, during the hyperpolarizing response (Fig. 10.5D) indicates that it is actually inhibitory and should probably restrict the excitation of a neuron produced by the M-current inhibition during slow e.p.s.p.

The lack of discharges during the slow e.p.s.p. in amphibian and mammalian sympathetic ganglion neurons, as well as in brain neurons, suggests that in these neurons the action potential generation is not a function of slow e.p.s.p. Detailed investigations showed that slow e.p.s.p. and depolarizations caused by muscarinic agonists in these neurons are accompanied by an essential increase in electrical excitability (Brown and Adams, 1980; Brown and Constanti, 1980; Adams et al., 1982b; Halliwell and Adams, 1982; Cole and Nicoll, 1983, 1984a; Brown and Selyanko, 1985a, b; Brown et al., 1986). This is explained by the membrane depolarization during slow e.p.s.p. and, what is more essential, by the increase in input resistance (see Table 9.1), which allows reaching threshold depolarization for spike generation using comparatively small currents. As a result, the depolarizing current pulse passed through the membrane during slow e.p.s.p. triggers a long train of spikes instead of one or two spikes elicited in control, and the amplitude of fast e.p.s.p. increases during muscarinic depolarization (Schulman and Weight, 1976). Therefore, it was assumed that with this mechanism fast e.p.s.p. can be transformed from subthreshold, or "inefficient," into suprathreshold for spike generation. On the other hand, spike generation does not affect slow e.p.s.p., as shown by the lack of any effect on slow e.p.s.p. of the antidromic spike (Kumamoto and Kuba, 1983).

Slow i.p.s.p. is most strong in C-neurons of amphibian sympathetic ganglion (Dodd and Horn, 1983). In some neurons, slow i.p.s.p. up to 30 mV can be elicited by single orthodromic stimuli , while in the rest a short train of orthodromic stimuli is needed for eliciting this response (Fig. 10.4B). The maximum amplitude of slow i.p.s.p. (about 40 mV) is probably restricted by its reversal potential equal to the potassium equilibrium potential. However, despite such a large amplitude, slow i.p.s.p. fails to inhibit action potentials elicited by antidromic (Fig. 10.4) or suprathreshold orthodromic stimulation (Dodd and Horn, 1983). These findings indicate that slow i.p.s.p. is inefficient in inhibiting fast (nicotinic) transmission through sympathetic ganglion. However, as Dodd and Horn have shown, slow i.p.s.p. can effectively inhibit the discharges elicited by noncholinergic slow e.p.s.p. (Fig. 10.6A). Continuous discharges accompanying noncholinergic slow e.p.s.p. are transformed by slow i.p.s.p. into bursts of impulses separated by short silent time intervals. Inhibition caused by slow i.p.s.p. can be imitated by passing a weak hyperpolarizing current through the membrane (Dodd and Horn, 1983).

In amphibian cardiac ganglion neurons, slow i.p.s.p. can also be evoked by a single orthodromic stimulus (Hartzell et al., 1977). By contrast, repetitive orthodromic stimulation is needed to evoke slow i.p.s.p. in mammalian parasympathetic neurons (Gallagher et al., 1982). The amplitude of slow i.p.s.p. in these neurons (6 mV, on the average) is much smaller than in amphibian parasympathetic neurons. Despite that, slow i.p.s.p. is effective in inhibiting discharges in spontaneously active neurons (Fig. 10.6B). This effect here, as in amphibian sympathetic neurons, is fully reproduced by passing a weak hyperpolarizing current through the membrane.

Thus, the results obtained from amphibian sympathetic C-neurons and mammalian parasympathetic neurons indicate that only spontaneous electrical activity or very weak excitatory action produced by slow e.p.s.p. can be inhibited by slow i.p.s.p. Since inhibition caused by slow i.p.s.p. is reproduced by equivalent artificial membrane hyperpolarization, the inhibitory effect of slow i.p.s.p. is apparently not due to membrane shunting.

An interesting feature of muscarinic hyperpolarization has been revealed in thalamic reticular neurons (McCormick and Prince, 1986). It appears that this hyperpolarization inhibits trans-synaptic spike initiation through rising threshold depolarization amplitude, but at the same time promotes burst discharges. The latter effect results from driving the membrane potential to a more negative range, where calcium current, being inactivated at the resting membrane potential level, becomes able to generate bursts of discharges. This result indicates that the generally accepted division of postsynaptic potentials into purely excitatory and purely inhibitory ones is an oversimplification (see Brown, 1986).

10.3. CONCLUSIONS

Activation of muscarinic acetylcholine receptors may affect neuronal electrical activity by changing either characteristics of the action potential or the resting membrane potential and conductance, which, in turn, modify neuronal electrical excitability.

In amphibian sympathetic ganglion neurons, muscarinic agonists decrease both the spike (due to suppression of active sodium and calcium conductances) and post-spike afterhyperpolarization (due to the suppression of the slow component of calcium-dependent potassium conductance and, probably, due to the delayed rectification inhibition). In myenteric and central neurons, muscarinic agonists decrease only the calcium-dependent component of post-spike afterhyperpolarization. It is suggested that inhibition of the spike, especially of its calcium component, can underlie muscarinic regulation of transmitter release from presynaptic nerve terminals. The muscarinic suppression of post-spike afterhyperpolarization has been shown to increase the rate of spontaneous discharges.

Probably only in myenteric neurons can slow (muscarinic) e.p.s.p. normally trigger spikes. In other types of neurons, slow e.p.s.p. is, as a rule, a subthreshold for spike generation and, therefore, can play here only an excitatory modulatory role since it is accompanied by an increase in electrical excitability.

Muscarinic slow i.p.s.p. does not inhibit action potentials elicited by suprathreshold orthodromic stimulation but, nevertheless, it effectively arrests spontaneous discharges as well as discharges caused by noncholinergic slow e.p.s.p. In thalamic reticular neurons, muscarinic hyperpolarization plays a dual role: it inhibits single spike generation and, at the same time, it promotes burst discharges.

Thus, activation of muscarinic acetylcholine receptors may modulate neuronal electrical activity. Another possible function of muscarinic acetylcholine receptors is synaptic control of the metabolism of nerve cells.

Chapter 11

EFFECTS OF NONCHOLINERGIC TRANSMITTERS ON NEURONAL ACETYLCHOLINE RECEPTORS

11.1. EFFECTS ON NICOTINIC ACETYLCHOLINE RECEPTORS

Noncholinergic transmitters essentially affect the activity of nicotinic acetylcholine receptors both in vertebrate and invertebrate neurons. The mechanisms of this action on two types of neurons are different.

A series of studies conducted by Japanese authors on amphibian sympathetic ganglion neurons showed that the responses to acetylcholine were decreased by serotonin, catecholamines, and neuropeptides (substance P and luteinizing hormone-releasing hormone) (Akasu et al., 1981a; Koketsu et al., 1982a, c; Akasu et al., 1983a, b, c), whereas in the presence of ATP they were increased (Akasu and Koketsu, 1985). Investigations carried out in parallel on the amphibian neuromuscular junction showed that the above transmitters act here in a similar way (Akasu et al., 1981a, b, 1983a, b, c; Koketsu et al., 1982c).

An example of acetylcholine response inhibition by substance P is shown in Fig. 11.1A, while an example of acetylcholine response potentiation by ATP is shown in Fig. 11.1B.

At least one of the noncholinergic transmitters, luteinizing hormone-releasing hormone, is supposed to be, in addition to acetylcholine, an excitatory transmitter in amphibian sympathetic ganglion (Jan et al., 1979, 1980a, b; Jan and Jan, 1982). Other transmitters, although not involved in synaptic transmission through sympathetic ganglion, also evoke responses in ganglion neurons.

Serotonin usually causes membrane depolarization with an increase in conductance (Watanabe and Koketsu, 1973), while substance P and ATP, as well as luteinizing hormone-releasing hormone, elicit membrane depolarization accompa-

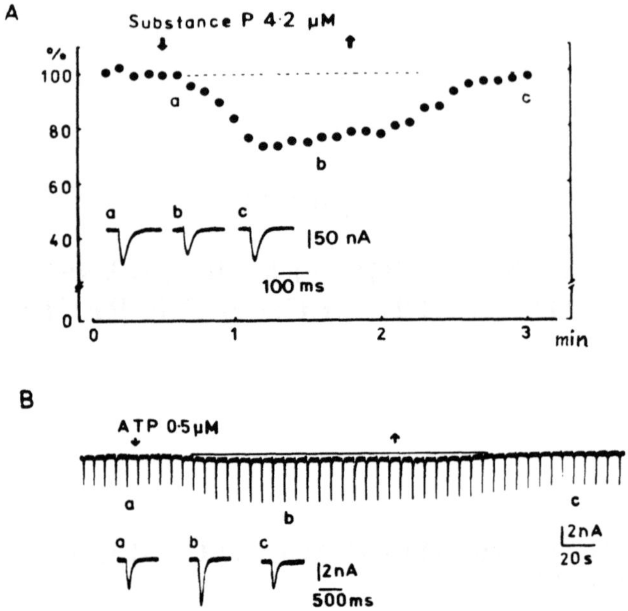

Fig. 11.1. Effects of noncholinergic transmitters on nicotinic acetylcholine receptors in frog sympathetic ganglion neurons. A) Change in time of the amplitude of current, elicited by iontophoretic application of acetylcholine on the neuron under the influence of substance P (4.2 µM). Control current amplitude is taken as 100%. Inset: examples of acetylcholine responses recorded before (a), during (b), and after (c) the action of substance P. Holding membrane potential was –55 mV (Akasu et al., 1983c). B) Effects of ATP (0.5 µM) on responses caused by iontophoretic applications of carbacholine on a neuron. Inset: examples of carbacholine responses recorded before (a), during (b), and after (c) the action of ATP. Holding membrane potential was –60 mV (Akasu and Koketsu, 1985).

nied by increased membrane conductance (Adams and Brown, 1980; Jan and Jan, 1982; Adams et al., 1983; Akasu et al., 1983b). Catecholamines produce both membrane depolarization and hyperpolarization, with varying changes in membrane conductance (see Skok, 1973; Kuba and Koketsu, 1978).

The modulatory actions of transmitters on acetylcholine response cannot be explained by changes in membrane potential, since these actions are retained under voltage-clamp conditions. Moreover, the modulation of acetylcholine response persists under conditions when responses to noncholinergic transmitters are not observed. This takes place at low concentrations of serotonin, insufficient for producing responses, or at hyperpolarized membrane potential levels, at which responses to certain transmitters (neuropeptides, ATP) do not occur. In the amphibian neuromuscular junction, noncholinergic transmitters modulate acetylcholine response without eliciting, as a rule, their own responses.

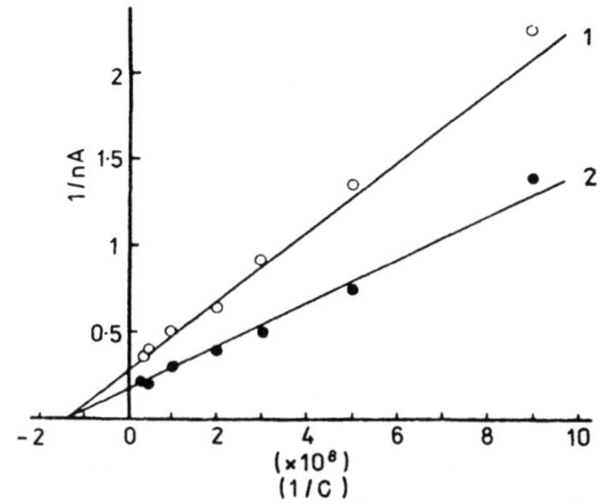

Fig. 11.2. Effect of ATP on concentration dependence of action of acetylcholine on frog sympathetic ganglion neuron. Double reciprocal plot (Lineweaver–Burk plot) for the dependence of the peak amplitude of the current elicited by iontophoretic application of acetylcholine on iontophoretic charge, obtained before (1) and during (2) the action of ATP (0.5 mM) (Akasu and Koketsu, 1985).

These data show that in amphibian sympathetic ganglion neurons and in neuromuscular junction there is no apparent link between acetylcholine response and responses to other transmitters. Thus, it has been concluded that noncholinergic transmitters exert their modulatory effect on acetylcholine responses by direct modulatory interaction with acetylcholine receptors. The possibility that these transmitters act on the recognition center of the acetylcholine receptor was verified by studying their effects on the concentration dependence of acetylcholine action. Special attention was paid to possible changes in two parameters of concentration dependence for acetylcholine, i.e., maximum response (V_{max}) and dissociation constant (K_D), characterizing the efficacy and affinity of acetylcholine for a receptor, respectively.

It has been found in this study that serotonin interacts with the recognition center of an acetylcholine receptor. This conclusion follows from the increase in K_D and from the constancy of V_{max} during the action of serotonin (Akasu et al., 1981a). By contrast, the other transmitters (catecholamines, neuropeptides, ATP) seem to act not on the recognition center of the acetylcholine receptor, but rather on its allosteric binding sites. This conclusion is based on the fact that the transmitters mentioned do not change the affinity of acetylcholine for a receptor (K_D) but affect its efficacy (V_{max}), either decreasing (catecholamines, neuropeptides: Koketsu et al., 1982a; Akasu et al., 1983b) or increasing it (ATP: Akasu and Koketsu, 1985). Figure 11.2 shows the effect of one of the transmitters, ATP, on the concentraton dependence of acetylcholine action. It should be pointed out that none of the transmitters studied changed the stoichiometry between acetylcholine and the nicotinic receptor, since the Hill coefficient characterizing this relationship remained unchanged.

The effects of serotonin and other transmitters on acetylcholine response, in particular their effects on concentration curves for acetylcholine, strikingly remind one of the action of some ganglion-blocking drugs. Based on this similarity as well as on the fact that these transmitters are usually present in cholinergic synapses, these substances are called biogenic, or endogenous, antagonists (Koketsu, 1984).

The conclusion that serotonin acts on the recognition center of the acetylcholine receptor and that other transmitters act on allosteric binding sites in the receptor was initially drawn from the effects of the transmitters on the concentration dependence of acetylcholine action. More direct and persuasive evidence was obtained in experiments with protection of the recognition center of the end-plate receptor by erabutoxin, a selective and irreversible blocker of the recognition center (Kato et al., 1978). A comparison of the action on nicotinic receptors of serotonin and catecholamines indicated that only serotonin, not catecholamines, prevented their irreversible blockade by erabutoxin (Koketsu et al., 1982b). This result is consistent with the competitive action of serotonin and noncompetitive actions of catecholamines on the nicotinic acetylcholine receptor.

The competition between acetylcholine and serotonin for the recognition center of a nicotinic receptor suggests that there is a similarity between the recognition centers of acetylcholine and serotonin receptors. This suggestion is supported by the fact that tubocurarine, a classical antagonist of nicotinic acetylcholine receptors, appears to be an effective blocker of serotonin receptors in various types of cells: in myenteric neurons (Hirst and Silinsky, 1975), mammalian sensory neurons (Higashi and Nishi, 1982), and molluscan ganglion neurons (Gerschenfeld and Paupardin-Tritsch, 1974). The action of tubocurarine on serotonin receptors is competitive only at low concentrations (lower than 5 μM), while at higher concentrations tubocurarine also exerts, as at the end-plate (see Chapter 5), an additional, noncompetitive action (Higashi and Nishi, 1982).

The similarity between serotonin and nicotinic receptors is supported by the fact that in about a half of the rabbit superior cervical ganglion neurons the iontophoretic application of serotonin elicits membrane depolarization and discharges, with the reversal potential for depolarization identical to that for fast e.p.s.p. (Skok and Selyanko, 1978a, b, 1979).

An attempt has also been made to clarify pharmacological properties of binding sites for two other types of transmitters – ATP and catecholamines – in nicotinic acetylcholine receptors. The binding site for ATP seems to be a purinergic receptor of P_2 subtype (see Burnstock, 1981). This is indicated by the observation that the action of ATP on acetylcholine response cannot be reproduced by adenosine and is not abolished by theophilline (Akasu and Koketsu, 1985). It is essential that other postsynaptic effects of ATP in amphibian sympathetic ganglion, on the resting membrane potential and on the calcium component of action potential, are also not reproduced by adenosine (Akasu et al., 1983a). Thus, all known postsynaptic effects of ATP in amphibian sympathetic ganglion neurons are mediated via purinergic receptors of P_2 subtype.

The binding site for catecholamines appears to be similar to the β-adrenoreceptor, since the effect on acetylcholine current of isoproterenol, an agonist selective for the β-adrenoreceptor, is much stronger than those of adrenaline and noradrenaline (Koketsu et al., 1982a).

The noncholinergic transmitters do not affect the characteristics of nicotinic acetylcholine receptor channels – ionic selectivity and kinetics (Akasu et al., 1981a; Koketsu et al., 1982a, c; Akasu et al., 1983b, c; Akasu and Koketsu, 1985). This is supported by the lack of changes in the reversal potential and in the decay of cholinergic postsynaptic current under the influence of these transmitters both in neurons and muscle fibers. No changes in end-plate current decay have been found during the action of adrenaline or noradrenaline (Snetkov, 1977). At the same time, some phenylalkylamines (tyramine, dopamine), as well as serotonin, make the end-plate current decay voltage independent and make the peak current nonmonotonically voltage dependent (Magazanik et al., 1976; Snetkov, 1977). This result is inconsistent with the above suggestion that serotonin acts on end-plate receptors competitively (Koketsu et al., 1982b).

Several attempts have been made to clarify the mechanism of the modulatory effect of noncholinergic transmitters on nicotinic acetylcholine receptors.

No involvement of second messengers in the modulatory actions of these transmitters on nicotinic receptors in vertebrate neurons has been shown. In the cultured PC 12 cell line, evidence has been obtained that one of the possible messengers, cAMP, is not involved in the inhibition of nicotinic receptor activity by substance P (McGee and Liepe, 1984). The elevation in the intracellular cAMP level via stimulation of adenylate cyclase by forskoline does not result in inhibition of the carbacholine response and does not affect the rate of desensitization of nicotinic receptors of the inhibitory action of substance P on carbacholine response. The involvement of the messengers is also inconsistent with the data obtained in sympathetic ganglion. Activation of muscarinic or adrenergic receptors produce in these ganglion cells an accumulation of essential amounts of possible messengers, including cyclic nucleotides (see Chapter 9). However, no changes in nicotinic sensitivity of ganglion neurons during activation of these receptors have been observed (Christ and Nishi, 1971; Koketsu and Yamada, 1982).

One of the possible explanations for the inhibitory actions of the transmitters on acetylcholine receptors is that they accelerate receptor desensitization. A study of chick sympathetic and parasympathetic ganglion neurons in culture has revealed that substance P shortens the decay of acetylcholine response without changing its peak amplitude (Role, 1984). It has been suggested, therefore, that acceleration of desensitization is the only effect that substance P induces in nicotinic acetylcholine receptors in neurons of this type. Other workers have shown the effect of substance P on desensitization of acetylcholine receptors in cultured sympathetic ganglion neurons (Stallcup and Patrick, 1980) and in isolated chromaffin cells (Clapham and Neher, 1984).

The acceleration of the desensitization of nicotinic acetylcholine receptors by neuropeptides has also been revealed in the amphibian neuromuscular junction (Akasu et al., 1984b). Since desensitization is a two-component process (see Magleby and Pallotta, 1981; Feltz and Trautmann, 1982; Andreev et al., 1984), the authors of this observation have suggested that neuropeptides change the ratio between contributions of the fast and slow components of desensitization in favor of the fast component.

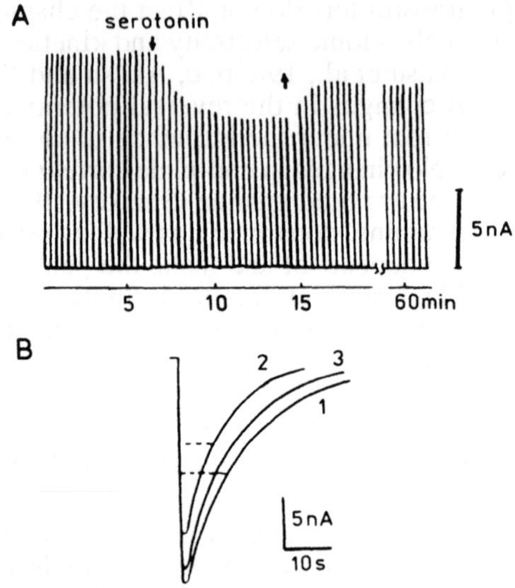

Fig. 11.3. Effect of serotonin on nicotinic responses in molluscan neuron. A) Change in time of the current evoked by iontophoretic application of acetylcholine (vertical lines) under the influence of serotonin (5-HT, 1 μM). B) Changes in amplitude and in time course of the response evoked by acetylcholine (5 μM), under the influence of 5-HT (1 μM). Acetylcholine responses are shown as obtained before (1), during (2), and after (3) the action of 5-HT. Interrupted lines denote the level of current corresponding to half of its peak value (Akopyan et al., 1980b).

Thus, the acceleration of desensitization of nicotinic acetylcholine receptors can indeed be one of the mechanisms responsible for suppression of their activity. However, this mechanism cannot explain the modulatory effects of transmitters on postsynaptic acetylcholine responses because desensitization apparently does not occur during normal synaptic transmission.

It should be kept in mind that noncholinergic transmitters can modulate cholinergic transmission not only postsynaptically, but also presynaptically, by modulating acetylcholine release from presynaptic terminals (Christ and Nishi, 1971; Koketsu et al., 1982b; Silinsky and Ginsborg, 1983). Combination of pre- and postsynaptic actions of noncholinergic transmitters can essentially strengthen their modulatory effects on cholinergic transmission.

The above-mentioned data show that the modulatory effects of noncholinergic transmitters on nicotinic acetylcholine receptors in vertebrate neurons and muscle fibers can be attributed to their direct actions on receptors. By contrast, this modulation in invertebrate neurons seems to occur indirectly and is mediated by second messengers.

Using the intracellular perfusion method (see Krishtal and Pidoplichko, 1975; Kostyuk et al., 1975), Akopyan et al. (1980b, 1981) studied the actions of serotonin and dopamine on acetylcholine-induced current in nonidentified snail neurons. They found that serotonin and dopamine both decreased the peak and shortened the decay of acetylcholine-induced current (Fig. 11.3). Shortening of current decay, marked by desensitization of nicotinic receptors and the decrease in current peak amplitude, seems to be mediated by different mechanisms, since at low concentrations serotonin decreases the current amplitude without shortening (or even slightly lengthening) current decay. Unlike in vertebrate neurons, no competition between serotonin and acetylcholine for the recognition center of the nicotinic acetylcholine receptor occurs in snail neurons. This is shown by removal of the serotonin effect on the acetylcholine response (but not the response itself) by the serotonin antagonist, lysergic acid diethylamide. Serotonin does not affect the reversal potential for the acetylcholine response.

Based on these results, it has been assumed that serotonin exerts its effect on nicotinic receptors in snail neurons indirectly, via intracellular messengers between serotonin and acetylcholine receptors. Among the possible messengers involved are intracellular Ca^{2+} ions which are known to inhibit acetylcholine-induced current in snail neurons from the inner side of the membrane (Chemeris et al., 1982). In order to verify the possible involvement of intracellular Ca^{2+} ions, the effect of intracellularly applied EGTA on inhibition of the nicotinic acetylcholine response by serotonin has been investigated (Akopyan et al., 1980b, 1981). It turns out that under these conditions the partial effect of serotonin on acetylcholine response is retained. Therefore, Ca^{2+} ions seem to participate in developing the serotonin effect, but together with other messengers rather than in isolation.

Another possible candidate for messenger is cAMP, which is accumulated in snail neurons during the action of serotonin (Cedar and Schwartz, 1972; Levitan, 1978). Akopyan and co-workers have shown that cAMP can indeed be a messenger between serotonin and nicotinic acetylcholine receptors. They have found that an increase in the intracellular concentration of cAMP, attained either by its direct introduction into the cell or by activation of adenylate cyclase, changes the acetylcholine response similarly to that produced by serotonin. Moreover, the decrease in hydrolysis of cAMP, due to phosphodiesterase inhibition, strengthens the effect of serotonin on acetylcholine response. On the other hand, the effect of serotonin on acetylcholine response decreases, following inhibition of cAMP-dependent proteinkinase. All these data indicate that interaction between serotonin and nicotinic acetylcholine receptors in snail neurons is mediated by cAMP participating in phosphorylation of a receptor protein via cAMP-dependent proteinkinase. The effect of dopamine on the acetylcholine response seems likewise to be mediated by cAMP, since in snail neurons cAMP is involved in dopamine-induced inhibition (Akopyan et al., 1980a).

Interestingly, the inhibition of the acetylcholine response by serotonin is not fully reversible (Akopyan et al., 1980b, 1981), which suggests that it may underlie the long-lasting inhibition of cholinergic transmission.

Turpaev and his co-workers, while studying the effect of serotonin on identified neurons of *Helix pomatia* L., have found that this effect is potentiating in some neurons (RPa3 and RPa2), whereas in other neurons (LPa3) it is inhibitory (Turpaev et al., 1983). The inhibitory, as well as potentiating effect of serotonin on acetylcholine response, is mediated by an elevation of intracellular cAMP level. This is indicated by the finding that serotonin-induced effects on acetylcholine responses in the above neurons are imitated by inhibition of phosphodiesterase using theophilline or by the GTP-induced activating of adenylate cyclase. The mechanism underlying the diverse actions of serotonin on various molluscan neurons is not yet known. One possible explanation is the heterogeneity of serotonin receptors, which indeed takes place in molluscan ganglia neurons (Gerschenfeld and Paupardin-Tritsch, 1974).

In some molluscan neurons, serotonin does not affect nicotinic acetylcholine receptors. According to Chad and Kerkut (1981), the response elicited in an *Aplysia* neuron by the simultaneous application of acetylcholine and serotonin results in arithmetic summation of individual responses to these transmitters. This result indicates that acetylcholine and serotonin receptors in *Aplysia* neurons are fully independent of one another.

It has recently been shown that biogenic amines (serotonin, dopamine, and adrenalin) affect acetylcholine responses in molluscan neurons when applied both extracellularly or intracellularly (Turpaev et al., 1986). The actions of intracellularly applied transmitters can be associated with their effects on either usual "external" receptors, due to diffusion of transmitters through the membrane, or with their interactions with special "internal" receptors of a cell. It is the former mechanism which is likely to be responsible for the responses of molluscan neurons to the intracellularly applied acetylcholine (Poskonova et al., 1981). By contrast, biogenic amines exert their actions from inside the cell via activation of internal receptors, as shown by the following observations. First, in some neurons the changes in acetylcholine response differ in sign and time course, depending on whether they are produced by extracellularly or intracellularly applied biogenic amines. Second, the responses produced by biogenic amines themselves differ in sign if applied extracellularly or intracellularly.

The functional significance of internal receptors for biogenic amines is not yet clear. However, insofar as biogenic amines are present in many types of neurons, they may mediate effective control of acetylcholine receptor activity.

In molluscan neurons, various changes in acetylcholine responses are observed following the intracellular introduction of morphine (Gromov et al., 1981). The action of morphine is much stronger when applied intracellularly than extracellularly. It is suggested that morphine affects nicotinic acetylcholine responses through internal receptors that normally interact with endogenous peptides – analgesics. It should be pointed out that the observed effects of numerous agents on the nicotinic acetylcholine receptor exerted from the inner side of the membrane are consistent with the presence of binding sites for some antagonists and Ca^{2+} ions in the cytoplasmic part of the acetylcholine receptor molecule (Skok, 1985; see Chapter 6).

11.2. EFFECTS ON MUSCARINIC ACETYLCHOLINE RECEPTORS

In mammalian sympathetic ganglion, dopamine causes an essential (50–100%) increase in the amplitude of muscarinic depolarization (Libet and Tosaka, 1970; Ashe and Libet, 1981). Dopamine also elicits here its own hyperpolarizing response, although its relation to the above potentiation of muscarinic response seems unlikely. First, the potentiation develops after the dopamine response or even in the absence of it (at low concentrations of dopamine). Second, the potentiation is not reproduced by other catecholamines (adrenalin or noradrenalin) which, nevertheless, are more effective than dopamine in eliciting membrane hyperpolarization. And, third, the effect of dopamine on muscarinic response seems to develop via activation of receptors different from those responsible for dopamine response, since dopamine response can be eliminated by α_2 adrenoreceptor antagonists (Brown and Caulfield, 1979), whereas potentiation of the muscarinic response by dopamine is blocked by D_1-receptor antagonists (Ashe and Libet, 1981).

It should be noted that the antagonists of D_1-receptors blocking the potentiation of the muscarinic response do not affect the muscarinic response itself. This indicates that the D_1-receptor is located apart from the muscarinic acetylcholine receptor.

Dopamine exerts its effect on muscarinic depolarization elicited either synaptically (slow e.p.s.p.) or by exogenous muscarinic agonists. Unlike muscarinic depolarization caused by exogenous agonists, slow e.p.s.p. is potentiated by dopamine only if it follows inhibition of the dopamine-destroying enzyme, catechol-O-methyltransferase (Ashe and Libet, 1981).

Potentiation of slow e.p.s.p. is produced by both dopamine and by conditional synaptic stimulation of ganglion neurons (Mochida and Libet, 1985). The latter potentiation is of two types: homosynaptic, produced by the application of conditional and testing stimuli to the same synaptic input, and heterosynaptic, produced by two types of stimuli applied to different synaptic inputs. Conditional potentiation of slow e.p.s.p. does not need the preliminary inhibition of catechol-O-methyltransferase for its occurrence (Mochida and Libet, 1985).

Potentiation of slow e.p.s.p. caused by conditional synaptic stimulation, as well as dopamine-induced potentiation, can be abolished by antagonists of D_1-receptors (see Libet, 1984). Based on these results, as well as on the fact that dopamine is present in sympathetic ganglion and is released by synaptic stimulation (Libet and Owman, 1974), it has been concluded that the synaptic potentiation of slow e.p.s.p. is associated with the action on ganglion neurons of dopamine released in the ganglion (Mochida and Libet, 1985).

Potentiation of slow e.p.s.p. by exogenous dopamine or by conditional synaptic stimulation is of long duration: After a transient application of dopamine or a short train of orthodromic stimuli, slow e.p.s.p. remains increased for several hours. Therefore, it is suggested that potentiation of slow e.p.s.p. may cause long-lasting changes in synaptic transmission efficiency, underlying such physiological phenomena as learning and memory (see Libet, 1984).

Modulation of slow e.p.s.p. by dopamine or by conditioning synaptic stimulation is mediated via intracellular messengers, i.e., cAMP. According to the criteria listed in Chapter 9, the possible role of cAMP as a messenger is supported by the following observations. First, dopamine markedly increases the cAMP level in the ganglion (see Chapter 9). The antagonists blocking the potentiation of slow e.p.s.p. by dopamine also abolish dopamine-induced accumulation of cAMP in the ganglion (Kobayashi and Tosaka, 1983). Second, cAMP increases slow e.p.s.p., as demonstrated by extracellular application of cAMP derivatives (Libet et al., 1975), as well as by intracellular injection of cAMP (Kobayashi et al., 1978). The increase in slow e.p.s.p. caused by cAMP is not abolished by antagonists of D_1-receptors, which prevent the potentiating action of dopamine. Third, stimulation of cAMP production caused by apomorphine (Lindl, 1979) is accompanied by an increase in muscarinic depolarization (see Libet, 1979).

It should be noted that an attempt to explain the potentiation of slow e.p.s.p. by an increase in the intracellular cAMP level faces some problems. It is not clear why potentiation of slow e.p.s.p. is not observed during the action of noradrenalin and adrenalin which, like dopamine, increase the cAMP level in the ganglion. Furthermore, it is known that the cAMP level is regulated via β-adrenoreceptors, whereas the antagonists for these receptors do not affect potentiation of slow e.p.s.p. These contradictions may result from the fact that cAMP stores are concentrated mostly in glial cells in the ganglion, not in the neurons, and it may be in glial cells where activation of β-adrenoreceptors leads to massive accumulation of cAMP (Briggs et al., 1982). It is suggested that dopamine, unlike other catecholamines, activates cAMP production in neurons via action on D_1-receptors (see Kobayashi and Tosaka, 1983; Libet, 1984). This suggestion is consistent with the fact that antagonists for D_1-receptors block both the potentiation of muscarinic response and an increase in cAMP synthesis by dopamine.

The above results indicate that dopamine may mediate metabolic control of the activity of muscarinic acetylcholine receptors, which seems to occur via cAMP-dependent phosphorylation of a receptor protein; the synthesis of new proteins during potentiation of muscarinic depolarization has not been revealed (Ashe and Libet, 1984).

One interesting observation was made during an investigation of the role of cyclic nucleotides in the potentiation of muscarinic response by dopamine: cGMP was able to completely prevent the potentiation of muscarinic response by dopamine if applied immediately (within 4–5 min) after the use of dopamine or cAMP (Libet et al., 1975). With longer time intervals (≥10–15 min) between the use of dopamine or cAMP, on the one hand, and cGMP, on the other hand, the potentiation of muscarinic response was not affected. It was suggested that cGMP affects the initial step of modulation of muscarinic acetylcholine receptor by cAMP. It was also shown that cGMP accumulated in the neurons during muscarinic depolarization (see Chapter 9) was able to diminish the modulatory action of dopamine (see Libet, 1979). Thus, the effect of cGMP on the modulatory action of dopamine on muscarinic response is actually a mechanism with negative feedback: the larger the muscarinic depolarization, the higher the cGMP production and the smaller the additional increase in muscarinic depolarization caused by dopamine.

As has been mentioned above, in sympathetic ganglion muscarinic response seems to be mediated by an intracellular messenger. Therefore, dopamine apparently exerts its effect on muscarinic response by changing the efficiency of coupling between the muscarinic receptor and its channel. Another transmitter, vasoactive intestinal polypeptide (VIP), apparently affects the binding step of the interaction between acetylcholine and muscarinic receptor. This polypeptide in mammalian sympathetic ganglion strongly (up to 15-fold) increases muscarinic depolarization (Mo and Dun, 1984; De Groat et al., 1985). From studies carried out earlier on cat submandibular salivary gland, it was known that VIP increases the affinity of muscarinic agonists to their receptors (Lundberg et al., 1982). One may thus suggest that an increase in efficiency of muscarinic transmission under the influence of neuropeptide occurs in sympathetic ganglion via the same mechanism.

A number of other transmitters may influence muscarinic response in addition to dopamine and VIP. Acetylcholine and three other transmitters (substance P, luteinizing hormone-releasing hormone, and ATP) cause a depolarizing response resulting from the closure of the same potassium channels, the M-channels (Adams and Brown, 1980; Brown and Adams, 1980; Adams et al., 1982b, 1983; Akasu and Koketsu, 1982; Jones, 1985). The identity of the M-channels closed by these three transmitters is shown by the fact that each of them, acting on its own receptor, is able to close the majority of M-channels. It is found, for instance, that at least 90% of the M-channels are controlled via muscarinic acetylcholine receptors (Adams et al., 1982b; Jones, 1985). The competition between various transmitters for common channels provides a basis for interactions between the corresponding responses.

Reciprocal interactions between the responses elicited in B-neurons of amphibian sympathetic ganglion by acetylcholine and neuropeptides have indeed been found (Jan and Jan, 1982; Kuffler and Sejnowski, 1983). An example of the interaction between slow peptidergic and muscarinic e.p.s.c.'s is shown in Fig. 11.4. One can see that at low amplitudes of peptidergic response muscarinic response is decreased, whereas at maximum amplitude of peptidergic response it disappears (occlusion effect). The effect in the opposite direction, i.e., that of the muscarinic response on the peptidergic one, has also been reported.

Interestingly, in C-neurons of amphibian sympathetic ganglion, unlike in B-neurons, the mechanisms of muscarinic and peptidergic responses are totally different: Peptides close M-channels, causing membrane depolarization, whereas muscarinic agonists open potassium channels, thus eliciting membrane hyperpolarization (Dodd and Horn, 1983; Jones, 1984).

In *Aplysia* neurons, reciprocal interactions between slow (muscarinic-like) outward current elicited by carbacholine and analogous currents caused by histamine and dopamine have been found (Ascher and Chesnoy-Marchais, 1982). Each of these three responses can be selectively eliminated by its own specific antagonist; therefore it is unlikely that any transmitter interacts with the other transmitter's receptor. On the other hand, reciprocal interaction between two responses

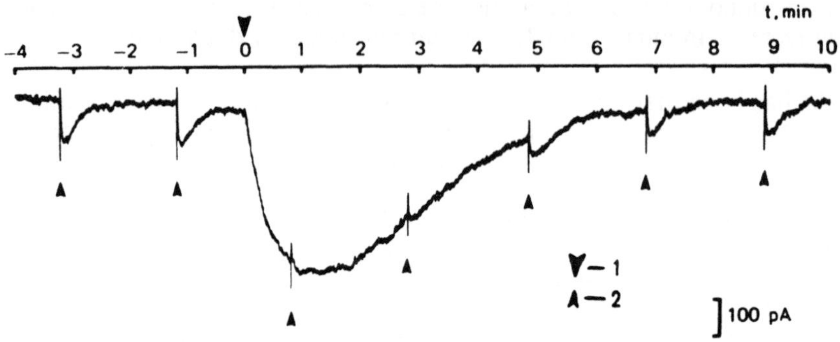

Fig. 11.4. Interaction between slow (muscarinic) and late slow (peptidergic) e.p.s.c.'s in frog sympathetic ganglion neuron. Late slow e.p.s.c. (1) was evoked by repetitive (20 Hz, 5 s) stimulation of the VIIth and VIIIth nerves, while slow e.p.s.c. (2) was elicited by repetitive (50 Hz, 200 ms) stimulation of cholinergic input. Holding membrane potential was –58 mV (Kuffler and Sejnowski, 1983).

elicited by the same transmitter does not differ from that of the responses elicited by two different transmitters. This allows one to conclude that there are certain common steps in the generation of all three responses following activation of the corresponding receptors, and it is these steps where interaction between the responses occurs.

Interaction between the responses elicited by the same or by different transmitters depends on their concentrations: At low transmitter concentrations it results in an increase of testing response (potentiation), whereas at higher concentrations the result is a decrease of the response (desensitization). Cross-potentiation and cross-desensitization of the responses caused by different transmitters have been observed during the actions of the transmitters either on the same membrane area or on different membrane areas. One might speculate that cross-potentiation and cross-desensitization may underlie homosynaptic and heterosynaptic modulation of synaptic transmission.

11.3. CONCLUSIONS

Neuronal responses to the activation of nicotinic acetylcholine receptors can be modulated by various noncholinergic transmitters; the mechanisms of this effect are different in vertebrate and in invertebrate neurons. In amphibian sympathetic ganglion, serotonin, catecholamines, and neuropeptides (substance P and luteinizing hormone-releasing hormone) cause a decrease, while ATP causes an increase, in nicotinic response. The effects of these transmitters on nicotinic response are not related to the generation by them of their own responses followed by a change in membrane potential and in membrane conductance, but are instead due to their direct actions on nicotinic acetylcholine receptors. The action of serotonin on nicotinic acetylcholine receptors in the ganglion is competitive, i.e., is associated with its action on the recognition center of the nicotinic receptor. By contrast, other transmitters appear to act noncompetitively, via interaction with the allosteric binding site of the receptor. The binding site for ATP in the nicotinic receptor possesses features typical for a β-adrenoreceptor. An inhibition of nicotinic response, as has been shown for substance P, may result from acceleration of receptor desensitization.

In many molluscan neurons, serotonin and dopamine cause a long-lasting decrease in nicotinic response. Unlike in sympathetic ganglion, this effect is not associated with the blockade of nicotinic acetylcholine receptors by these transmitters and results from an increase in the intracellular cAMP level due to activation of receptors different from nicotinic ones. In some molluscan neurons, serotonin elicits an increase, instead of a decrease, in nicotinic response, although this action is also mediated by the intracellular accumulation of cAMP. The effect of biogenic amines on nicotinic response occurs when they are applied from either the outer or inner side of the membrane, implying that the receptors for biogenic amines are located on both sides of the membrane. Nicotinic response can also be modulated by peptide-analgesics, and their effects are much stronger if induced from the inner side of the membrane than if induced from its outer side.

Noncholinergic transmitters also affect muscarinic responses. In mammalian sympathetic ganglion neurons muscarinic response increases under the influence of dopamine or vasoactive intestinal polypeptide (VIP). Dopamine does not act on muscarinic receptors directly, and its unusually long effect on muscarinic response results from the accumulation of cAMP in ganglion cells due to the activation of D_1-receptors. A dopamine-induced increase in muscarinic response can be prevented by elevation of the intracellular cGMP level. However, cGMP is effective only when used before or immediately after dopamine is applied. Unlike dopamine, VIP seems to exert a more direct action on muscarinic acetylcholine receptors, causing an increase in their affinity for acetylcholine.

In mammalian sympathetic ganglion neurons, as well as in molluscan ganglion neurons, muscarinic response and responses to some other transmitters reveal reciprocal interactions between each other. In sympathetic ganglion, such an interaction occurs between muscarinic and peptidergic responses, and it is attributed to closure of the same potassium channels (M-channels). In molluscan neurons the reciprocal interactions between the responses mediated via muscarinic-like, dopamine, and histamine receptors have been observed. It has been suggested that there are common steps in the development of these responses, which are responsible for their interaction.

REFERENCES

Adams, D. J., Gage, P. W., and Hamill, O. P. (1976). Voltage sensitivity of inhibitory postsynaptic currents in *Aplysia* buccal ganglia, *Brain Res.* **115**:506–511.

Adams, D. J., Dwyer, T. M., and Hille, B. (1980). The permeability of end-plate channels to monovalent and divalent metal cations, *J. Gen. Physiol.*, **75**:493–510.

Adams, D. J., Nonner, W., Dwyer, T. M., and Hille, B. (1981). Block of end-plate channels by permeant cations in frog skeletal muscle, *J. Gen. Physiol.*, **78**:593–615.

Adams, P. R. (1975). Kinetics of agonist conductance changes during hyperpolarization at frog end-plate, *Br. J. Pharmacol.* **53**:308–310.

Adams, P. R. (1976). Drug blockade of open end-plate channels, *J. Physiol. (London)*. **260**:531–552.

Adams, P. R. (1977). Voltage jump analysis of procaine action at frog end-plate, *J. Physiol . (London)* **268**:291–318.

Adams, P. R. (1981). Biophysical aspects of transmitter action, *Fed. Proc.* **40**:2611–2613.

Adams, P. R., and Brown, D. A. (1975). Actions of γ-aminobutyric acid on sympathetic ganglion cells, *J. Physiol. (London)* **250**:85–120.

Adams, P. R., and Brown, D. A. (1980). Luteinizing hormone-releasing factor and muscarinic agonists act on the same voltage-sensitive K^+-current in bullfrog sympathetic neurons, *Br. J. Pharmacol.* **68**:353–355.

Adams, P. R., and Brown, D. A. (1982). Synaptic inhibition of the M-current: slow excitatory post-synaptic potential mechanism in bullfrog sympathetic neurons, *J. Physiol . (London)* **332**:263–272.

Adams, P. R., and Feltz, A. (1980a). Quinacrine (mepacrine) action at frog end-plate, *J. Physiol . (London)* **306**:261–281.

Adams, P. R., and Feltz, A. (1980b). End-plate channel opening and the kinetics of quinacrine (mepacrine) block, *J. Physiol. (London)* **306**:283–306.

Adams, P. R., and Sakmann, B. (1978). A comparison of current-voltage relations for full and partial agonists, *J. Physiol.* (*London*) **283**:621–644.

Adams, P. R., Brown, D. A., and Constanti, A. (1982a). M-currents and other potassium currents in bullfrog sympathetic neurons, *J. Physiol.* (*London*) **330**:537–572.

Adams, P. R., Brown, D. A., and Constanti, A. (1982b). Pharmacological inhibition of the M-current, *J. Physiol.* (*London*) **332**:223–262.

Adams, P. R., Constanti, A., Brown, D. A., and Clark, R. B. (1982c). Intracellular Ca^{2+} activates a fast voltage-sensitive K^+ current in vertebrate sympathetic neurons, *Nature* (*London*) **296**:746–749.

Adams, P. R., Brown, D. A., and Jones, S. W. (1983). Substance P inhibits the M-current in bullfrog sympathetic neurons, *Br. J. Pharmacol.* **79**:330–333.

Adler, M., Oliveira, A. C., Albuquerque, E. X., Mansour, N. A., and Eldefrawi, A. T. (1979). Reaction of tetraethylammonium with the open and closed conformations of the acetylcholine receptor ionic channel complex, *J. Gen. Physiol.* **74**:129–152.

Akasu, T., and Koketsu, K. (1976). Ionic mechanism of the slow inhibitory postsynaptic potential of sympathetic ganglion cells in bullfrogs, *Kurume Med. J.* **23**:183–188.

Akasu, T., and Koketsu, K. (1980). Desensitization of the muscarinic receptor controlling action potentials of sympathetic ganglion cells in bullfrogs, *Life Sci.* **27**:2261–2267.

Akasu, T., and Koketsu, K. (1982). Modulation of voltage-dependent current by muscarinic receptor in sympathetic neurons of bullfrog, *Neurosci. Lett.* **29**:41–45.

Akasu, T., and Koketsu, K. (1983). Electrogenesis of the slow inhibitory postsynaptic potential in bullfrog sympathetic ganglia, *Jpn. J. Physiol.* **33**:279–300.

Akasu, T., and Koketsu, K. (1985). Effect of adenosine triphosphate on the sensitivity of the nicotinic acetylchole-receptor in the bullfrog sympathetic ganglion cells, *Br. J. Pharmacol.* **84**:525–531.

Akasu, T., Ohta, Y., and Koketsu, K. (1978). The effect of adrenaline on the electrogenic Na^+ pump in cardiac muscle cells, *Experientia* **34**:488–490.

Akasu, T., Hirai, K., and Koketsu, K. (1981a). 5-Hydroxytryptamine controls ACh-receptor sensitivity of bullfrog sympathetic ganglion cells, *Brain Res.* **211**:217–220.

Akasu, T., Hirai, K., and Koketsu, K. (1981b). Increase of acetylcholine receptor sensitivity by adenosine triphosphate: a novel action of ATP on ACh sensitivity, *Br. J. Pharmacol.* **74**:505–507.

Akasu, T., Hirai, K., and Koketsu, K. (1983a). Modulatory actions of ATP on membrane potentials of bullfrog sympathetic ganglion cells. *Brain Res.* **258**:313–317.

Akasu, T., Kojima, M., and Koketsu, K.(1983b). Luteinizing hormone-releasing hormone modulates nicotinic ACh-receptor sensitivity in amphibian cholinergic transmission, *Brain Res.* **279**:347–351.

Akasu, T., Kojima, M., and Koketsu, K. (1983c). Substance P modulates the sensitivity of the nicotinic receptor in amphibian cholinergic transmission, *Br. J. Pharmacol.* **80**:123–131.

Akasu, T., Nishimura, T., and Koketsu, K. (1983d). Modulation of action potential during the late slow excitatory postsynaptic potential in bullfrog sympathetic ganglia, *Brain Res.* **280**:349–354.

Akasu, T., Gallagher, J. P., Koketsu, K., and Shinnick-Gallagher, P. (1984a). Slow excitatory postsynaptic currents in bullfrog sympathetic neurons, *J. Physiol. (London)* **351**:583–593.

Akasu, T., Ohta, Y., and Koketsu, K. (1984b). Neuropeptides facilitate the desensitization of nicotinic acetylcholine receptors in frog skeletal muscle endplate, *Brain Res.* **290**:342–347.

Akopyan, A. R., Bocharova, L. S., and Chemeris, N. K. (1980a). Cyclic AMP-dependent dopamine responses in snail neurons, *Comp. Biochem. Physiol.* **67C**:211–214.

Akopyan, A. R., Chemeris, N. K., Iljin, V. I., and Veprintsev, B. N. (1980b). Serotonin, dopamine, and intracellular cyclic AMP inhibit the responses of nicotinic cholinergic membrane in snail neurons, *Brain Res.* **201**:480–484.

Akopyan, A. R., Veprintsev, B. N., Iljin, V. I., and Chemeris, N. K. (1981). Serotonin-induced biochemical modulation of the functional state of cholinoreceptors as a possible mechanism for synaptic plasticity, *Dokl. Akad. Nauk SSSR* **259**:753–756.

Akopyan, A. R., Iljin, V. I., and Chemeris, N. K. (1984). Modulation by dopamine of Ca-component of action potential in soma of *L. stagnalis* mollusc neurons, *Biofizika* **29**:284–288.

Akopyan, A. R., Chemeris, N. K., and Iljin, V. I. (1985). Neurotransmitter-induced modulation of neuronal Ca current is not mediated by intracellular Ca^{2+} or cAMP, *Brain Res.* **326**:145–148.

Albuquerque, E. X., Barnard, E. A., Chin, T. H., Lapa, A. J., Dolly, J. O., Jansson, S.-E., Daly, J., and Witkop, B. (1973). Acetylcholine receptor and ion conductance modulator sites at the murine neuromuscular junction: evidence for specific toxin reactions, *Proc. Natl. Acad. Sci. USA* **70**:949–953.

Albuquerque, E. X., Eldefrawi, A. T., Eldefrawi, M. E., Mansour, N. A., and Tsai, M. C. (1978). Amantadine: neuromuscular blockade by suppression of ionic conductance of the acetylcholine receptor, *Science* **199**:788–790.

Alger, B. E., and Nicoll, R. A. (1980). Epileptiform burst after hyperpolarization: calcium-dependent potassium potential in hippocampal CA1 pyramidal cells, *Science* **210**:1122–1124.

Allen, G., Trinnaman, B. J., and Green, N. M. (1980). The primary structure of the calcium ion-transporting adenosine triphosphatase protein of rabbit skeletal sarcoplasmic reticulum, *Biochem. J.* **187**:591–616.

Altman, J. (1969). Autoradiographic and histologic studies of post-natal neurogenesis, *J. Comp. Neurol.* **136**:269–294.

Ambache, N. (1949). The nicotinic action of substances supposed to be purely smooth-muscle stimulating. (B) Effect of $BaCl_2$ and pilocarpine on the superior cervical ganglion, *J. Physiol. (London)* **110**:164–172.

Anderson, C. R., and Stevens, C. F. (1973). Voltage clamp analysis of acetylcholine-produced end-plate current fluctuations at frog neuromuscular junction, *J. Physiol. (London)* **235**:655–691.

Anderson, S., Bankier, A. T., Barell, B. G., de Bruijn, S. M. H. L., Coulson, A. R., Drouin, J., Eperon, I. C., Nierlich, D. P., Roe, B. A., Sanger, F., Schreier, P. H., Smith, A. J. H., Staden, R., and Young, I. G. (1981). Sequence and organization of the human mitochondrial genome, *Nature* (*London*) **290**:457–465.

Andreev, A. A., Veprintsev, B. N., and Vulfius, C. A. (1984). Two-component desensitization of nicotinic receptors induced by acetylcholine agonists in *Lymnaea stagnalis* neurons, *J. Physiol.* (*London*) **353**:375–391.

Anichkov, S. V. (1974). *Selective Action of Mediator Drugs* [in Russian], Meditsina, Moscow.

Ariano, M. A. (1985). Muscarinic acetylcholine receptors are localized on striatal cyclic GMP-containing neurons, *Brain Res.* **334**:152–156.

Ariyoshi, M., Hasuo, H., Koketsu, K., Ohta, Y., and Tokimasa, T. (1985). Histamine is an antagonist of the acetylcholine receptor at the frog end-plate, *Br. J. Pharmacol.* **85**:65–73.

Aronstam, R. S., Hoss, W., and Abood, L. G. (1977). Conversion between configurational states of the muscarinic receptor in rat brain, *Eur. J. Pharmacol.* **46**:279–282.

Aronstam, R. S., Abood, L. G., and Hoss, W. (1978). Influence of sulfhydryl reagents and heavy metals on the functional state of the muscarinic acetylcholine receptor in rat brain, *Mol. Pharmacol.* **14**:575–586.

Ascher, P., and Chesnoy-Marchais, D. (1982). Interactions between three slow potassium responses controlled by three distinct receptors in *Aplysia* neurons, *J. Physiol.* (*London*) **324**:67–92.

Ascher, P., and Erulkar, S. (1983). Cholinergic chloride channels in snail neurons, in: *Single-Channel Recording* (E. Neher and B. Sakmann, eds.), Plenum Press, New York, pp. 401–407.

Ascher, P., Marty, A., and Neild, T. O. (1978a). Life time and elementary conductance of the channels mediating the excitatory effects of acetylcholine in *Aplysia* neurons, *J. Physiol.* (*London*) **278**:177–206.

Ascher, P., Marty, A., and Neild, T. O. (1978b). The mode of action of antagonists of the excitatory response to acetylcholine in *Aplysia* neurons, *J. Physiol.* (*London*) **278**:207–235.

Ascher, P., Large, W. A., and Rang, H. (1979). Studies on the mechanism of action of acetylcholine antagonists on rat parasympathetic ganglion cells, *J. Physiol.* (*London*) **295**:139–170.

Ashe, J. H., and Libet, B. (1981). Modulation of slow postsynaptic potentials by dopamine in rabbit sympathetic ganglion, *Brain Res.* **217**:93–106.

Ashe, J. H., and Libet, B. (1982). Pharmacological properties and monoaminergic mediation of the slow IPSP in mammalian sympathetic ganglion, *Brain Res.* **242**:345–349.

Ashe, J. H., and Libet, B. (1984). Effect of inhibitors of protein synthesis on dopamine modulation of the slow EPSP in rabbit superior cervical ganglion, *Brain Res.* **290**:170–173.

Ashe, J. H., and Yarosh, C. A. (1984). Differential and selective antagonism of the slow-inhibitory postsynaptic potential and slow-excitatory postsynaptic potential by gallamine and pirenzepine in the superior cervical ganglion of the rabbit, *Neuropharmacology* **11**:1321–1329.

Auerbach, A., Del Castillo, J., Specht, P. C., and Titmus, M. (1983). Correlation of agonist structure with acetylcholine receptor kinetics: studies on the frog end-plate and on chick embryo muscle, *J. Physiol.* (*London*) **343**:551–568.

Avissar, S., Amitai, G., and Sokolovsky, M. (1983). Oligomeric structure of muscarinic receptors is shown by photo affinity labeling: subunit assembly may explain high and low affinity agonist states, *Proc. Natl. Acad. Sci. USA* **80**:156–159.

Axelrod, J., and Hirata, F. (1981). Phospholipid methylation and receptor-mediated transmission of biological signals through membranes, in: *Drug Receptors and Their Effectors* (N. J. M. Birdsall, ed.), Macmillan, London, pp. 51–57.

Babmindra, V. P., and Dyachkova, L. N. (1968). On the structure of synapses in the extramural ganglia of autonomic nervous system (data obtained by light and electron microscopy), *Tsitologiya* **10**:413–421.

Baker, P. F., Hodgkin, A. L., and Ridgeway, E. B. (1971). Depolarization and calcium entry in squid giant axons, *J. Physiol.* (*London*) **218**:709–755.

Ballanyi, K., Grafe, P., Reddy, M. M., and ten Bruggencate, G. (1984). Different types of potassium transport linked to carbachol and GABA actions in rat sympathetic neurons, *Neuroscience* **12**:917–927.

Banks, B. E. C., Miledi, R., and Shipolini, R. A. (1974). The primary sequences and neuromuscular effects of three neurotoxin polypeptides from the venom of *Dendroaspis viridis*, *Eur. J. Biochem.* **45**:457-468.

Barker, J. L. (1975). CNS depressants: effects on post-synaptic pharmacology, *Brain Res.* **92**:35–55.

Barlow, R. B. (1955), *Introduction to Chemical Pharmacology*, Wiley, New York.

Barlow, R. B., and Ing, H. R. (1948). Curare-like action of polymethylene bis-quaternary ammonium salts, *Br. J. Pharmacol.* **3**:298–304.

Barlow, R. B., and Tubby, J. H. (1974). Actions of some esters of 3,3-dimethylbutan-1-ol (the carbone analog of choline) on the guinea-pig ileum, *Br. J. Pharmacol.* **51**:95–100.

Barlow, R. B., and Zoller, A. (1964). Some effects of long chain polymethylene bis-onium salts on junctional transmission in the peripheral nervous system, *Br. J. Pharmacol.* **23**:131–150.

Barlow, R. B., Berry, R. J., Glenton, P. A. M., Nikolaou, N. M., and Soh, R. S. (1976). A comparison of affinity constants for muscarine-sensitive acetylcholine receptors in guinea-pig atrial pacemaker cells at 29°C and in ileum at 29°C and 37°C, *Br. J. Pharmacol.* **58**:613–620.

Beam, K. G. (1976a). A voltage clamp study of the effects of two lidocaine derivatives on the time course of end-plate currents, *J. Physiol.* (*London*) **258**:279–300.

Beam, K. G. (1976b). A quantitative description of end-plate currents in the presence of two lidocaine derivatives, *J. Physiol.* (*London*) **258**:301–322.

Beckett, A. H., Harper, N. J., Clitherow, J. W., and Lesser, E. (1961). Muscarinic receptors, *Nature* (*London*) **189**:671–673.

Beckett, A. H., Harper, N. J., and Clitherow, J. W. (1963). The absolute configurations of the α- and β-methylcholine isomers and their acetyl and succinyl esters, *J. Pharm. Pharmacol.* **15**:349–361.

Beers, W. H., and Reich, E. (1970). Structure and activity of acetylcholine, *Nature* (London) **228**:917.

Beleslin, D. B., and Malobabic, Z. S. (1974). Cholinoceptive sites in the ganglia of myenteric plexus subserving the peristaltic reflex of guinea-pig isolated ileum, *Neuropharmacology* **13**:1091–1094.

Belluzzi, O., Sacchi, O., and Wanke, E. (1985). Identification of delayed potassium and calcium currents in the rat sympathetic neuron under voltage clamp, *J. Physiol. (London)* **358**:109–129.

Benardo, L., and Prince, D. A. (1981). Acetylcholine-induced modulation of hippocampal pyramidal neurons, *Brain Res.* **211**:227–234.

Benardo, L. S., and Prince, D. A. (1982a). Cholinergic excitation of mammalian hippocampal pyramidal cells, *Brain Res.* **249**:315–333.

Benardo, L. S., and Prince, D. A. (1982b). Ionic mechanisms of cholinergic excitation in mammalian hippocampal pyramidal cells, *Brain Res.* **249**:333–344.

Ben-Barak, J., and Dudai, Y. (1979). Cholinergic binding sites in rat hippocampal formation: properties and ontogenesis, *Brain Res.* **166**:245–257.

Benham, C. D., Bolton, T. B., and Lang, R. J. (1985). Acetylcholine activates an inward current in single mammalian smooth muscle cells, *Nature (London)* **316**:345–347.

Bernhard, S. A. (1968). *The Structure and Functions of Enzymes*, W. A. Benjamin, New York–Amsterdam.

Berrie, C. P., Birdsall, N. J. M., Burgen, A. S. V., and Hulme, E. C. (1979). Guanine nucleotides modulate muscarinic receptor binding in the heart, *Biochem. Biophys. Res. Commun.* **87**:1000–1005.

Bird, S. J., and Aghajanian, G. K. (1975). Denervation supersensitivity in the cholinergic septo-hippocampal pathway: a microiontophoretic study, *Brain Res.* **100**:355–370.

Bird, S. J., and Aghajanian, G. K. (1976). The cholinergic pharmacology of hippocampal pyramidal cells: a microiontophoretic study, *Neuropharmacology* **15**:273–282.

Birdsall, N. J. M., and Hulme, E. C. (1976). Biochemical studies on muscarinic acetylcholine receptors, *J. Neurochem.*, **27**:7–16.

Birdsall, N. J. M., Burgen, A. S. V., Hiley, C. R., and Hulme, E. C. (1976). Binding of agonists and antagonists to muscarinic receptors, *J. Supramol. Struct.* **4**:367–371.

Birdsall, N. J. M., Burgen, A. S. V., and Hulme, E. C. (1978). The binding of agonists to brain muscarinic receptors, *Mol. Pharmacol.* **14**:723–726.

Birdsall, N. J. M., Burgen, A. S. V., and Hulme, E. C. (1979). A study of the muscarinic receptor by gel electrophoresis, *Br. J. Pharmacol.* **66**:337–342.

Birdsall, N. J. M., Hulme, E. C., and Burgen, A. S. V. (1980). The character of the muscarinic receptor in different regions of the rat brain, *Proc. R. Soc. London*, **207**:1–12.

Birdsall, N. J. M., Hulme, E. C., and Stockton, J. M. (1983). Muscarinic receptor subclasses: allosteric interactions, in: Molecular Neurobiology, *Cold Spring Harbor Symp. Quant. Biol.* **48**:53–56.

Birks, R. I., and Fitch, S. J. G. (1974), Storage and release of acetylcholine in a sympathetic ganglion, *J. Physiol. (London)* **240**:125–134.

Biscoe, T. J., and Straughan, D. W. (1966). Micro-electrophoretic studies of neurons in cat hippocampus, *J. Physiol. (London)* **183**:341–359.

Bjorklund, A., Cegrell, L., Falck, B., Ritzen, M., and Rosengren, E. (1970). Dopamine-containing cells in sympathetic ganglia, *Acta Physiol. Scand.* **78**:334–338.

Blackman, S. J. (1970). Dependence on membrane potential of the blocking action of hexamethonium at a sympathetic ganglionic synapse, *Proc. Univ. Otago Med. Sch.* **48**:4–5.

Blackman, J. G., Ginsborg, B. L., and Ray, C. (1963). Synaptic transmission in the sympathetic ganglion of the frog, *J. Physiol.* (*London*) **167**:355–373.

Bobrishev, A. Yu., and Skok, V. I. (1987). Action of arfonad on excitatory postsynaptic current in sympathetic ganglion nerve cell, *Dokl. Akad. Nauk Ukr. SSR*. No. 5:60–61.

Bolton, T. B. (1979). Mechanisms of action of transmitters and other substances on smooth muscle, *Physiol. Rev.* **59**:606–718.

Bon, C., and Changeux, J.-P. (1977). Ceruleotoxin: a possible marker of the cholinergic ionophore, *Eur. J. Biochem.* **74**:43–51.

Boulter, J., Evans, K., Goldman, D., Martin, G., Treco, D., Heinemann, S., and Patrick, J. (1986). Isolation of a cDNA clone coding for a possible neural nicotinic acetylcholine receptor α-subunit, *Nature (London)* **319**:368–374.

Bowman, W. C., and Webb, S. N. (1972). Neuromuscular blocking and ganglion blocking activities of some acetylcholine antagonists in the cat, *J. Pharm. Pharmacol.* **24**:762–772.

Bregestovski, P. D. (1980). Analysis of membrane noise in molluscan nerve cells, in: *Physiology and Biochemistry of Mediator Processes* [in Russian], Moscow, pp. 30–31.

Bregestovski, P. D., and Iljin, V. I. (1980). Effect of calcium antagonist D-600 on the postsynaptic membrane, *J. Physiol.* (*Paris*) **76**:515–522.

Bregestovski, P. D., Miledi, R., and Parker, I. (1979). Calcium conductance of acetylcholine-induced end-plate channels, *Nature (London)* **279**:638–639.

Breitwieser, G. E., and Szabo, G. (1985). Uncoupling of cardiac muscarinic and β-adrenergic receptors from ion channels by a guanine nucleotide analog, *Nature (London)* **317**:538–540.

Briggs, C. A., Whiting, G. J., Ariano, M. A., and McAfee, D. A. (1982). Cyclic nucleotide metabolism in the sympathetic ganglion, *Cell. Molec. Neurobiol.* **2**:129–141.

Brisson, A., and Unwin, P. N. T. (1985). Quaternary structure of the acetylcholine receptor, *Nature (London)* **315**:474–477.

Brown, D. A. (1969). Responses of normal and denervated cat superior cervical ganglion to some stimulant compounds, *J. Physiol.* (*London*) **201**:225–236.

Brown, D. A. (1980). Locus and mechanism of action of ganglion-blocking agents, in: *Pharmacology of Ganglionic Transmission* (D. A. Kharkevich, ed.), Springer-Verlag, Berlin, pp. 185–235.

Brown, D. A. (1983). Slow cholinergic excitation – a mechanism for increasing neuronal excitability, *Trends Neurosci.* **6**:302–307.

Brown, D. A. (1986). Acetylcholine and brain cells, *Nature* (*London*) **319**:358–359.

Brown, D. A., and Adams, P. R. (1980). Muscarinic suppression of a novel voltage-sensitive K^+-current in a vertebrate neuron, *Nature* (*London*) **283**:673–676.

Brown, D. A., and Caulfield, M. P. (1979). Hyperpolarizing "α_2"-adrenoceptors in rat sympathetic ganglia, *Br. J. Pharmacol.* **65**:435–445.

Brown, D. A., and Constanti, A. (1980). Intracellular observations on the effects of muscarinic agonists on rat sympathetic neurons, *Br. J. Pharmacol.* **71**:362–364.

Brown, D. A., and Fumagalli, L. (1977). Dissociation of α-bungarotoxin binding and receptor block in the rat superior cervical ganglion, *Brain Res.* **129**:165–168.

Brown, D. A., and Griffith, W. H. (1983). Calcium-activated outward current in voltage-clamped hippocampal neurons of the guinea-pig, *J. Physiol.* (*London*) **337**:287–302.

Brown, D. A., and Kwiatkowski, D. (1976). A note on the effects of dithiothreitol (DTT) in the depolarization of isolated sympathetic ganglia by carbachol and bromoacetylcholine, *Br. J. Pharmacol.* **56**:128–130.

Brown, D. A., and Scholfield, C. N. (1974). Changes of intracellular sodium and potassium ion concentrations in isolated rat superior cervical ganglia induced by depolarizing agents, *J. Physiol.* (*London*) **242**:307–319.

Brown, D. A., and Selyanko, A. A. (1985a). Two components of muscarine-sensitive membrane current in rat sympathetic neuron, *J. Physiol.* (*London*)) **358**:335–363.

Brown, D. A., and Selyanko, A. A. (1985b). Membrane currents underlying the cholinergic slow excitatory postsynaptic potential in the rat sympathetic ganglion, *J. Physiol.* (*London*) **365**:365–387.

Brown, D. A., Garthwaite, J., Hayashi, E., and Yamada, S. (1976). Action of surugatoxin on nicotinic receptors in the superior cervical ganglion of the rat, *Br. J. Pharmacol.* **58**:157–159.

Brown, D. A., Fatherazi, S., Garthwaite, J., and White, R. D. (1980a). Muscarinic receptors in rat sympathetic ganglia, *Br. J. Pharmacol.* **70**:577–592.

Brown, D. A., Forward, A., and Marsh, S. (1980b). Antagonist discrimination between ganglionic and ileal muscarinic receptors, *Br. J. Pharmacol.* **71**:362–364.

Brown, D. A., Gähwiler, B. H., Marsh, S. J., and Selyanko, A. A. (1986). Mechanisms of muscarinic excitatory synaptic transmission in ganglia and brain, in: *Subtypes of Muscarinic Receptors II*, TIPS Suppl., pp. 66–71.

Buckley, N. J., and Burnstock, G. (1984). Distribution of muscarinic receptors on cultured myenteric neurons, *Brain Res.* **310**:133–137.

Buckley, N. J., and Burnstock, G. (1986). Localization of muscarinic receptors on cultured myenteric neurons: a combined autoradiographic and immunocytochemical approach, *J. Neurosci.* **6**:531–540.

Bülbring, E. (1944). The action of adrenaline on transmission in the superior cervical ganglion, *J. Physiol.* (*London*) **103**:55–67.

Bülbring, E., and Depierre, F. (1949). The action of synthetic curarizing compounds on skeletal muscle and sympathetic ganglia both normal and denervated, *Br. J. Pharmacol.* **4**:22–32.

Bulygin, I. A. (1964). *Closing Function and Receptor Function of Autonomic Ganglia* [in Russian], Nauka i Tekhnika, Minsk.

Bulygin, I. A. (1983). A consideration of the general principles of organization of sympathetic ganglia, *J. Auton. Nerv. Syst.* **8**:303–330.

Burden, S. J., Hartzell, H. C., and Yoshikami, D. (1975). Acetylcholine receptors at neuromuscular synapses: phylogenetic differences detected by snake α-neurotoxins, *Proc. Natl. Acad. Sci. USA* **72**:3245–3249.

Burgen, A. S. V. (1965). The role of ionic interaction at the muscarinic receptor, *Br. J. Pharmacol.* **25**:4–17.

Burnstock, G. (1981). Neurotransmitters and trophic factors in the autonomic nervous system, *J. Physiol. (London)* **313**:1–35.

Burnstock, G., and Hoyle, C. H. V. (1985). Actions of adenine dinucleotides in the guinea-pig taenia coli: NAD acts indirectly on P_1-purinoceptors, NADP acts like a P_2-purinoceptor agonist, *Br. J. Pharmacol.* **84**:825–831.

Bursztajn, S., and Gerson, D. D. (1977). Discrimination between nicotinic receptors in vertebrate ganglia and skeletal muscle by α-bungarotoxin and cobra venoms, *J. Physiol. (London)* **269**:17–31.

Busis, N. A., Weight, F. F., and Smith, P. A. (1978). Synaptic potentials in sympathetic ganglia: are they mediated by cyclic nucleotides? *Science* **200**:1079–1081.

Capaldi, R. A. (1982). Structure of intrinsic membrane proteins, *TIBS* **7**:292–295.

Carbonetto, S. T., Fambrough, D. M., and Muller, K. J. (1978). Nonequivalence of α-bungarotoxin receptors and acetylcholine receptors in chick sympathetic neurons, *Proc. Natl. Acad. Sci. USA* **75**:1018–1020.

Carpenter, D. O., Greene, L. A., Shain, W., and Vogel, Z. (1976). Effects of eserine and neostigmine on the interaction of α-bungarotoxin with *Aplysia* acetylcholine receptors, *Mol. Pharmacol.* **12**:999–1006.

Carroll, R. C., Eldefrawi, M. E., and Edelstein, S. J. (1973). Studies on the structure of the acetylcholine receptor from *Torpedo marmorata*, *Biochem. Biophys. Res. Commun.* **55**:864–872.

Case, R., Creese, R., Dixon, W. J., Massey, F. J., and Taylor, D. B. (1977). Movement of labeled decamethonium in muscle fibers of the rat, *J. Physiol. (London)* **272**:283–294.

Cassell, J. F., Clark, A. L., and McLachlan, E. M. (1986). Characteristics of phasic and tonic sympathetic ganglion cells of the guinea-pig, *J. Physiol. (London)* **372**:457–483.

Cedar, H., and Schwartz, J. (1972). Cyclic AMP in the nervous system of *Aplysia californica*. Effect of serotonin and dopamine, *J. Gen. Physiol.* **60**:570–587.

Chad, J. E., and Kerkut, G. A. (1981). Summation of membrane current conductance responses as evidence for independent ionophore populations, *Comp. Biochem. Physiol.* **69C**:61–65.

Chang, C. C., and Lee, C. Y. (1963). Isolation of neurotoxins from the venom of *Bungarus multicinctus* and their modes of neuromuscular blocking action, *Arch. Int. Pharmacodyn.* **144**:241–257.

Chang, H. W., and Bock, E. (1977). Molecular forms of acetylcholine receptor. Effects of calcium ions and a sulfhydryl reagent on the occurrence of oligomers, *Biochemistry* **16**:4513.

Chang, H. W., and Neumann, E. (1976). Dynamic properties of isolated acetylcholine receptor proteins: release of calcium ions caused by acetylcholine binding, *Proc. Natl. Acad. Sci. USA* **74**:3364–3368.

Changeux, J. P., Kasai, M., and Lee, C. Y. (1970). The use of a snake venom toxin to characterize the cholinergic receptor protein, *Proc. Natl. Acad. Sci. USA* **67**:1241–1247.

Changeux, J. P., Benedetti, L., Bourgeois, J. P., Brission, A., Cartaud, J., Devaux, P., Grünhagen, H., Moreau, M. Popot, J. L., Sobel, A., and Weber, M. (1976). Some structural properties of the cholinergic receptor protein in its membrane environment relevant to its function as a pharmacological receptor, *Cold Spring Harbor Symp. Quant. Biol.* **40**:211–230.

Changeux, J. P., Bon, F., Cartaud, J., Devillers-Thiery, A., Giraudat, J., Heidmann, T., Holton, B., Nghiem, H.-O., Popot, J. L., Van Rapenbush, R., and Tzartos, S. (1983). Allosteric properties of the acetylcholine receptor protein from *Torpedo marmorata*, *Cold Spring Harbor Symp. Quant. Biol.* **48**:35–52.

Changeux, J. P., Devillers-Thiery, A., and Chemouilli, P. (1984). Acetylcholine receptor: an allosteric protein, *Science* **225**:1335–1345.

Chao, Y., Vandlen, R. L., and Raftery, M. A. (1975). Preferential chemical modification of a binding subsite on the acetylcholine receptor, *Biochem. Biophys. Res. Commun.* **63**:300–307.

Chassaing, C., Dureng, G., Baissat, J., and Duchene-Marullaz, P. (1984). Pharmacological evidence for cardiac muscarinic receptor subtypes, *Life Sci.* **35**:1739–1747.

Chemeris, N. K., Kazachenko, V. N., Kislov, A. N., and Kurchikov, A. L. (1982). Inhibition of acetylcholine responses by intracellular calcium in *Limnaea stagnalis* neurons, *J. Physiol. (London)* **323**:1–19.

Chen, R., Krämer, C., Schmidmayer, W., and Henning, U. (1979). Primary structure of major outer membrane protein I of *Escherichia coli* B/r, *Proc. Natl. Acad. Sci. USA* **76**:5014–5017.

Cherubini, E., and North, R. A. (1984). Inhibition of calcium spikes and transmitter release by γ-aminobutyric acid in the guinea-pig myenteric plexus, *Br. J. Pharmacol.* **82**:101–105.

Chesnoy-Marchais, D., and Ascher, P. (1983). Effects of various cations on the slow K^+ conductance increases induced by carbachol, histamine, and dopamine in *Aplysia* neurons, *Brain Res.* **259**:57–67.

Chesnut, T. J. (1983). Two-component densitization of the neuromuscular junction of the frog, *J. Physiol (London)* **336**:229–241.

Chiappinelli, V. A., (1983). ϰ-Bungarotoxin: a probe for the neuronal nicotinic receptor in the avian ciliary ganglion, *Brain Res.* **277**:9–21.

Chiappinelli, V. A., and Giacobini, E. (1978). Time course of appearance of α-binding sites during development of chick ciliary ganglion and iris, *Neurochem. Res.* **3**:465–478.

Chiappinelli, V. A., and Zigmond, R. E. (1978), α-Bungarotoxin blocks nicotinic transmission in the avian ciliary ganglion, *Proc. Natl. Acad. Sci. USA* **75**:2999–3003.

Chiappinelli, V. A., Cohen, J. B., and Zigmond, R. E. (1981). The effects of α- and β-neurotoxins from the venoms of various snakes on transmission in autonomic ganglia, *Brain Res.* **211**:107–126.

Chizmadjev, Yu. A., Khodorov, B. I., and Aityan, S. H. (1973). Theory of ionic transport through sodium channel in biological membranes, *Dokl. Akad. Nauk SSSR* **213**:722–725.

Chothia, C. (1970). Interaction of acetylcholine with different cholinergic nerve receptors, *Nature* (*London*) **225**:36–38.
Chothia, C., and Pauling, P. (1970). The conformation of cholinergic molecules of nicotinic nerve receptors, *Proc. Natl. Acad. Sci. USA* **65**:477–482.
Chou, T. C., and de Elio, F. J. (1948). The anticurare activity of eserine on the superior cervical ganglion of cat, *Br. J. Pharmacol.* **3**:113–115.
Chou, T. C., and Lee, C. Y. (1969). Effect of whole and fractionated cobra venom on sympathetic ganglionic transmission, *Eur. J. Pharmacol.* **8**:326–330.
Christ, D. D., and Nishi, S. (1971). Site of adrenaline blockade in the superior cervical ganglion of the rabbit, *J. Physiol.* (*London*) **213**:107–117.
Clapham, D. E., and Neher, E. (1984). Substance P reduces acetylcholine-induced currents in isolated bovine chromaffin cells, *J. Physiol.* (*London*) **347**:255–277.
Clark, R. B., Donaldson, P. L., Gration, K. A. F., Lambert, J. J., Pick, T., Ramsey, R., Spanjer, W., and Usherwood, P. N. R. (1982). Block of locust muscle glutamate receptors by δ-philanthotoxin occurs after receptor activations, *Brain Res.* **241**:105–114.
Clarke, P. B. S., Pert, C. B., and Pert, A. (1984). Autoradiographic distribution of nicotine receptors in rat brain, *Brain Res.* **323**:390–395.
Clarke, P. B. S., Schwartz, R. D., Paul, S. M., Pert, C. B., and Pert, A. (1985). Nicotinic binding in rat brain: autoradiographic comparison of ^{3}H-acetylcholine, ^{3}H-nicotine, and ^{125}I-α-bungarotoxin, *J. Neurosci.* **5**:1307–1315.
Claudio, T., Ballivet, M., Patrick, J., and Heinemann, S. (1983). Nucleotide and deduced amino acid sequence of *Torpedo californica* acetylcholine receptor α–subunit, *Proc. Natl. Acad. Sci. USA* **80**:1111–1115.
Cocchia, D., and Fumagalli, L. (1981). Immunocytochemical localization of α-bungarotoxin receptors in the chick ciliary ganglion: synaptic and extrasynaptic sites? *Neurochem. Int.* **3**:123–128.
Cohen, J. B., and Changeux, J. P. (1973). Interaction of a fluorescent ligand with membrane bound cholinergic receptor from *Torpedo marmorata*, *Biochemistry* **12**:4855–4864.
Cole, A. E., and Nicoll, R. A. (1983). Acetylcholine mediates a slow synaptic potential in hippocampal pyramidal cells, *Science* **221**:1299–1301.
Cole, A. E., and Nicoll, R. A. (1984a). Characterization of a slow cholinergic postsynaptic potential recorded in vitro from rat hippocampal pyramidal cells, *J. Physiol.* (*London*) **352**:173–188.
Cole, A. E., and Nicoll, R. A. (1984b). The pharmacology of cholinergic excitatory responses in hippocampal pyramidal cells, *Brain Res.* **305**:283–290.
Cole, A. E., and Shinnick-Gallagher, P. (1981). Comparison of the receptors mediating the catecholamine hyperpolarization and slow inhibitory postsynaptic potential in sympathetic ganglia, *J. Pharmacol. Exp. Ther.* **217**:440–444.
Cole, A. E., and Shinnick-Gallagher, P. (1984). Muscarinic inhibitory transmission in mammalian sympathetic ganglia mediated by increased potassium conductance, *Nature* (*London*) **307**:270–271.
Colquhoun, D. (1979). The link between drug binding and response: theories and observations, in: *The Receptors: A Comprehensive Treatise* (R. D. O'Brien, ed.), Plenum Press, New York, Vol. 1:93–142.

Colquhoun, D., and Hawkes, A. G. (1977). Relaxation and fluctuations of membrane currents that flow through drug-operated ion channels, *Proc. R. Soc. London* **199**:231–262.

Colquhoun, D., and Hawkes, A. G. (1983). The principles of the stochastic interpretation of ion-channel mechanisms, in: *Single-Channel Recording* (E. Neher and B. Sakmann, eds.), Plenum Press, New York, pp. 135–174.

Colquhoun, D., and Ogden, D. C. (1984). Evidence from single-channel recording of channel block by nicotinic agonists at the frog neuromuscular junction, *J. Physiol. (London)* **353**:90.

Colquhoun, D., and Sakmann, B. (1983). Bursts of openings in transmitter-activated ion channels, in: *Single-Channel Recording* (E. Neher and B. Sakmann, eds.), Plenum Press, New York, pp. 345–364.

Colquhoun, D., and Sakmann, B. (1985). Fast events in single-channel currents activated by acetylcholine and its analogs at the frog muscle end-plate, *J. Physiol. (London)* **369**:501–557.

Colquhoun, D., and Sheridan, R. E. (1982). The effect of tubocurarine competition on the kinetics of agonist action on the nicotinic receptor, *Br. J. Pharmacol.* **75**:77–86.

Colquhoun, D., Dreyer, F., and Sheridan, R. E. (1979). The actions of tubocurarine at the frog neuromuscular junction, *J. Physiol. (London)* **293**:247–284.

Connor, E. A., Levy, S. M., and Parsons, R. L. (1983). Kinetic analysis of atropine-induced alterations in bullfrog ganglionic fast synaptic current, *J. Physiol. (London)* **337**:137–158.

Constanti, A., and Brown, D. A. (1981). M-current in voltage-clamped mammalian sympathetic neurons, *Neurosci. Lett.* **24**:289–294.

Constanti, A., and Galvan, M. (1983). M-current in voltage-clamped olfactory cortex neurons, *Neurosci. Lett.* **39**:65–70.

Constanti, A., and Sim, J. A. (1985). A slow muscarine-sensitive Ca-dependent K current in guinea-pig olfactory cortex neurons in vitro, *J. Physiol. (London)* **365**:47P.

Constanti, A., and Sim, J. A. (1987). Muscarinic receptors mediating suppression of the M-current in guinea-pig olfactory cortex neurons may be of the M_2-subtype, *Br. J. Pharmacol.* **90**:3–6.

Constanti, A., Adams, P. R., and Brown, D. A. (1981). Why do barium ions imitate acetylcholine? *Brain Res.* **206**:244–250.

Conti-Tronconi, B., Gotti, C., Paggi, P., and Rossi, A. (1979). Acetylcholine receptors in the ciliary ganglion and in the iris muscle of the chick: specific binding and effect on the synaptic transmission of the neurotoxin from *Naja naja siamensis*, *Br. J. Pharmacol.* **66**:33–38.

Courtney, K. R. (1975). Mechanism of frequency-dependent inhibition of sodium currents in frog myelinated nerve by the lidocaine derivative GEA 968, *J. Pharmacol. Exp. Ther.* **195**:225–236.

Coyle, J. T. (1974). Biochemical aspects of catecholaminergic neurons in the brain of fetal and neonatal rat, in: *Degeneration and Growth in Neurons* (K. Fuxe, L. Olson, and Y. Zotterman, eds.), Pergamon Press, New York, pp. 425–434.

Coyle, J. T., and Yamamura, H. I. (1976). Neurochemical aspects of the ontogenesis of cholinergic neurons in the rat brain, *Brain Res.* **118**:429–440.

Creese, R., Franklin, G. I., and Mitchell, L. D. (1977). Sodium entry in rat diaphragm induced by depolarizing drugs, *J. Physiol.* (*London*) **272**:295–316.

Criado, M., Hochschwender, S., Sarin, V., Fox, J. L., and Lindstrom, J. (1985a). Evidence for unpredicted transmembrane domains in acetylcholine receptor subunits, *Proc. Natl. Acad. Sci. USA* **82**:2004–2008.

Criago, M., Sarin, V., Fox, L., and Lindstrom J. (1985b). Structural localization of the sequence α 235–242 of the nicotinic acetylcholine receptor, *Biochem. Biophys. Res. Commun.* **128**:864–871.

Crouch, M. F., Morris, C. P., Cheah, T. B., and Rush, R. A. (1982). Developmental loss of a smooth muscle muscarinic receptor population, *Brain Res.* **232**:212–215.

Cull-Candy, S. G., and Miledi, R. (1982). Properties of miniature excitatory junctional currents at the locust nerve-muscle junction, *J. Physiol.* (*London*) **326**:527–551.

Cull-Candy, S. G., Miledi, R., and Parker, L. (1981). Single glutamate-activated channels recorded from locust muscle fibers with perfused patch-clamp electrodes, *J. Physiol.* (*London*) **321**:195–210.

Curtis, D. R., and Ryall, R. W. (1966). The excitation of Renshaw cells of cholinomimetics, *Exp. Brain Res.* **2**:49–65.

Damle, V. N., and Karlin, A. (1978). Affinity labeling of one of two α-neurotoxin binding sites in acetylcholine receptor from *Torpedo californica*, *Biochemistry* **17**:2039–2045.

Daniels, M. P., and Vogal, Z. (1980). Localization of α-bungarotoxin binding sites in synapses of the developing chick retina, *Brain Res.* **201**:45–56.

Dascal, N., Landau, E. M., and Lass, Y. (1984). *Xenopus* oocyte resting potential, muscarinic responses, and the role of calcium and guanosine 3',5'-cyclic monophosphate, *J. Physiol.* (*London*) **352**:551–574.

De Groat, W. C., Kawatani, M., and Rutigliano, M. (1985). Ganglionic depolarization and enhancement of muscarinic ganglionic excitatory mechanisms in the cat by vasoactive intestinal polypeptide, *J. Physiol.* (*London*) **364**:41P.

Del Castillo, J., and Katz, B. (1957). Interaction at end-plate receptors between different choline derivatives, *Proc. R. Soc. London* **146**:369–381.

Dennis, M. J., and Sargent, P. B. (1979). Loss of extrasynaptic acetylcholine sensitivity upon reinnervation of parasympathetic ganglion cells, *J. Physiol.* (*London*) **289**:263–275.

Dennis, M. J., Harris, A. J., and Kuffler, S. W. (1971). Synaptic transmission and its duplication by focally applied acetylcholine in parasympathetic neurons in the heart of the frog, *Proc. R. Soc. London* **177**:509–539.

Derkach, V. A. (1986). Relaxations of acetylcholine current in sympathetic ganglion neurons, *Dokl. Akad. Nauk Ukr. SSR*, No. 4:60–62.

Derkach, V. A., Selyanko, A. A., and Skok, V. I. (1983). Acetylcholine-induced current fluctuations and fast excitatory postsynaptic currents in rabbit sympathetic neurons, *J. Physiol.* (*London*) **336**:511–526.

Derkach, V. A., Selyanko, A. A., and Skok, V. I. (1984). Localization of sites for interaction with bis-quaternary ammonium compounds in cholinoreceptors, *Dokl. Akad. Nauk Ukr. SSR*, No. 10, 70–72.

Derkach, V. A., North, R. A., Selyanko, A. A., and Skok, V. I. (1985). Activity of single ionic channels activated by acetylcholine in the sympathetic neuron, *Dokl. Akad. Nauk SSSR* **284**:495–498.

Derkach, V. A., North, R. A., Selyanko, A. A., and Skok, V. I. (1987). Single channels activated by acetylcholine in rat superior cervical ganglion, *J. Physiol. (London)* **388**:141–151.

Devillers-Thiery, A., Giraudat, S., Bintaboulet, M., and Changeux, J.-P. (1983). Complete mRNA coding sequence of the acetylcholine binding α subunit of *Torpedo marmorata* acetylcholine receptor: a model for the transmembrane organization of the polypeptide chain, *Proc. Natl. Acad. Sci. USA* **80**:2067–2071.

Dingledine, R., Dodd, J., and Kelly, J. S. (1977). ACh-evoked excitation of cortical neuron, *J. Physiol. (London)* **273**:79–80P.

Dionne, V. E. (1981). The kinetics of slow muscle acetylcholine-operated channels in the garter snake, *J. Physiol. (London)* **310**:159–190.

Dionne, V. E., and Stevens, C. F. (1975). Voltage dependence of agonist effectiveness of the frog neuromuscular junctions: resolution of paradox, *J. Physiol. (London)* **251**:345–270.

Dneprovski, A. S., and Temnikova, T. I. (1979). *Theoretical Basis of Organic Chemistry* [in Russian], Khimiya, Leningrad.

Dodd, J., and Horn, J. (1983). Muscarinic inhibition of sympathetic C neurons in the bullfrog, *J. Physiol. (London)* **334**:271–291.

Dodd, J., Dingledine, R., and Kelly, J. S. (1981). The excitatory action of acetylcholine on hippocampal neurons of the guinea pig and rat maintained in vitro, *Brain Res.* **207**:109–127.

Dreyer, F., and Peper, K. (1974). The acetylcholine sensitivity in the vicinity of the neuromuscular junction of the frog, *Pflügers Arch.* **348**:273–286.

Dreyer, F., Müller, K. D., Peper, K., and Sterz, R. (1976). The M. omohyoideus of the mouse as a convenient mammalian muscle preparation, *Pflügers Arch.* **367**:115–122.

Duggan, A. W., Hall, J. G., and Lee, C. Y. (1976). α-Bungarotoxin, cobra neurotoxin, and excitation of Renshaw cells by acetylcholine, *Brain Res.* **107**:166–170.

Dun, N. J. (1980). Ganglionic transmission: electrophysiology and pharmacology, *Fed. Proc.* **39**:2982–2989.

Dun, N. J., and Karczmar, A. G. (1978). Involvement of an interneuron in the generation of the slow inhibitory postsynaptic potential in mammalian sympathetic ganglia, *Proc. Natl. Acad. Sci. USA* **75**:4029–4032.

Dun, N. J., and Karczmar, A. G. (1980). Blockade of ACh potentials by α-bungarotoxin in rat superior cervical ganglion cells, *Brain Res.* **196**:536–540.

Dun, N. J., and Karczmar, G. (1977). The presynaptic site of action of norepinephrine in the superior cervical ganglion of guinea pig, *J. Pharmacol. Exp. Ther.* **200**:328–335.

Dun, N. J., , Nishi, S., and Karczmar, A. G. (1976). Alterations in nicotinic responses of rabbit superior cervical ganglion cells after chronic preganglionic denervation, *Neuropharmacology* **15**:211–218.

Dun, N. J., Kaibara, K., and Karczmar, A. G. (1977). Dopamine and adenosine 3',5'-monophosphate responses of single mammalian sympathetic neurons, *Science* **197**:778–780.

Dun, N. J., Kaibara, K., and Karczmar, A. G. (1978). Muscarinic and cGMP-induced membrane potential changes: differences in electrogenic mechanisms, *Brain Res.* **150**:658–661.

Dunant, Y., and Dolivo, M. (1967). Rélations entre les potentiels synaptiques lents et l'excitabilité du ganglion sympathique chez le rat, *J. Physiol.* (*Paris*) **59**:281–294.

Dunlap, K., and Fischbach, G. D. (1978). Neurotransmitters decrease the calcium component of sensory neuron action potentials, *Nature* (*London*) **276**:837–839.

Dunlap, K., and Fischbach, G. D. (1981). Neurotransmitters decrease the calcium conductance activated by depolarization of embryonic chick sensory neurons, *J. Physiol.* (*London*) **317**:519–535.

Dunn, P. M., and Marshall, L. M. (1985). Lack of nicotinic supersensitivity in frog sympathetic neurons following denervation, *J. Physiol.* (*London*) **365**:211–225.

Dwyer, T. M., Adams, D. J., and Hille, B. (1980). The permeability of the endplate channel to organic cations in frog muscle, *J. Gen. Physiol.* **75**:469–492.

Dyadyusha, G. G., Kornilov, M. Yu., Zamkovoy, V. I., Dehtyar, M. L., and Skok, V. I. (1985). Conformational analysis of bis-quaternary ammonium compounds which are able to block cholinoreceptor selectively, *Dokl. Akad. Nauk Ukr. SSR*, No. 2, 64–67.

Eccles, R. M. (1952). Responses of isolated curarized sympathetic ganglia, *J. Physiol.* (*London*) **117**:196–217.

Eccles, R. M., and Libet, B. (1961). Origin and blockade of the synaptic responses of curarized sympathetic ganglia, *J. Physiol.* (*London*) **157**:484–503.

Egan, T. M., and North, R. A. (1985a). Acetylcholine acts on m_2-muscarinic receptors to excite rat locus coeruleus neurons, *Br. J. Pharmacol.* **85**:733–735.

Egan, T. M., and North, R. A. (1985b). Muscarinic receptor mediating potassium conductance increase of parabrachial neurons has a low affinity for pirenzepine, in: *Subtypes of Muscarinic Receptors*, Second Intern. Symposium, Boston University, p. 24.

Egan, T. M., and North, R. A. (1986). Acetylcholine hyperpolarizes central neurons by acting on an m_2-muscarinic receptor, *Nature* (*London*) **319**:405–407.

Ehlert, F. J., Roeske, W. R., and Yamamura, H. J. (1981). Muscarinic receptor: regulation by guanine nucleotides, ions, and N-ethylmaleimide, *Fed. Proc.* **40**:153–159.

Eldefrawi, M. E. (1974). Neuromuscular transmission – the transmitter-receptor combination, in: *The Peripheral Nervous System* (J. Hubbard, ed.), Plenum Press, New York–London, pp. 181–200.

Eldefrawi, M. E., and Eldefrawi, A. T. (1973). Cooperativities in the binding of acetylcholine to its receptor, *Biochem. Pharmacol.* **22**:3145–3150.

Eldefrawi, M. E., Eldefrawi, A. T., and O'Brien, R. D. (1971). Binding sites for cholinergic ligands in a particulate fraction of *Electrophorus* electroplax, *Proc. Natl. Acad. Sci. USA* **68**:1047–1050.

Eldefrawi, M. E., Eldefrawi, A. T., Penfield, L. A., O'Brien, R. D., and Campen, E. (1975). Binding of calcium and zinc to the acetylcholine receptor purified from *Torpedo californica*, *Life Sci.* **16**:925–936.

Ellenbroeck, J. W., Nivard, R. J. F., Van Rossum, J. M., and Ariens, E. J. (1965). Absolute configuration and parasympathetic action: Pharmacodynamics of enantiomorphic and diastereoisomeric esters of betamethylcholine, *J. Pharm. Pharmacol.* **17**:393–404.

Engelman, D. M., Menderson, R., McLachlan, A. D., and Wallace, B. A. (1980). Path of the polypeptide in bacteriorhodopsin, *Proc. Natl. Acad. Sci. USA* **77**:2023–2027.

Eusebi, F., Pasetto, N., and Siracusa, G. (1984). Acetylcholine receptors in human oocytes, *J. Physiol.* (*London*) **346**:321–330.

Eyzaguirre, C., and Monti-Bloch, L. (1982). Nicotinic and muscarinic reactive sites in mammalian glomus cells, *Brain Res.* **252**:181–184.

Fairclough, R. H., Finer-Moore, J., Love, R. A., Kristofferson, K., Desmeules, P. J., and Stround, R. M. (1983). Subunit organization and structure of an acetylcholine receptor, in: Molecular Neurobiology, *Cold Spring Harbor Symp. Quant. Biol.* **48**:9–20.

Fakstorp, J. T., Pedersen, J. G. A., Poulsen, E., and Schilling, M. (1957). Ganglionic blocking activity of bis-trialkylammoniumethyl ether salts and their branched homologs, *Acta Pharmacol. Toxicol. (Kbh)* **13**:52–58.

Fambrough, D. E. (1979). Control of acetylcholine receptors in skeletal muscle, *Physiol. Rev.* **59**:165–227.

Farley, S. M., Yeh, J. Z., Watanabe, S., and Narahashi, T. (1981). End-plate channel block by guanidine derivatives, *J. Gen. Physiol.* **77**:273–293.

Fedorov, A. N., Kurski, M. D., and Skok, V. I. (1985). Effect of bis-quaternary ammonium compounds on Mg^{2+}, Ca^{2+}-ATPase activity and transport of Ca^{2+} in sarcoplasmic reticulum of skeletal muscle, *Dokl. Akad. Nauk Ukr. SSR*, No. 9:78–80.

Fehlhammer, A., Bode, W., and Huber, R. (1977). Crystal structure of bovine trypsinogen at 1.8 Å resolution, *J. Mol. Biol.* **111**:415–438.

Feltz, A., and Trautmann, A. (1982). Desensitization at the frog neuromuscular junction: a biphasic process, *J. Physiol.* (*London*) **322**:257–272.

Fertuck, H. C., and Salpeter, M. M. (1976). Quantitation of junctional and extrajunctional acetylcholine receptors by electron microscope autoradiography after ^{125}I–α-bungarotoxin binding at mouse neuromuscular junctions, *J. Cell. Biol.* **69**:144–158.

Fex, J., and Adams, J. C. (1978). α-Bungarotoxin blocks reversibly cholinergic inhibition in the cochlea, *Brain Res.* **159**:440–444.

Fiekers, J. F., Spannbauer, P. M., Scubon-Mulieri, B., and Parsons, R. L. (1980). Voltage dependence of desensitization. Influence of calcium and activation kinetics, *J. Gen. Physiol.* **75**:511–529.

Fields, S., Winter, G., and Brownlee, G. G. (1981), Structure of the neuraminidase gene in human influenza virus A/Pr/8/34, *Nature* (*London*) **290**:213–217.

Finer-Moore, J., and Stroud, R. M. (1984). Amphipathic analysis and possible formation of the ion channel in an acetylcholine receptor, *Proc. Natl. Acad. Sci. USA* **81**:155–159.

Finger, W. (1983). Effects of glycine on the crayfish neuromuscular junction. I. Glycine-operated inhibitory postsynaptic channels and a glycine-effected decrease in membrane conductance, *Pflügers Arch.* **397**:121–127.

Finkel, A. S. (1983). A cholinergic chloride conductance in neurons of *Helix aspersa*, *J. Physiol. (London)* **344**:119–135.

Fischmeister, R., and Hartzell, H. C. (1986). Mechanism of action of acetylcholine on calcium current in single cells from frog ventricle, *J. Physiol. (London)* **376**:183–202.

Flood, J. F., Landry, D. W., and Jarvik, M. E. (1981). Cholinergic receptor interactions and their effects on long-term memory processing, *Brain Res.* **215**:177–185.

Forsbracy, P., and Johnson, E. S. (1980). Release modulating acetylcholine receptors on cholinergic neurons of the guinea-pig ileum, *Br. J. Pharmacol.* **68**:289–300.

Freeman, J. A. (1977). Possible regulatory function of acetylcholine receptor in maintenance of retinotectal synapses, *Nature (London)* **269**:218–222.

Freeman, J. A., Schmidt, J. T., and Oswald, R. E. (1980). Effect of α-bungarotoxin in retino-tectal synaptic transmission in goldfish and toad, *Neuroscience* **5**:929–942.

Freschi, J. E. (1983). Membrane currents of cultured rat sympathetic neurons under voltage clamp, *J. Neurophysiol.* **50**:1460–1478.

Freschi, J. E., and Shain, W. G. (1980). Slow muscarinic depolarization in neurons of dissociated rat superior cervical ganglia can be evoked by iontophoresis of acetylcholine, *Brain Res.* **185**:429–434.

Fuder, H., Rink, D., and Muscholl, E. (1982). Sympathetic nerve stimulation on the perfused rat heart. Affinities of N-methylatropine and pirenzepine at pre- and postsynaptic muscarine receptors, *Naunyn-Schmiedeberg's Arch. Pharmacol.* **318**:210–219.

Fumagalli, L., and De Renzis, G. (1980). α-Bungarotoxin binding sites in the rat superior cervical ganglion are influenced by postganglionic axotomy, *Neuroscience* **5**:611–616.

Fumagalli, L., De Renzis, G., and Miani, N. (1976). Acetylcholine receptors: number and distribution in intact and deafferented superior cervical ganglion of the rat, *J. Neurochem.* **27**:47–52.

Gabella, G. (1976). *Structure of the Autonomic Nervous System*, Chapman and Hall, London.

Gage, P. W. (1976). Generation of end-plate potentials, *Physiol. Rev.*, **56**:177-247.

Gage, P. W., and Armstrong, C. M. (1968). Miniature end-plate currents in voltage-clamped muscle fibers, *Nature (London)* **218**:363–365.

Gage, P. W., and Hamill, O. P. (1975). General anesthetics: synaptic depression consistent with increased membrane fluidity, *Neurosci. Lett.* **1**:61–65.

Gage, P. W., and Van Helden, D. (1979). Effects of permeant monovalent cations on end-plate channels, *J. Physiol. (London)* **288**:509–528.

Gage, P. W., Hamill, O. P., and Wachtel, R. E. (1983). Sites of action of procaine at the motor end-plate, *J. Physiol. (London)* **335**:123–137.

Gahwiler, B. H., and Brown, D. A. (1985). Functional innervation of cultured hippocampal neurons by cholinergic afferents from co-cultured septal explants, *Nature* (*London*) **313**:577–579.

Gallagher, J. P., and Shinnick-Gallagher, P. (1977). Cyclic nucleotides injected intracellularly into rat superior cervical ganglion cells, *Science* **189**:851–853.

Gallagher, J. P., Griffith, W. H., and Shinnick-Gallagher, P. (1982). Cholinergic transmission in cat parasympathetic ganglia, *J. Physiol.* (*London*) **332**:473–486.

Galvan, M., and Adams, P. R. (1982). Control of calcium current in rat sympathetic neurons by norepinephrine, *Brain Res.* **244**:135–144.

Galvan, M., Satin, L. S., and Adams, P. R. (1984). Comparison of conventional microelectrode and whole-cell patch clamp recordings from cultured bullfrog ganglion cells, *Soc. Neurosci. Abstr.* **10**:146.

Gangitano, C., Fumagalli, L., De Renzis, G., and Sangiacomo, C. O. (1978). α-Bungarotoxin-acetylcholine receptors in the chick ciliary ganglion during development, *Neuroscience* **3**:1101–1108.

Gardner, D., and Stevens, C. F. (1980). Rate-limiting step of inhibitory postsynaptic current decay in *Aplysia* buccal ganglia, *J. Physiol.* (*London*) **304**:145–164.

George, W. J., Wilkerson, D. R., and Kadowitz, P. J. (1973). Influence of acetylcholine on contractile force and nucleotide levels in the isolated perfused rat heart, *J. Pharmacol. Exp. Ther.* **184**:228–235.

Ger, B. A., Zeimal, E. V., Michelson, M. J., Sokolov, G. P., Protas, L. L., and Shelkovnikov, S. A. (1976). Comparative data on stereoselectivity of cholinoreceptors, *Zh. Évol. Biokhim. Fiziol.* **12**:473–475.

Ger, B. A., Zeimal, E. V., and Katchman, A. N. (1979a). Revealing of different types of cholinoreceptors on the membrane of isolated *Planorbarius corneus* neuron, *Zh. Évol. Biokhim. Fiziol.* **15**:520–526.

Ger, B. A., Katchman, A. N., and Zeimal, E. V. (1979b). The slow potassium-dependent acetylcholine current in isolated neurons: its time course and temperature dependence, *Brain Res.* **171**:355–359.

Ger, B. A., Zeimal, E. V., and Katchman, A. N. (1980). Ionic mechanisms of fast (nicotinic) phase of acetylcholine response in identified gastropoda neurons, *Neirofiziologiya* **12**:533–540.

Gerschenfeld, H. M. (1973). Chemical transmission in invertebrate central nervous systems and neuromuscular junctions, *Physiol. Rev.* **53**:1–119.

Gerschenfeld, H. M., and Paupardin-Tritsch, D. (1974). Ionic mechanisms and receptor properties underlying the response of molluscan neurons to 5-hydroxytryptamine, *J. Physiol.* (*London*) **243**:427–456.

Gerschenfeld, H. M., and Piccolino, M. (1977). Muscarinic antagonists block cone to horizontal cell transmission in turtle retina, *Nature* (*London*) **268**:257–259.

Gill, E. W. (1959). Interquaternary distance and ganglion-blocking activity in bisquaternary compounds, *Proc. R. Soc. London* **150**:381–402.

Gillo, B., and Lass, G. (1984). The mechanism of steroid anesthetic (alphaxalone) block of acetylcholine-induced ionic currents, *Br. J. Pharmacol.* **82**:783–789.

Ginsborg, B. L. (1965). The actions of McN-A-343, pilocarpine and acetyl-β-methylcholine on sympathetic ganglion cells of the frog, *J. Pharmacol. Exp. Ther.* **150**:216–219.

Ginsborg, B. L. (1971). On the presynaptic acetylcholine receptors in sympathetic ganglia of the frog, *J. Physiol.* (*London*) **216**:237–246.

Glitsch, H. G., and Pott, L. (1978). Effects of acetylcholine and parasympathetic nerve stimulation on membrane potential in quiescent guinea-pig atria, *J. Physiol.* (*London*) **279**:655–668.

Gloge, H., Lüllman, H., and Mutschler, E. (1966). The action of tertiary and quaternary arecaidine and dihydroarecaidine esters on the guinea-pig isolated ileum, *Br. J. Pharmacol.* **27**:185–195.

Gmiro, V. E., Lukomskaya, N. Y., and Serdyuk, S. E. (1984). Problem of selectivity in action of drugs blocking transmission through sympathetic and parasympathetic ganglia, in: *Physiology of Transmitters, The Peripheral Synapse* [in Russian], Kazan, pp. 71–73.

Gmiro, V. E., Groisman, S. D., Lukomskaya, N. Y., Melnichenko, L. V., Serdyuk, S. E., and Skok, V. I. (1987). Selective blockers of parasympathetic ganglia, *Dokl. Akad. Nauk SSSR* **292**:497–501.

Godovnikov, N. N., Dorofeyeva, N. A., Kabachnik, M. I., Kardanov, N. A., Trifonova, S. A., and Shelkovnikov, S. A. (1985). The difference in muscarinolytic activities of antagonists relative to muscarinic cholinoreceptors of various organs of the same animal species, *Dokl. Akad. Nauk SSSR* **284**:503–505.

Goldman, D. E. (1943). Potential, impedance, and rectification in membranes, *J. Gen. Physiol.* **27**:37–60.

Golovanov, I. B., Ivkov, V. G., and Gagloyev, V. N. (1966). Notes on electron structure of the ether part of the acetylcholine molecule, *Biofizika* **11**:1071–1076.

Gotti, C., Omini, C., Berti, F., and Clementi, F. (1985a). Isolation of a polypeptide from the venom of *Bungarus multicinctus* that binds to ganglia and blocks the ganglionic transmission in mammals, *Neuroscience* **15**:563–575.

Gotti, C., Spagnoli, D., Omini, C., and Clementi, F. (1985b). α-Bungarotoxin and P15 toxin binding site in mammalian sympathetic ganglia, *Neuroscience Lett.* **57**:227–231.

Goyal, R. K., and Rattan, S. (1978). Neurohumoral, hormonal and drug receptors for the lower esophageal sphincter, *Gastroenterology* **74**:598–618.

Grafe, P., Mayer, C. J., and Wood, J. D. (1980). Synaptic modulation of calcium-dependent potassium conductance in myenteric neurons in the guinea-pig, *J. Physiol.* (*London*) **305**:235–248.

Gregor, I., and Tsugita, A. (1982). The amino acid sequence of cytochrome C oxidase subunit VI from *Saccharomyces cerevisiae*, *J. Biol. Chem.* **257**:13081–13083.

Greene, L. A., Sytokowski, A. J., Vogel, Z., and Nirenberg, M. (1973). Bungarotoxin used as a probe for acetylcholine receptors in cultured neurons, *Nature* (*London*) **243**:163–166.

Greengard, P., and Kebabian, J. W. (1974). Role of cyclic AMP in synaptic transmission in the mammalian peripheral nervous system, *Fed. Proc.* **33**:1059–1067.

Griffith, W. H., and Gallagher, J. P. (1980). An intracellular investigation of the cat vesicle pelvic ganglia, *J. Neurophysiol.* **43**:343–354.

Griffith, W. H., III, Gallagher, J. P., and Shinnick-Gallagher, P. (1981). Sucrose-gap recordings of nerve-evoked potentials in mammalian parasympathetic ganglia, *Brain Res.* **209**:446–451.

Grob, D. (1967). Neuromuscular blocking drugs, in: *Physiological Pharmacology* (W. Root and F. G. Hofman, eds.), Academic Press, New York, pp. 389–460.

Gromov, L. A., Kryvorotov, S. V., and Skryma, R. N. (1981). Effect of morphine on dialyzed snail neurons, *Neirofiziologiya* **13**:332–333.

Gullick, W. J., and Lindstrom, J. M. (1982). Structural similarities between acetylcholine receptors from fish electric organs and mammalian muscle, *Biochemistry* **21**:4563–4569.

Gurney, A. M., and Rang, H. P. (1984). The channel-blocking action of methonium compounds on rat submandibular ganglion cells, *Br. J. Pharmacol.* **82**:623–642.

Gustafsson, B., and Wigstrom, H. (1981). Evidence for two types of afterhyperpolarization in CA1 pyramidal cells in the hippocampus, *Brain Res.* **206**:462–468.

Guy, H. R. (1984). A structural model of the acetylcholine receptor channel based on partition energy and helix packing calculations, *Biophys. J.* **45**:249–261.

Guyton, A. C., and Reeder, R. C. (1950). Quantitative studies on the autonomic actions of curare, *J. Pharmacol. Exp. Ther.* **98**:188–194.

Gyermek, L., and Unna, K. R. (1958). Relation of structure of synthetic muscarines and muscarones to their pharmacological action, *Proc. Soc. Exp. Biol. Med.* **98**:882–885.

Haas, H. L., and Konnerth, A. (1983). Histamine and noradrenalin decrease calcium-activated potassium conductance in hippocampal pyramidal cells, *Nature (London)* **302**:432–434.

Haefely, W. E. (1980). Nonnicotinic chemical stimulation of autonomic ganglia, in: *Pharmacology of Ganglionic Transmission* (D. A. Kharkevich, ed.), Springer-Verlag, Berlin, pp. 313–357.

Halliwell, J. V., and Adams, P. R. (1982). Voltage-clamp analysis of muscarinic excitation in hippocampal neurons, *Brain Res.* **250**:71–92.

Hamill, O. P., Marty, A., Neher, E., Sakmann, B., and Sigworth, F. J. (1981). Improved patch-clamp techniques for high resolution current recording from cells and cell-free membrane patches, *Pflügers Arch.* **391**:85–100.

Hamilton, S. L., McLaughlin, M., and Karlin, A. (1979). Formation of disulfide-linked oligomers of acetylcholine receptor in membrane from *Torpedo* electric tissue, *Biochemistry* **18**:155–163.

Hamilton, S. L., Pratt, D. R., and Eaton, D. C. (1985). Arrangement of the subunits of the nicotinic acetylcholine receptor of *Torpedo californica* as determined by α-neurotoxin cross-linking, *Biochemistry* **24**:2210–2219.

Hammer, R., Berrie, C. P., Birdsall, N. J. M., Burgen, A. S. V., and Hulme, E. C. (1980). Pirenzepine distinguishes between different subclasses of muscarinic receptors, *Nature (London)* **283**:90–92.

Hancock, J. C., and Volle, R. L. (1970). Cholinoceptive sites in the rat superior cervical ganglion, *Arch. Int. Pharmacodyn.* **184**:111–120.

Hanin, J., Jenden, D. J., and Cho, A. K. (1966). The influence of pH on the muscarinic action of oxotremorine, arecoline, pilocarpine, and their quaternary ammonium analogs, *Mol. Pharmacol.* **2**:352–359.

Harris, A. J., Kuffler, S. W., and Dennis, M. J. (1971). Differential chemosensitivity of synaptic and extrasynaptic areas on the neuronal surface membrane in parasympathetic neurons of the frog tested by micro-application of acetylcholine, *Proc. R. Soc. London* **177**:541–553.

Hartzell, H. C. (1980). Distribution of muscarinic acetylcholine receptors and presynaptic nerve terminals in amphibian heart, *J. Cell Biol.* **86**:6–20.

Hartzell, H. C. (1981). Mechanisms of slow postsynaptic potentials, *Nature (London)* **291**:539–544.

Hartzell, H. C., Kuffler, S. W., Stickgold, R., and Yoshikami, D. (1977). Synaptic excitation and inhibition resulting from direct action of acetylcholine on two types of chemoreceptors on individual amphibian para-sympathetic neurons, *J. Physiol. (London)* **271**:817–846.

Hashiguchi, T., Ushiyama, N. S., Kobayashi, H., and Libet, B. (1978). Does cyclic GMP mediate the slow excitatory synaptic potential in sympathetic ganglia? *Nature (London)* **271**:267–268.

Hashiguchi, T., Kobayashi, H., Tosaka, T., and Libet, B. (1982). Two muscarinic depolarizing mechanisms in mammalian sympathetic neurons, *Brain Res.* **242**:378–382.

Hayashi, E., and Yamada, S. (1975). Pharmacological studies on surugatoxin, the toxic principle from Japanese ivory mollusc (*Babylonia japonica*), *Br. J. Pharmacol.* **53**:207–215.

Head, S. D. (1983). Temperature and end-plate currents in rat diaphragm, *J. Physiol. (London)* **334**:441–459.

Hedlund, B., and Bartfai, T. (1979). The importance of thiol- and disulfide groups in agonist and antagonist binding to the muscarinic receptor, *Mol. Pharmacol.* **15**:531–544.

Heilbronn, E., and Bartfai, T. (1978). Muscarinic acetylcholine receptor, *Prog. Neurobiol.* **11**:171–188.

Herrling, P. L. (1981). The membrane potential of cat hippocampal neurons recorded in vivo displays four different reaction-mechanisms to iontophoretically applied transmitter agonists, *Brain Res.* **212**:331–343.

Heschler, I., Kameyama, M., and Trautwein, W. (1986). On the mechanism of muscarinic inhibition of the cardiac Ca current, *Pflügers Arch.* **407**:182–189.

Hey, P. (1952). On relationships between structure and nicotine-like stimulant activity in choline esters and ethers, *Br. J. Pharmacol.* **7**:117–129.

Higashi, H., and Nishi, S. (1982). 5-Hydroxytryptamine receptors of visceral primary afferent neurons on rabbit nodose ganglia, *J. Physiol. (London)* **323**:543–567.

Hille, B. (1970). Ionic channels in nerve membranes, *Prog. Biophys. Mol. Biol.* **21**:3–32.

Hille, B. (1984). *Ionic Channels in Excitable Membranes*, Sinauer Associates, Sunderland, Massachusetts.

Hille, B., Sanchez, J., Siemen, D., and Dani, J. (1983). Opening of acetylcholine receptor channels without acetylcholine, Proc. Int. Union Physiol. Sci. 29th Congr., Sydney, Australia, **15**:413.

Hirst, G. D. S., and McLachlan, E. M. (1984). Post-natal development of ganglia in the lower lumbar sympathetic chain of the rat, *J. Physiol.* (*London*) **349**:119–134.

Hirst, G. D. S., and Silinsky, E. M. (1975). Some effects of 5-hydroxytryptamine, dopamine, and noradrenalin on neurons in the submucous plexus of guinea-pig small intestine, *J. Physiol.* (*London*) **251**:817–832.

Hirst, G. D. S., Johnson, S. M., and van Helden, D. F. (1985a). The calcium current in a myenteric neuron of the guinea-pig ileum, *J. Physiol.* (*London*) **361**:297–314.

Hirst, G. D. S., Johnson, S. M., and van Helden, D. F. (1985b). The slow calcium-dependent potassium current in a myenteric neuron of the guinea-pig ileum, *J. Physiol.* (*London*) **361**:315–337.

Hodgkin, A. L., and Huxley, A. F. (1952). A quantitative description of membrane current and its application to conduction and excitation in nerve. *J. Physiol.* (*London*) **117**:500–544.

Hodgkin, A. L., and Katz, B. (1949). The effect of sodium ions on the electrical activity of the giant axon of the squid, *J. Physiol.* (*London*) **108**:37–77.

Holton, P., and Ing, H. R. (1959). The specificity of the trimethylammonium group in acetylcholine, *Br. J. Pharmacol.* **4**:190–196.

Holz, R. W., and Senter, R. A. (1981). Choline stimulates nicotinic receptors on adrenal medullary chromaffin cells to induce catecholamine secretion, *Science* **214**:466–468.

Holtz, G. G., IV, Rane, S. G., and Dunlap, K. (1986). GTP-binding proteins mediate transmitter inhibition of voltage-dependent calcium channels, *Nature* (*London*) **319**:670–672.

Horn, J. P., and McAfee, D. A. (1980). Alpha-adrenergic inhibition of calcium-dependent potentials in rat sympathetic neurons, *J. Physiol.* (*London*) **301**:191–204.

Hotson, J. R., and Prince, D. A. (1980). A calcium-activated hyperpolarization follows repetitive firing in hippocampal neurons, *J. Neurophysiol.* **43**:409–419.

Hotson, J. R., Prince, D. A., and Swartzkroin, P. A. (1979). Anomalous inward rectification in hippocampal neurons, *J. Neurophysiol.* **42**:889–895.

Hounsgaard, J. (1978). Presynaptic inhibitory action of acetylcholine in area CA1 of the hippocampus, *Exp. Neurol.* **62**:787-797.

Huang, L. M., Catterall, W. A., and Ehrenstein, G. (1978). Selectivity of cations and nonelectrolytes for acetylcholine-activated channels in cultured muscle cells, *J. Gen. Physiol.* **71**:397–410.

Hucho, F. (1986). The nicotinic acetylcholine receptor ion channel, *Eur. J. Biochem.* **158**:211–226.

Hulme, E., and Birdsall, N. (1986). Distinctions in acetylcholine receptor activity, *Nature* (*London*) **323**:396–397.

Hulme, E. C., Birdsall, N. J. M., Burgen, A. S. V., and Mehta, P. (1978). The binding of antagonists to brain muscarinic receptors, *Mol. Pharmacol.* **14**:737–750.

Hunt, S., and Schmidt, J. (1979). The relationship of α-bungarotoxin binding activity and cholinergic termination within the rat hippocampus, *Neuroscience* **4**:585–592.

Ing, H. R. (1949). The structure–action relationships of the choline group, *Science* **109**:264–266.

Ing, H. R. (1957). Structure–action relationship of hypotensive drugs, in: *Hypotensive Drugs* (M. Harington, ed.), Pergamon Press, London–New York, pp. 7–22.

Ito, H., and Nishi, K. (1982). Effects of organic Ca^{2+}-antagonists on sympathetic ganglion, *Jpn. J. Pharmacol.* **32**, Suppl.:48.

Ivanov, A. Ya., and Melnichenko, L. V. (1971). Action of d-tubocurarine on synaptic transmission in cat ciliary ganglion, *Fiziol. Zh.*, No. 1:94–95.

Ivanov, A. Ya., and Skok, V. I. (1971). On the origin of slow positive potential in cat superior sympathetic ganglion, *Neirofiziologiya* **3**:76–83.

Ivanov, A. Ya., and Skok, V. I. (1972). Slow synaptic process in sympathetic ganglion, in: *Interneuronal Transmission in the Autonomic Nervous System* (P. G. Kostyuk, ed.) [in Russian], Naukova Dumka, Kiev, pp. 136–145.

Ivanov, A. Ya., and Skok, V. I. (1980). Slow inhibitory postsynaptic potentials and hyperpolarization evoked by noradrenalin in the neurons of mammalian sympathetic ganglion, *J. Auton. Nerv. Syst.* **1**:255–268.

Ivanov, A. Ya., and Skok, V. I. (1981). Properties of slow excitatory postsynaptic potential in mammalian sympathetic ganglion neurons, *Neirofiziologiya* **13**:371–378.

Jacob, M. H., and Berg, D. K. (1983). The ultrastructural localization of α-bungarotoxin binding sites in relation to synapses on chick ciliary ganglion neurons, *J. Neurosci.* **3**:260–271.

Jan, L. N., and Jan, Y. N. (1982). Peptidergic transmission in sympathetic ganglia of the frog, *J. Physiol.* (*London*) **327**:219–246.

Jan, Y. N., Jan, L. Y., and Kuffler, S. W. (1979). A peptide as possible transmitter in sympathetic ganglia of the frog, *Proc. Natl. Acad. Sci. USA* **76**:1501–1505.

Jan, L. Y., Jan, Y. N., and Brownfield, M. S. (1980a). Peptidergic transmitters in synaptic boutons of sympathetic ganglia, *Nature* (*London*) **288**:380–382.

Jan, Y. N., Jan, L. Y., and Kuffler, S. W. (1980b). Further evidence for peptidergic transmission in sympathetic ganglia, *Proc. Natl. Acad. Sci. USA* **77**:5008–5012.

Jänig, W., Krauspe, R., and Wiedersatz, G. (1983). Reflex activation of postganglionic vasoconstrictor neurons supplying skeletal muscle by stimulation of arterial chemoreceptors via nonnicotinic synaptic mechanisms in sympathetic ganglia, *Pflügers Arch.* **396**:95–100.

Jaup, B. H., Stockbrügger, R., and Dotevall, G. (1979). The effect of pirenzepine on basal, pentagastrine- and insulin-stimulated gastric acid secretion in patients with peptic ulcer disease, *Scand. J. Gastroenterol.* **14**:621–624.

Johnson, S. M., Katayama, Y., and North, R. A. (1980). Slow synaptic potentials in neurons of the myenteric plexus, *J. Physiol.* (*London*) **301**:505–516.

Jones, S. W. (1984). Muscarinic and peptidergic actions on C cells of bullfrog sympathetic ganglia, *Soc. Neurosci. Abstr.* **10**:207.

Jones, S. W. (1985). Muscarinic and peptidergic excitation of bullfrog sympathetic neurons, *J. Physiol.* (*London*) **366**:63–87.

Kaivarainen, A. I. (1980). *Dynamic Behavior of Proteins in Aqueous Medium and Their Functions* [Russian translation], Nauka, Leningrad.

Kandel, E. R. (1976). *Cellular Basis of Behavior*, W. H. Freeman and Company, San Francisco.

Kao, P., Dwork, A., Kaldany, R., Silver, M., Wideman, J., Stein, S., and Karlin, A. (1984). Identification of α subunit half-cystine specifically labeled by an affinity reagent for the acetylcholine receptor binding site, *J. Biol. Chem.* **259**:1662–1665.

Karlin, A. (1969). Chemical modification of the active site of the acetylcholine receptor, *J. Gen. Physiol.* **54**:245–254.

Karlin, A. (1973). Molecular interactions of the acetylcholine receptor, *Fed. Proc.* **32**:1847–1853.

Karlin, A., and Winnik, M. (1968). Reduction and specific alkylation of the receptor for acetylcholine, *Proc. Natl. Acad. Sci. USA* **60**:668–671.

Karlin, A., Weill, M., McNamee, M. G., and Valderrama, R. (1976). Facets of the structures of acetylcholine receptors from *Electrophorus* and *Torpedo*, *Cold Spring Harbor Symp. Quant. Biol.* **40**:203–210.

Karlin, A., Cox, R., Kaldany, R. R., Lobel, P., and Holtzman, E. (1983). The arrangement and functions of the chains of the acetylcholine receptor of *Torpedo* electric tissue, *Cold Spring Harbor Symp. Quant. Biol.* **48**:1–8.

Katchman, A. N., and Zeimal, E. V. (1982). Ionic mechanisms of the rapid (nicotinic) phase of acetylcholine response in identified *Planorbarius corneus* neurons, *Brain Res.* **241**:95–103.

Katchman, A. N., Ger, B. A., and Zeimal, E. V. (1980). The slow phase of the acetylcholine response in isolated neurons of the gastropod mollusc *Planorbarius corneus*, *Gen. Pharmacol.* **11**:55–64.

Kato, E., and Kuba, K. (1980). Inhibition of transmitter release in bullfrog sympathetic ganglia induced by γ-aminobutyric acid, *J. Physiol. (London)* **298**:271–283.

Kato, E., Kuba, K., and Koketsu, K. (1978). Effects of erabutoxins on neuromuscular transmission in frog skeletal muscles, *J. Pharmacol. Exp. Ther.* **204**:446–453.

Kato, E., Kuba, K., and Koketsu, K. (1980). Effects of erabutoxins on the cholinergic receptors of bullfrog sympathetic ganglion cells, *Brain Res.* **191**:294–298.

Katz, B., and Miledi, R. (1964). Further observations on the distribution of acetylcholine-reactive sites in skeletal muscle, *J. Physiol. (London)* **170**:379–388.

Katz, B., and Miledi, R. (1972). The statistical nature of the acetylcholine potential and its molecular components, *J. Physiol. (London)* **224**:665–699.

Kebabian, J. W., Steiner, A. L., and Greengard, P. (1975). Muscarinic cholinergic regulation of cyclic guanosine-3',5'-monophosphate in autonomic ganglia: possible role in synaptic transmission, *J. Pharmacol. Exp. Ther.* **193**:474–488.

Kehoe, J. S. (1972). Three acetylcholine receptors in *Aplysia* neurons, *J. Physiol. (London)* **225**:115–146.

Kehoe, J. S., and Marty, A. (1980). Certain slow synaptic responses: their properties and possible underlying mechanisms, *Annu. Rev. Biophys. Bioenerg.* **9**:437–465.

Kehoe, J., Sealock, R., and Bon, C. (1976). Effects of α-toxins from *Bungarus multicinctus* and *Bungarus caeruleus* on cholinergic responses in *Aplysia* neurons, *Brain Res.* **107**:527–540.

Kemp, G., and Morley, B. J. (1986). Ganglionic nAChRs and high-affinity nicotinic binding sites are not equivalent, *FEBS Lett.* **205**:265–268.

Kerkut, G. A., and Walker, R. J. (1967). Action of acetylcholine, dopamine, and 5-hydroxytryptamine on the spontaneous activity of the cells of Retzius of the leech *Hirudo medicinalis*, *Br. J. Pharmacol. Chemother.* **30**:644–654.

Kharkevich, D. A. (1967). *Ganglion-Blocking and Ganglion-Stimulating Agents*, Pergamon Press, Oxford.

Kharkevich, D. A., Skoldinov, A. P., Samoilov, D. N., and Shorr, V. A. (1980). On the possibility of selective drug action on M-cholinoreceptors of definite localization, *Farmakol. Toksikol.* **63**:645–652.

Kilbinger, H. (1977). Modulation by oxotremorine and atropine of acetylcholine release evoked by electrical stimulation of the myenteric plexus of guinea-pig ileum, *Naunyn-Schmiedeberg's Arch. Pharmacol.* **300**:145–151.

Kilbinger, H., and Wagner, P. (1975). Inhibition by oxotremorine of acetylcholine resting release from guinea-pig ileum longitudinal muscle strips, *Naunyn-Schmiedeberg's Arch. Pharmacol.* **287**:47–60.

Kilbinger, H., and Wessler, I. (1980). Inhibition by acetylcholine of the stimulation-evoked release of [^{3}H] acetylcholine from the guinea-pig myenteric plexus, *Neuroscience* **5**:1331–1340.

Kilbinger, H., Halim, S., Lambrecht, G., Weiler, W., and Wessler, I. (1984). Comparison of affinities of muscarinic antagonists to pre- and post-junctional receptors to guinea-pig ileum, *Eur. J. Pharmacol.* **103**:313–320.

Kistler, J., Stround, R. M., Klymkowsky, M. W., Ladanette, R. A., and Flairclough, R. H. (1982). Structure and function of an acetylcholine receptor, *Biophys. J.* **37**:371–383.

Klymkowsky, M. W., and Stroud, R. M. (1979). Immunospecific identification and three-dimensional structure of a membrane-bound acetylcholine receptor from *Torpedo californica*, *J. Mol. Biol.* **128**:319–334.

Kobayashi, H., and Libet, B. (1968). Generation of slow postsynaptic potentials without increase in ionic conductance, *Proc. Natl. Acad. Sci. USA* **60**:1304–1311.

Kobayashi, H., and Libet, B. (1970). Actions of noradrenalin and acetylcholine on sympathetic ganglion cells, *J. Physiol.* (*London*) **208**:353–372.

Kobayashi, H., and Libet, B. (1974). Is inactivation of potassium conductance involved in slow postsynaptic excitation of sympathetic ganglion cells? Effects of nicotine, *Life Sci.* **14**:1871–1883.

Kobayashi, H., and Tosaka, T. (1983). Slow synaptic actions in mammalian sympathetic ganglia, with special reference to the possible roles played by cyclic nucleotides, in: *Autonomic Ganglia* (L. G. Elfin, ed.), Wiley, New York, pp. 281–307.

Kobayashi, H., Ushiyama, N. S., and Hashiguchi, T. (1977). Intracellular study of the effects of cyclic nucleotides in the mammalian sympathetic ganglion cells, in: *Proc. 27th Int. Congr. Physiol. Sci.*, Paris, Vol. 13, p. 394.

Kobayashi, H., Hashiguchi, T., and Ushiyama, N. S. (1978). Postsynaptic modulation by cyclic AMP, intra- or extracellularly applied, or by stimulation of preganglionic nerve, in mammalian sympathetic ganglion cells, *Nature* (*London*) **271**:268–270.

Koketsu, K. (1969). Cholinergic synaptic potentials and underlying ionic mechanisms, *Fed. Proc.* **28**:101–112.

Koketsu, K. (1984). Modulation of receptor sensitivity and action potentials by transmitters in vertebrate neurons, *Jpn. J. Physiol.* **34**:945–960.

Koketsu, K., and Nakamura, M. (1976). The electrogenesis of adrenaline-hyperpolarization of sympathetic ganglion cells in bullfrogs, *Jpn. J. Physiol.* **26**:63–77.

Koketsu, K., and Yamada, M. (1982). Presynaptic muscarinic receptors inhibiting active acetylcholine release in the bullfrog sympathetic ganglion, *Br. J. Pharmacol.* **77**:75–82.

Koketsu, K., and Yamamoto, K. (1975). Unusual cholinergic response of bullfrog sympathetic ganglion cells, *Eur. J. Pharmacol.* **31**:281–286.

Koketsu, K., Nishi, S., and Soeda, H. (1968). Calcium and acetylcholine-potential of bullfrog sympathetic ganglion cell membrane, *Life Sci.* **7**:955–963.

Koketsu, K., Akasu, T., Miyagawa, M., and Hirai, R. (1982a). Modulation of nicotinic transmission by biogenic amines in bullfrog sympathetic ganglia, *J. Auton. Nerv. Syst.* **6**:47–53.

Koketsu, K., Akasu, T., Miyagawa, M., and Hirai, R. (1982b). Biogenic antagonists of the nicotinic receptor: their interactions with erabutoxin, *Brain Res.* **250**:391–393.

Koketsu, K., Miyagawa, M., and Akasu, T. (1982c). Catecholamine modulates nicotinic ACh-receptor sensitivity, *Brain Res.* **236**:487–491.

Komalahiranyan, A., and Volle, R. L. (1963). Alterations of transmission in sympathetic ganglia treated with a sulfhydryl-group inhibitor, N-ethylmaleimide (NEM), *J. Pharmacol. Exp. Ther.* **139**:304–311.

Kononenko, N. I., and Scherbatko, A. P. (1985). Effect of serotonin on inward calcium current in snail neurons, *Dokl. Akad. Nauk SSSR* **281**:1494–1497.

Kordas, M. (1969). The effect of membrane polarization on the time course of the end-plate current in frog sartorius muscle, *J. Physiol. (London)* **204**:493–502.

Kordas, M. (1970). The effect of procaine on neuromuscular transmission, *J. Physiol. (London)* **209**:689–699.

Kordas, M. (1972a). An attempt at an analysis of the factors determining the time course of the end-plate current. I. The effects of prostigmine and the ratio of Mg^{2+} to Ca^{2+}, *J. Physiol. (London)* **224**:317–322.

Kordas, M. (1972b). An attempt at an analysis of the factors determining the time course of the end-plate current. II. Temperature, *J. Physiol. (London)* **224**:333–348.

Kosower, E. M. (1983). A molecular model for the exobilayer portion of the α subunit of the acetylcholine receptor with binding sites for acetylcholine and noncompetitive antagonists, *Biochem. Biophys. Res. Commun.* **116**:17–22.

Kostyuk, P. G. (1984). Ionic channels in the membrane of the nerve cell and their metabolic control, *Usp. Fiz. Nauk* **15**:7–22.

Kostyuk, P. G., and Doroshenko, P. A. (1983). Calcium binding and calcium channels, in: *Calcium-Binding Proteins* (B. De Bernard, Y. L. Sottocasa, Y. Sandri, et al., eds.), Elsevier, Amsterdam–New York–Oxford, pp. 331–338.

Kostyuk, P. G., and Krishtal, O. A. (1981). *Mechanisms of Electrical Excitability of the Nerve Cell* [in Russian], Nauka, Moscow.

Kostyuk, P. G., Krishtal, O. A.,, and Pidoplichko, V. I. (1975). Intracellular dialysis of nerve cells: effect of intracellular fluoride and phosphate on the inward current, *Nature* (*London*) **257**:691–693.

Kouvelas, E. D., Dichter, M. A., and Greene, L. A. (1978). Chick sympathetic neurons develop receptors for α-bungarotoxin in vitro but the toxin does not block nicotinic receptors, *Brain Res.* **154**:83–93.

Koval, L. M., Derkach, V. A., Selyanko, A. A., Skok, V. I., and Ivanov, A. Ya. (1982). Distribution of muscarinic receptors in mammalian sympathetic ganglion: autoradiographic and electrophysiological studies, *J. Auton. Nerv. Syst.* **6**:37–46.

Kretsinger, R. H. (1976a). Calcium-binding proteins, *Annu. Rev. Biochem.* **45**:239–266.

Kretsinger, R. H. (1976b). Calcium in biological systems, *Coord. Chem. Rev.* **18**:29–124.

Krishtal, O. A., and Pidoplichko, V. I. (1975). Intracellular perfusion of giant *Helix* neurons, *Neirofiziologiya* **7**:327–329.

Kristic, M. K. (1972). The action of choline on the superior cervical ganglion of the cat, *Eur. J. Pharmacol.* **17**:87–96.

Krnjevic, K. (1978). Acetylcholine and cyclic GMP, in: *Cholinergic Mechanisms and Psychopharmacology*, New York–London, pp. 261–266.

Krnjevic, K., and Phillis, J. W. (1963). Pharmacological properties of acetylcholine-sensitive cells in the cerebral cortex, *J. Physiol.* (*London*) **166**:328–350.

Krnjevic, K., and Van Meter, W. G. (1976). Cyclic nucleotides in spinal cells, *Can. J. Physiol. Pharmacol.* **54**:416–421.

Krnjevic, K., Pumain, R., and Renaud, L. (1971a). Effects of Ba^{2+} and tetraethylammonium on cortical neurons, *J. Physiol.* (*London*) **215**:233–245.

Krnjevic, K., Pumain, R., and Renaud, L. (1971b). The mechanism of excitation by acetylcholine in the cerebral cortex, *J. Physiol.* (*London*) **215**:247–268.

Krnjevic, K., Puil, E., and Werman, R. (1976). Is cyclic guanosine monophosphate the internal "second messenger" for cholinergic actions on central neurons? *Can. J. Physiol. Pharmacol.* **54**:172–176.

Kuba, K., and Koketsu, K. (1974). Ionic mechanism of the slow excitatory postsynaptic potential in bullfrog ganglion cells, *Brain Res.* **81**:338–342.

Kuba, K., and Koketsu, K. (1975). Direct control of action potentials by acetylcholine in bullfrog sympathetic ganglion cells, *Brain Res.* **89**:166–169.

Kuba, K., and Koketsu, K. (1976a). Analysis of the slow excitatory postsynaptic potential in bullfrog sympathetic ganglion cells, *Jpn. J. Physiol.* **26**:647–664.

Kuba, K., and Koketsu, K. (1976b). Muscarinic effects of acetylcholine on the action potential of bullfrog sympathetic ganglion cells, *Jpn. J. Physiol.* **26**:703–716.

Kuba, K., and Koketsu, K. (1978). Synaptic events in sympathetic ganglia, *Prog. Neurobiol.* **11**:77–169.

Kuba, K., and Nishi, S. (1979). Characteristics of the fast excitatory postsynaptic current in bullfrog cells, *Pflügers Arch.* **378**:205–212.

Kuba, K., and Takeshita, S. (1983). On the mechanism of calcium action on the acetylcholine receptor-channel complex at the frog end-plate membrane, *Jpn. J. Physiol.*. **33**:931–944.

Kubo, T., Fukuda, K., Mikami, A., Maeda, A., Takahashi, H., Mishina, M., Haga, T., Haga, K., Ichiyama, A., Kaugawa, K., Kojima, M., Matsuo, H., Hirose, T., and Numa, S. (1986). Cloning, sequencing, and expression of complementary DNA encoding the muscarinic acetylcholine receptor, *Nature* (*London*), **323**:411–416.

Kuffler, S. W., and Sejnowski, T. J. (1983). Peptidergic and muscarinic excitation at amphibian sympathetic synapses, *J. Physiol.* (*London*) **341**:257–278.

Kuffler, S. W., and Yoshikami, D. (1975). The distribution of acetylcholine sensitivity at the postsynaptic membrane of vertebrate skeletal twitch muscle: ionophoretic mapping in micron range, *J. Physiol.* (*London*) **244**:703–730.

Kuffler, S. W., Dennis, M. J., and Harris, A. J. (1971). The development of chemosensitivity in extrasynaptic areas of the neuronal surface after denervation of parasympathetic ganglion cells in the heart of the frog, *Proc. R. Soc. London* **177**:555–563.

Kuhar, M. J., Birdsall, N. J. M., Burgen, A. S. V., and Hulme, E. C. (1980). Ontogeny of muscarinic receptors in rat brain, *Brain Res.* **184**:375–383.

Kumamoto, E., and Kuba, K. (1983). Sustained rise in ACh sensitivity of a sympathetic ganglion cell induced by postsynaptic electrical activities, *Nature* (*London*) **305**:145–146.

Kurachi, Y., Nakajima, T., and Sugimoto, T. (1986). On the mechanism of activation of muscarinic K^+ channels by adenosine in isolated atrial cells: involvement of GTP-binding proteins, *Pflügers Arch.* **407**:264–274.

Kurchikov, A. L., and Kazachenko, V. N. (1979). *Two Components in Current Relaxations of Cholinoreceptive Membrane in Lymnea stagnalis Neurons* [in Russian], VINITI, No 429-79, pp. 1–16.

Kurchikov, A. L., and Kazachenko, V. N. (1984a). Effect of membrane hyperpolarization on the conductance of cholinoreceptive membrane, *Biol. Membr.* **1**:289–293.

Kurchikov, A. L., and Kazachenko, V. N. (1984b). Kinetics of acetylcholine interaction with cholinoreceptors in molluscan neurons, *Biol. Membr.* **1**:384–389.

Kurchikov, A. L., Kazachenko, V. N., and Veprintsev, B. N. (1985). Interaction of molluscan cholinoreceptors with suberyldicholine and other agonists, *Biol. Membr.* **2**:525–533.

Kusano, K., Miledi, R., and Stinnakre, J. (1982). Cholinergic and catecholaminergic receptors in the *Xenopus* oocyte membrane, *J. Physiol.* (*London*) **328**:143–170.

Laduron, P. (1980). Axonal transport of muscarinic receptors, *Nature* (*London*), **286**:287–288.

Lambrecht, G., Moser, U., Wess, J., Rotter, J., Fuder, H., Kilbinger, H., Müller, H., Linoh, H., Tacke, R., Zilch, H., and Mutschler, E. (1985). Hexahydrosiladifenidol: a selective antagonist at muscarinic M2 receptor subtypes, in: *Subtypes of Muscarinic Receptors, Second Intern. Symp.*, Boston University, p. 38.

Lapetina, E. G., Brown, W. E., and Michell, R. H. (1976). Muscarinic cholinergic stimulation of phosphatidylinositol turnover in isolated rat superior cervical sympathetic ganglia, *J. Neurochem.* **26**:649–651.

Laporte, Y., and Lorente de No, R. (1950). Potential changes evoked in a curarized sympathetic ganglion by presynaptic volleys of impulses, *J. Cell. Comp. Physiol.* **35**, Suppl. 2:61–106.

Large, W. A., and Rang, H. P. (1978). Factors affecting the rate of incorporation of a false transmitter into mammalian motor nerve terminals, *J. Physiol. (London)* **285**:1–24.

Large, W. A., and Sim, J. A. (1986). A comparison between mechanisms of action of different nicotinic blocking agents on rat submandibular ganglia, *Br. J. Pharmacol.* **89**:583–592.

Larrabee, M. G., and Leicht, W. S. (1965). Metabolism of phosphatidylinositol and other lipids in active neurons of sympathetic ganglia and other peripheral nervous tissues. The site of the inositide effect, *J. Neurochem.* **12**:1–13.

Laüger, P. (1985). Structural fluctuations and current noise of ionic channels, *Biophys. J.* **48** :369–373.

Lee, C.-Y., and Chen, Y.-M. (1976). Species differences in reversibility of neuromuscular blockade by elapid and sea snake neurotoxins, in: *Animal, Plant, and Microbial Toxins* (A. Ohsaka, K. Hayashi, and Y. Sawai, eds.), Plenum Press, New York, Vol. 2, pp. 193–204.

Lee, C.-Y., Chang, C. C., and Chen, Y.-M. (1972a). Reversibility of neuromuscular blockade by neurotoxins from elapid and sea snake venoms, *J. Formosan Med. Assoc.* **71**:344–349.

Lee, C.-Y., Chang, S. L., Kau, S. T., and Luh, S.-H. (1972b). Chromatographic separation of the venom of *Bungarus multicinctus* and characterization of its components, *J. Chromatogr.* **72**:71–82.

Lees, G. M., and Nishi, S. (1972). Analysis of the mechanism of action of some ganglion-blocking drugs in the rabbit superior cervical ganglion, *Br. J. Pharmacol.* **46**:78–88.

Lesser, E. (1965). The stereospecificity of acetyl-alpha-methylcholine, *Br. J. Pharmacol.* **25**:213–216.

Levitan, H., and Tauc, L. (1972). Acetylcholine receptors: topographic distribution and pharmacological properties of two receptor types on a single molluscan neuron, *J. Physiol. (London)* **222**:537–558.

Levitan, J. B. (1978). Adenylate cyclase in isolated *Helix* and *Aplysia* neuronal cell bodies: stimulation by serotonin and peptide-containing extract, *Brain Res.* **154**:404–408.

Lewis, C. A. (1979). Ion-concentration dependence of the reversal potential and the single channel conductance of ion channels at the frog neuromuscular junction, *J. Physiol. (London)* **286**:417–445.

Libet, B. (1964). Slow synaptic responses and excitatory changes in sympathetic ganglia, *J. Physiol. (London)* **174**:1–25.

Libet, B. (1967). Long latent periods and further analysis of slow synaptic responses in sympathetic ganglia, *J. Neurophysiol.* **30**:494–514.

Libet, B. (1970). Generation of slow inhibitory and excitatory postsynaptic potentials, *Fed. Proc.* **29**:1945–1956.

Libet, B. (1971). Inactivation of potassium conductance in slow postsynaptic excitation, *Science* **172**:503–504.

Libet, B. (1979). Dopaminergic synaptic processes in the superior cervical ganglion: models for synaptic actions, in: *The Neurobiology of Dopamine* (A. Horn, J. Korf, and B. H. C. Westerink, eds.), Academic Press, London, pp. 453–474.

Libet, B. (1984). Heterosynaptic interaction at a sympathetic neuron as a model for induction and storage of a postsynaptic memory trace, in: *Neurobiology of Learning and Memory* (G. Lynch, G. L. McGaugh, and N. M. Weinberger, eds.), Guilford Press, New York, pp. 405–430.

Libet, B., and Kobayashi, H. (1969). Generation of adrenergic and cholinergic potentials in sympathetic ganglion cells, *Science* **164**:1530–1532.

Libet, B., and Kobayashi, H. (1974). Adrenergic mediation of slow inhibitory postsynaptic potential in sympathetic ganglia of the frog, *J. Neurophysiol.* **37**:805–814.

Libet, B., and Owman, Ch. (1974). Concomitant change in formaldehyde-induced fluorescence of dopamine interneurons and in slow inhibitory postsynaptic potentials of the rabbit superior cervical ganglion, induced by stimulation of the preganglionic nerve or by a muscarinic agent, *J. Physiol. (London)* **237**:635–662.

Libet, B., and Tosaka, T. (1969). Slow inhibitory and excitatory postsynaptic responses in single cells of mammalian sympathetic ganglia, *J. Neurophysiol.* **32**:43–50.

Libet, B., and Tosaka, T. (1970). Dopamine as a synaptic transmitter and modulator in sympathetic ganglia: different mode of synaptic action, *Proc. Natl. Acad. Sci. USA* **67**:667–673.

Libet, B., Chichibu, S., and Tosaka, T. (1968). Slow synaptic responses and excitability in sympathetic ganglia of the bullfrog, *J. Neurophysiol.* **31**:383–395.

Libet, B., Kobayashi, H., and Tanaka, T. (1975). Synaptic coupling into the production and storage of a neuronal memory trace, *Nature (London)* **258**:155–157.

Linder, T. M., and Quastel, D. M. (1978). A voltage-clamp study of the permeability change induced by quanta of transmitter at the mouse endplate, *J. Physiol. (London)* **281**:535–556.

Lindl, T. (1979). cAMP and its relation to ganglionic transmission. A combined biochemical and electrophysiological study of the rat superior cervical ganglion in vitro, *Neuropharmacology* **18**:227–235.

Lindstrom, J. (1980). Antibodies as probes of acetylcholine receptor structure and function, in: *Psychopharmacology and Biochemistry of Neurotransmitter Receptors* (H. Yamamura, W. Olsen, and E. Usdin, eds.), Elsevier–North-Holland, Amsterdam, pp. 63–84.

Lingle, Ch. (1983a). Blockade of cholinergic channels by chlorisondamine on a crustacean muscle, *J. Physiol. (London)* **339**:395–417.

Lingle, Ch. (1983b). Different types of blockade of crustacean acetylcholine-induced currents, *J. Physiol. (London)* **339**:419–437.

Loring, R. H., Dahm, L. M., and Zigmond, R. E. (1985). Localization of α-bungarotoxin binding sites in the ciliary ganglion of the embryonic chick: an autoradiographic study at the light and electron microscopic level, *Neuroscience* **14**:645–660.

Lukomskaya, N. Y. (1972). Study of muscarine- and nicotine-sensitive cholinoreceptors in cat superior cervical ganglion neurons, in: *Interneuronal Trans-*

mission in Autonomic Nervous System (P. G. Kostyuk, ed.) [in Russian], Naukova Dumka, Kiev, pp. 146–155.

Lukomskaya, N. Y., and Gmiro, V. E. (1982). Study of cholinoreceptive membrane in sympathetic ganglion by analysis of structure–activity relationship, *J. Auton. Nerv. Syst.* **6**:363–371.

Lukomskaya, N. Y., Magazanik, L. G., and Potapjeva, N. N. (1981). Effect of bis-cationic ganglion-blockers on activation kinetics of synaptic ionic channels, in: *Physiology of Autonomic Ganglia* (V. I. Skok, ed.) [in Russian], Naukova Dumka, pp. 26–27.

Lundberg, M. J., Hedlund, B., and Bartfrai, T. (1982). Vasoactive intestinal polypeptide enhances muscarinic ligand binding in cat submandibular salivary gland, *Nature (London)* **295**:147–149.

MacDermott, A. B., Connor, E. A., Dionne, V. E., and Parsons, R. L. (1980), Voltage clamp study of fast excitatory synaptic currents in bullfrog sympathetic ganglion cells, *J. Gen. Physiol.* **75**:39–60.

MacLennan, D. H., and Reitmeier, R. A. F. (1982). The structure of the Ca^{2+}, Mg^{2+}-ATPase of sarcoplasmic reticulum, *Membr. Transport* **1**:567–571.

Madison, D. V., and Nicoll, R. A. (1982). Noradrenalin blocks accommodation of pyramidal cell discharge in the hippocampus, *Nature (London)* **229**:636–638.

Magazanik, L. G. (1968). On the mechanism of desensitization of postsynaptic membrane in muscle fiber, *Biofizika* **13**:199–203.

Magazanik, L. G., Ivanov, A. Ya., and Lukomskaya, N. Y. (1974). Study of action of snake venom polypeptides on cholinoreception of isolated rabbit superior cervical ganglion, *Neirofiziologiya* **6**:652–654.

Magazanik, L. G., Illesh, P., and Snetkov, V. A. (1976). Effect of serotonin on voltage-dependence of end-plate current in muscle fiber, *Dokl. Akad. Nauk SSSR* **227**:1268–1271.

Magazanik, L. G., Antonov, S. M., and Gmiro, V. E. (1984). Mechanisms of activation and blockade of glutamate-sensitive postsynaptic membrane, *Biol. Membr.* **1**:130–140.

Magleby, K. L., and Pallotta, B. S. (1981). A study of desensitization of acetylcholine receptors using nerve-released transmitter in the frog, *J. Physiol. (London)* **316**:225–250.

Magleby, K. L., and Stevens, C. F. (1972a). The effect of voltage on the time course of end-plate currents, *J. Physiol. (London)* **223**:151–171.

Magleby, K. L., and Stevens, C. F. (1972b). A quantitative description of end-plate currents, *J. Physiol. (London)* **223**:173–197.

Maleque, M. A., Takahashi, K., Witkop, B., Brossi, A., and Albuquerque, E. X. (1984). A study of the novel synthetic analog (±)-dipentylperhydrohistrionicotoxin on the nicotinic receptor-ion channel complex, *J. Pharmacol. Exp. Ther.* **230**:619–626.

Marchais, D., and Marty, A. (1979). Interaction of permeant ions with channels activated by acetylcholine in *Aplysia* neurons, *J. Physiol. (London)* **297**:9–45.

Marshall, L. M. (1981). Synaptic localization of α-bungarotoxin binding which blocks nicotinic transmission at frog sympathetic neurons, *Proc. Natl. Acad. Sci. USA* **78**:1948–1952.

Marshall, L. M. (1985). Presynaptic control of synaptic channel kinetics in sympathetic neurons, *Nature (London)* **317**:621–623.

Marty, A. (1980). Action of calcium ions on acetylcholine-sensitive channels in *Aplysia* neurons, *J. Physiol. (Paris)* **76**:523–527.

Marty, A., Neild, T. O., and Ascher, P. (1976). Voltage sensitivity of acetylcholine currents in *Aplysia* neurons in the presence of curare, *Nature (London)* **261**:501–503.

Maruyama, Y., and Petersen, O. H. (1982a). Single-channel currents in isolated patches of plasma membrane from basal surface of pancreatic acini, *Nature (London)* **299**:159–161.

Maruyama, Y., and Petersen, O. H. (1982b). Cholecystokinin activation of single channel currents as mediated by internal messenger in pancreatic acinar cell, *Nature (London)* **300**:61–63.

Maruyama, Y., Gallacher, D. V., and Petersen, O. H. (1983). Voltage- and Ca^{2+}-activated K^+ channel in baso-lateral acinar cell membranes of mammalian salivary glands, *Nature (London)* **302**:827–829.

Masukawa, L. M., and Albuquerque, E. X. (1978). Voltage- and time-dependent action of histrionicotoxin on the end-plate current of the frog muscle, *J. Gen. Physiol.* **72**:351–367.

Matthews, M. R. (1983). The ultrastructure of junctions in sympathetic ganglia of mammals, in: *Autonomic Ganglia* (L.-G. Elfvin, ed.), Wiley, New York, pp. 27–66.

Matthews, M. R., and Nelson, V. H. (1975). Detachment of structurally intact nerve endings from chromatolytic neurons of rat superior cervical ganglion during the depression of synaptic transmission induced by postganglionic axotomy, *J. Physiol. (London)* **245**:91–135.

Matthews-Bellinger, J. A., and Salpeter, M. M. (1983). Fine structural distribution of acetylcholine receptors at developing mouse neuromuscular junctions, *J. Neurosci.* **3**:644–657.

Mayer, M. L., Westbrook, G. L., and Guthrie, P. B. (1984). Voltage-dependent block by Mg^{2+} of NMDA responses in spinal cord neurons, *Nature (London)* **309**:261–263.

McAfee, D. A., and Greengard, P. (1972). Adenosine 3',5'-monophosphate: electrophysiological evidence for a role in synaptic transmission, *Science* **178**:310–312.

McAfee, D. A., and Yarowsky, P. J. (1979). Calcium-dependent potentials in the mammalian sympathetic neuron, *J. Physiol. (London)*) **290**:507–523.

McAfee, D. A., Schorderet, M., and Greengard, P. (1971). Adenosine 3',5'-monophosphate in nervous tissue: increase associated with synaptic transmission, *Science* **171**:1156–1158.

McCormick, A., and Prince, D. A. (1985). Two types of muscarinic responses to acetylcholine in mammalian cortical neurons, *Proc. Natl. Acad. Sci. USA* **82**:6344–6348.

McCormick, D. V., and Atassi, M. Z. (1984). Localization and synthesis of the acetylcholine-binding site in the α-chain of the *Torpedo californica* acetylcholine receptor, *Biochem. J.* **224**:995–1000.

McCormick, D. A., and Prince, D. A. (1986a). Acetylcholine induces burst firing in thalamic reticular neurons by activating a potassium conductance, *Nature (London)* **319**:402–405.

McCormick, D. A., and Prince, D. A. (1986b). Pirenzepine discriminates among ionic responses to acetylcholine in guinea-pig cerebral cortex and reticular nucleus of thalamus, in: Subtypes of muscarinic receptors II, *Trends Pharmacol. Sci. Suppl.*, pp. 72–77.

McGee, T., and Liepe, B. (1984). Acute elevation of cyclic AMP does not alter the ion-conducting properties of the neuronal nicotinic acetylcholine receptor of PC 12 cells, *Mol. Pharmacol.* **26**:51–56.

McLachlan, E. M. (1977). The effect of strontium and barium ions at synapses in sympathetic ganglia, *J. Physiol.* (*London*) **267**:497–518.

McLennan, H., and York, D. H. (1966). Cholinergic mechanisms in the caudate nucleus, *J. Physiol.* (*London*) **187**:163–175.

McNamee, M. G., and Ochoa, E. L. M. (1982). Reconstitution of acetylcholine receptor function in model membranes, *Neuroscience* **7**:2305–2319.

McQuarrie, C., Salvaterra, P. M., De Blas, A., Routes, J., and Mahler, H. R. (1976). Studies on nicotinic acetylcholine receptors in mammalian brain, *J. Biol. Chem.* **251**:6335–6339.

Messing, A. (1982). Cholinergic agonist-induced down regulation of neuronal α-bungarotoxin receptors, *Brain Res.* **232**:479–484.

Messing, A., and Gonatas, N. K. (1983). Extrasynaptic localization of α-bungarotoxin receptors in cultured chick ciliary ganglion neurons, *Brain Res.* **269**:172–176.

Messing, A., and Kim, S. U. (1981). Development of α-bungarotoxin receptors in cultured chick ciliary ganglion neurons, *Brain Res.* **208**:479–486.

Michelson, M. J., and Zeimal, E. V. (1973). *Acetylcholine. An Approach to the Molecular Mechanism of Action*, Pergamon Press, Oxford–New York.

Michelson, M. J., Kolganova, O. I., Lukomskaya, N. Y., Protas, L. L., and Sokolov, G. A. (1979). Comparative study of stereoselectivity in cholinoreceptors, *Zh. Évol. Biokhim. Fiziol.* **15**:42–47.

Miledi, R. (1960). Junctional and extrajunctional acetylcholine receptors in skeletal muscle fibers, *J. Physiol.* (*London*) **151**:24–30.

Miledi, R., and Szepaniak, A. C. (1975). Effect of *Dendroaspis* neurotoxins on synaptic transmission in the spinal cord of the frog, *Proc. R. Soc. London* **190**:267–274.

Miller, C. (1982). Bis-quaternary ammonium blockers as structural probes of the sarcoplasmic reticulum K channel, *J. Gen. Physiol.* **79**:869–891.

Mishina, M., Kurosaki, T., Tobimatsu, T., and Noda, M. (1984). Expression of functional acetylcholine receptor from cloned cDNAs, *Nature* (*London*) **307**:604–608.

Mishina, M., Tobimatsu, T., Imoto, K., Tanaka, K., Fujita, Y., Fukuda, K., Kurasaki, M., Takahashi, H., Morimoto, Y., Hirose, T., Inayama, S., Takahashi, T., Kuno, M., and Numa, Sh. (1985). Location of functional regions of acetylcholine receptor α subunit by site-directed mutagenesis, *Nature* (*London*) **313**:364–369.

Mitsuka, M., and Hatanaka, H. (1983). Selective loss of acetylcholine sensitivity in a nerve cell line cultured in hormone supplanted serum-free medium, *J. Neurosci.* **3**:1785–1790.

Mo, N., and Dun, N. J. (1984). Vasoactive intestinal polypeptide facilitates muscarinic transmission in mammalian sympathetic ganglia, *Neurosci. Lett.* **52**:19–23.

Mochida, S., and Kobayashi, H. (1986). Activation of M_2 muscarinic receptors causes an alteration of action potentials by modulation of Ca entry in isolated sympathetic neurons of rabbits, *Neurosci. Lett.* **72**:199–204.

Mochida, S., and Libet, B. (1985). Synaptic long-term enhancement (LTE) induced by a heterosynaptic neural input, *Brain Res.* **329**:360–363.

Morita, K., and Katayama, Y. (1984). Two types of acetylcholine receptors on the soma of primary afferent neurons, *Brain Res.* **290**:348–352.

Morita, K., and North, R. A. (1985). Significance of slow synaptic potentials for transmission of excitation in guinea-pig myenterix plexus, *Neuroscience* **14**:661–672.

Morita, K., North, R. A., and Tokimasa, T. (1982). Muscarinic presynaptic inhibition of synaptic transmission in myenteric plexus of guinea-pig ileum, *J. Physiol.* (*London*) **333**:141–149.

Morley, B. J. (1981). The properties of brain nicotine receptors, *Pharmacol. Ther.* **15**:111–122.

Morley, B. J., Dwyer, D. S., Strang-Brown, P. F., and Bradley, R. J. (1983). Evidence that certain peripheral anti-acetylcholine receptor antibodies do not interact with brain BuTX binding sites, *Brain Res.* **262**:109–116.

Muller, W., and Misgeld, U. (1986). Slow cholinergic excitation of guinea-pig hippocampal neurons is mediated by two muscarinic receptor subtypes, *Neurosci. Lett.* **67**:107–112.

Nador, K. (1960). Ganglion blocker, *Prog. Drug Res.* **2**:303–411.

Neher, E. (1983). The charge carried by single-channel currents of rat cultured muscle cells in the presence of local anesthetics, *J. Physiol.* (*London*)) **339**:663–678.

Neher, E., and Sakmann, B. (1975). Voltage-dependence of drug-induced conductance in frog neuromuscular junction, *Proc. Natl. Acad. Sci. USA* **72**:2140–2144.

Neher, E., and Sakmann, B. (1976). Single channel currents recorded from membrane of denervated frog muscle fibers, *Nature* (*London*), **260**:799–802.

Neher, E., and Steinbach, J. H. (1978). Local anesthetics transiently block currents through single acetylcholine-receptor channels, *J. Physiol.* (*London*) **277**:153–176.

Neher, E., and Stevens, C. F. (1977). Conductance fluctuations and ionic pores in membranes, *Annu. Rev. Biophys. Bioeng.* **6**:345–381.

Neumann, D., Fridkin, M., and Fuchs, S. (1984). Anti-acetylcholine receptor response achieved by immunization with a synthetic peptide from the receptor sequence, *Biochem. Biophys. Res. Commun.* **121**:673–679.

Neumann, D., Gershoni, J. M., Fridkin, M., and Fuchs, S. (1985). Antibodies to synthetic peptides as probes for the binding site on the α subunit of the acetylcholine receptor, *Proc. Natl. Acad. Sci. USA* **82**:3490–3493.

Neumann, D., Barchan, D., Fridkin, M., and Fuchs, S. (1986). Analysis of ligand binding to the synthetic dodecapeptide 185-196 of the acetylcholine receptor α subunit, *Neurobiology* **83**:9250–9353.

Newberry, N. R., Priestly, T., and Woodruff, G. N. (1985). Pharmacological distinction between two muscarinic responses on the isolated superior cervical ganglion of the rat, *Br. J. Pharmacol.* **116**:191–192.

Nishi, S. (1974). Ganglionic transmission, in: *The Peripheral Nervous System* (J. I. Hubbard, ed.), Plenum Press, New York–London, pp. 225–255.

Nishi, S. (1979). The catecholamine-mediated inhibition in ganglionic transmission, in: *Integrative Functions of the Autonomic Nervous System* (C. M. Brooks, K. Koisumi, and A. Sato, eds.), Univ. Tokyo Press and Elsevier–North-Holland Biomed. Press, Amsterdam, pp. 223–233.

Nishi, S., and Koketsu, K. (1960). Electrical properties and activities of single sympathetic neurons in frogs, *J. Cell. Comp. Physiol.* **55**:15–30.

Nishi, S., and Koketsu, K. (1967). Origin of ganglionic inhibitory postsynaptic potential, *Life Sci.* **6**:2049–2056.

Nishi, S., and Koketsu, K. (1968). Analysis of slow inhibitory postsynaptic potential of bullfrog sympathetic ganglion, *J. Neurophysiol* .**31**:717–728.

Nishi, S., Soeda, H., and Koketsu, K. (1967). Release of acetylcholine from sympathetic preganglionic nerve terminals, *J. Neurophysiol.* **30**:114–134.

Nishi, S., Soeda, H., and Koketsu, K. (1969). Unusual nature of ganglionic slow EPSP studied by a voltage-clamp method, *Life Sci.* **8**:33–42.

Nishiyama, A., and Petersen, O. H. (1974). Pancreatic acinar cells: membrane potential and resistance change evoked by acetylcholine, *J. Physiol. (London)* **238**:145–158.

Noda, M., Takahashi, H., Tanabe, T., Toyosato, M., Furutani, Y., Hirose, T., Asai, M., Inayama, S., Migata, T., and Numa, S. (1982). Primary structure of α subunit precursor of *Torpedo californica* acetylcholine receptor deduced from cDNA sequence, *Nature (London)* **299**:793–797.

Noda, M., Furutani, Y., Takahashi, H., Toyasato, M., Tanabe, T., Shimizu, S., Kikyotani, S., Kayano, T., Hirose, T., Inayama, S., and Numa, S. (1983a). Cloning and sequence analysis of calf cDNA and human genomic DNa encoding α subunit precursor of molecular acetylcholine receptor, *Nature (London)* **305**:818–823.

Noda, M., Takahashi, H., Tanabe, T., Toyosato, M., Furutani, Y., Hirose, T., Asai, M., Inayama, S., Migata, T., and Numa, S. (1983b). Primary structure of β- and δ-subunit precursors of *Torpedo californica* acetylcholine receptor deduced from cDNA sequence, *Nature (London)* **301**:251–255.

Noda, M., Takahashi, H., Tanabe, T., Toyosato, M., Kikyotani, S., Furutani, Y., Hirose, T., Takashima, H., Inayama, S., Migata, T., and Numa, S. (1983c). Structural homology of *Torpedo californica* acetylcholine receptor subunits, *Nature (London)* **302**:528–532.

Nohmi, M., and Kuba, K. (1984). (+)-Tubocurarine blocks the Ca^{2+}-dependent K^+-channel of the bullfrog sympathetic ganglion cell, *Brain Res.* **301**:146–148.

Noma, A., and Trautwein, W. (1978). Relaxation of the ACh-induced potassium current in the rabbit sinoatrial node cell, *Pflügers Arch.* **377**:193–200.

Noma, A., Di Francesco, D., and Trautwein, W. (1978). Dose–response curve for acetylcholine (ACh) and carbamylcholine (CCh) in the rabbit S-A node, *Pflügers Arch.* **377**:R4.

Noon, J. P. (1975). Norepinephrine release from nerve terminals within the rabbit superior cervical ganglion, *Naunyn-Schmiedeberg's Arch. Pharmacol.* **291**:139–162.

North, R. A. (1986). Muscarinic receptors and their associated ion conductance, *Trends Pharmacol. Sci. Suppl.*: 19–22.

North, R. A., and Tokimasa, T. (1982). Muscarinic synaptic potentials in guinea-pig myenteric plexus neurons, *J. Physiol. (London)* **333**:151–156.

North, R. A., and Tokimasa, T. (1983). Depression of calcium-dependent potassium conductance of guinea-pig myenteric neurons by muscarinic agonists, *J. Physiol. (London)* **342**:253–266.

North, R. A., and Tokimasa, T. (1984). The time course of muscarinic depolarization of guinea-pig myenteric neurons, *Brain Res.* **82**:85–91.

North, R. A., and Tokimasa, T. (1987). Persistent calcium-sensitive potassium current and the resting properties of guinea-pig myenteric neurons, *J. Physiol. (London)* **386**:333–353.

North, R. A., Slack, B. E., and Surprenant, A. (1985). Muscarinic m_1 and m_2 receptors mediate depolarization and presynaptic inhibition in guinea-pig enteric nervous system, *J. Physiol. (London)* **368**:435–452.

Nowak, L. M., and MacDonald, R. L. (1983a). Ionic mechanism of muscarinic cholinergic depolarization of mouse spinal cord neurons in cell culture, *J. Neurophysiol.* **49**:792–803.

Nowak, L. M., and MacDonald, R. L. (1983b). Muscarine-sensitive voltage-dependent potassium current in cultured murine spinal cord neurons, *Neurosci. Lett.* **35**:85–91.

Nowak, L., Bregestowski, P., Ascher, P., and Herbert, A. (1984). Magnesium gated glutamate-activated channels in mouse central neurons, *Nature (London)* **307**:462–465.

Nurse, C. A., and O'Lague, P. H. (1975). Formation of cholinergic synapses between dissociated sympathetic neurons and skeletal myotubes of the rat in cell culture, *Proc. Natl. Acad. Sci. USA* **72**:1955–1959.

Obata, K. (1974). Transmitter sensitivities of some nerve and muscle cells in culture, *Brain Res.* **73**:71–88.

Ogden, D. C., Siegelbaum, S. A., and Colquhoun, D. (1981). Block of acetylcholine-activated ion channels by an unchanged local anesthetic, *Nature (London)* **289**:596–598.

Ogden, D. C., Gray, P. T. A., Colquhoun, D., and Rang, H. P. (1984). Kinetics of acetylcholine activated ion channels in chick ciliary ganglion neurons grown in tissue culture, *Pflügers Arch.* **400**:44–50.

Ono, J. K., and Salvaterra, P. M. (1981). α-Toxin effects on cholinergic and noncholinergic responses of *Aplysia californica* neurons, *J. Neurosci.* **1**:259–270.

Osterrieder, W., Noma, A., and Trautwein, W. (1980). On the kinetics of the potassium channel activated by acetylcholine in the S–A node of the rabbit heart, *Pflügers Arch.* **386**:101–109.

Osterrieder, W., Yang, Q. F., and Trautwein, W. (1981). The time course of the muscarinic response to ionophoretic acetylcholine application to the S–A node of the rabbit heart, *Pflügers Arch.* **389**:283–291.

Ovchinnikov, Y. A., Abdulaev, N. G., Feigina, M. Y., Kiselev, A., and Lobanov, N. (1979). The structure basis of the functioning of bacteriorhodopsin: an overview, *FEBS Lett.*, **100**:219–224.

Ovchinnikov, Y. A., Abdulaev, N. G., Feigina, M. Y., Artamonov, I. D., Bogachuk, A. S., Zolotarjev, A. S., Yeganyan, E. P., and Kostetski, P. V. (1983a). Visual rhodopsin. II. Total amino acid sequence and arrangement in the membrane, *Bioorg. Khim.* **9**:1331–1340.

Ovchinnikov, Y. A., Tsetlin, V. I., Pluzhnikov, K. A., and Karelin, A. A. (1983b). Topography of the neurotoxin–acetylcholine receptor interaction: binding of *Naja naja oxiana* neurotoxin. II. Photoactive derivatives, in:

Physical Chemistry of Transmembrane Ion Motions [in Russian], pp. 487–490.

Pappano, A. J., and Volle, R. L. (1966). Observations on the role of calcium ions in ganglionic responses to acetylcholine, *J. Pharmacol. Exp. Ther.* **152**:171–180.

Parsons, R. L., Neel, D. S., and Connor, E. A. (1984). Characteristics of fast synaptic currents are altered by a reduction in external sodium, *Soc. Neurosci. Abstr.* **10**, Pt. I:202.

Paton, W. D. M., and Perry, W. L. M. (1953). The relationship between depolarization and block in the cat's superior cervical ganglion, *J. Physiol. (London)* **119**:43–57.

Paton, W. D. M., and Zaimis, E. J. (1949). The pharmacological action of polymethylene bis-trimethylammonium salts, *Br. J. Pharmacol.* **4**:381–400.

Patrick, J., and Stallcup, W. B. (1977). Immunological distinction between acetylcholine receptor and the α-bungarotoxin binding component on sympathetic neurons, *Proc. Natl. Acad. Sci. USA* **74**:4689–4692.

Patrick, J., Stallcup, W. B., Zavenelli, M., and Ravdin, P. (1980). Binding properties of a neurotoxin from the venom of the Green mamba, *Dendroaspis viridis*, *J. Biol. Chem.* **225**:526–533.

Patterson, B. A., and Volle, R. L. (1984). Muscarinic receptors and [^{3}H]inositol incorporation in a rat sympathetic ganglion, *J. Auton. Nerv. Syst.* **10**:69–72.

Pearson, G. T., and Petersen, O. H. (1984). Nervous control of membrane conductance in mouse lacrimal acinar cells, *Pflügers Arch.* **400**:51–59.

Pearson, G. T., Elanagan, P. M., and Petersen, O. H. (1984). Neural and hormonal control of membrane conductance in the pig pancreatic acinar cell, *Am. J. Physiol.* **247**:520–526.

Pellegrino, M., and Simonneau, M. (1984). Distribution of receptors for acetylcholine and 5-hydroxytryptamine on identified leech neurons growing in culture, *J. Physiol. (London)* **352**:669–684.

Pennefather, P., Lancaster, B., Adams, P. R., and Nicoll, R. A. (1985). Two distinct Ca-dependent K currents in bullfrog sympathetic ganglion cells, *Proc. Natl. Acad. Sci. USA* **82**:3040–3044.

Peper, K., Bradley, R. J., and Dreyer, F. (1982). The acetylcholine receptor at the neuromuscular junction, *Physiol. Rev.* **62**:1271–1340.

Petersen, O. H. (1976). Electrophysiology of mammalian gland cells, *Physiol. Rev.* **56**:535–577.

Petersen, O. H., and Maruyama, Y. (1984). Calcium-activated potassium channels and their role in secretion, *Nature (London)* **307**:693–697.

Petersen, O., Nishiyama, A., Laugier, R., and Philpott, H. G. (1981). Hormonal control of ion permeability of the pancreatic acinar cell membrane mediated by intracellular calcium, in: *Drug Receptors and Their Effectors* (N. J. M. Birdsall, ed.), Macmillan, London, pp. 63–73.

Pfaffinger, P. J., Martin, J. M., Hunter, D. D., Nathanson, N. M., and Hille, B. (1985). GTP-binding proteins couple cardiac muscarinic receptors to a K channel, *Nature (London)* **317**:536–538.

Pickard, M. R., Hawthorne, J. N., Hayashi, E., and Yamada, S. (1977). Effects of surugatoxin and other nicotinic and muscarinic antagonists on phosphatidylinositol metabolism in active sympathetic ganglia, *Biochem. Pharmacol.* **26**:448–450.

Pitman, R. M. (1971). Transmitter substances in insects: a review, *Comp. Gen. Pharmacol.* **2**:347–371.

Plummer, A. J., Barrett, W. E., and Rutledge, R. (1956). Some actions of chlorisondamine dimethylchloride (Su 3088) (Ecolid) on the gastrointestinal tract, *J. Pharmacol. Exp. Ther.* **116**:46.

Polz-Tejera, G., Hunt, S. P., and Schmidt, J. (1980). Nicotinic receptors in sensory ganglia, *Brain Res.* **195**:223–230.

Popot, J., and Changeux, J. (1984). Nicotinic receptor of acetylcholine: structure of an oligomeric integral membrane protein, *Physiol. Rev.* **64**:1162–1239.

Poskonova, M. A., Akopyan, A. R., and Chemeris, N. K.(1981). Diffusion of acetylcholine through the membrane of a neuron, *Dokl. Akad. Nauk SSSR* **257**:1496–1499.

Pott, L. (1979). On the time course of the acetylcholine-induced hyperpolarization in quiescent guinea-pig atria, *Pflügers Arch.* **380**:71–77.

Pott, L. (1983). Effects of fast and slowly acting antagonists on the time course of the muscarinic response in guinea-pig atria muscle, *Pflügers Arch.* **399**:35–41.

Pott, L., and Pusch, H. (1979). A kinetic model for the muscarinic action of acetylcholine, *Pflügers Arch.* **383**:75–77.

Pratesi, P., Villa, L., and Grana, E. (1968). Affinity and intrinsic activity of choline ethers and of compounds related to formethide. A contribution to the interpretation of the 5-atom rule, *Farm. Ed. Sci.* **23**:1213–1223.

Ptitsyn, O. B., and Finkelstein, A. V. (1983). Theory of protein secondary structure and algorithm of its prediction, *Biopolymers* **22**:15–25.

Pullman, B., and Pullman, A. (1963). *Quantum Biochemistry*, Wiley, New York–London.

Purves, R. D. (1976). Function of muscarinic and nicotinic acetylcholine receptors, *Nature (London)* **261**:149–151.

Putney, J. W. (1978). Stimulus-permeability coupling role of calcium in the receptor regulation of membrane permeability, *Pharmacol. Rev.* **30**:209–245.

Quik, M. (1982). Presence of an endogenous factor which inhibits binding of α-bungarotoxin 2.2 to its receptor, *Brain Res.* **245**:57–67.

Quik, M. (1985a). Inhibition of nicotinic receptor mediated ion fluxes in rat sympathetic ganglia by BGT-S1, a potent phospholipase, *Brain Res.* **325**:79–88.

Quik, M. (1985b). Binding characteristics of the bungarotoxin fraction II-S1 to rat brain membranes, *Eur. J. Pharmacol.* **111**:239–244.

Quik, M., and Lamarca, M. V. (1982). Blockade of transmission in rat sympathetic ganglia by a toxin which co-purifies with α-bungarotoxin, *Brain Res.* **238**:385–399.

Quik, M., Smith, P. A., Padjen, A. L., and Collier, B. (1981). A critical evaluation of the use of toxins from *Dendroaspis viridis* to block nicotinic responses at central and ganglionic synapses, *Brain Res.* **209**:129–143.

Raftery, M. A., Vandlen, R. L., Reed, K. L., and Lee, T. (1976). Characterization of *Torpedo californica* acetylcholine receptor: its subunit composition and ligand-binding properties, *Cold Spring Harbor Symp. Quant. Biol.* **40**:193–202.

Raftery, M. A., Dunn, S. M. J., Conti-Tronconi, B. M., Middlemas, D. S., and Crawford, R. D. (1983). The nicotinic acetylcholine receptor: subunit structure, functional binding sites, and ion transport properties, *Cold Spring Harbor Symp. Quant. Biol.* **48**:21–34.

Raisman, G., Field, P. M., Ostberg, A.-J., Iversen, L., and Zigmond, R. E. (1974). A quantitative ultrastructural and biochemical analysis of the process of reinnervation of the superior cervical ganglion in the adult rat, *Brain Res.* **71**:1–16.

Rang, H. P. (1981). The characteristics of synaptic currents and responses to acetylcholine of rat submandibular ganglion cells, *J. Physiol. (London)* **311**:23–55.

Rang, H. P. (1982). The action of ganglionic blocking drugs on the synaptic responses of rat submandibular ganglion cells, *Br. J. Pharmacol.* **75**:151–168.

Rang, H. P., and Gurney, A. M. (1982). Components of the synaptic current of rat submandibular ganglion cells, *J. Physiol. (Paris)* **78**:426–432.

Rang, H. P., and Ritter, J. M. (1971). The effect of disulfide bond reduction on the properties of cholinergic receptors in chick muscle, *Mol. Pharmacol.* **7**:620–631.

Rang, H. P., and Rylett, R. J. (1984). The interaction between hexamethonium and tubocurarine on the rat neuromuscular junction, *Br. J. Pharmacol.* **81**:519–531.

Rapier, C., Harrison, R., Lunt, G. G., and Wonnacott, S. (1985). Neosurugatoxin blocks nicotinic acetylcholine receptors in the brain, *Neurochem. Int.* **7**:389–396.

Ratnam, M., and Lindstrom, J. (1984). Structural features of the nicotinic acetylcholine receptor revealed by antibodies to synthetic peptides, *Biochem. Biophys. Res. Commun.* **122**:1225–1233.

Ratnam, M., Le Nguyen, D., Rivier, J., Surgent, P. B., and Lindstrom, J. (1986), Transmembrane topography of nicotinic acetylcholine receptor: immunochemical tests contradict theoretical predictions based on hydrophobicity profiles, *Biochemistry* **25**:2633–2643.

Ravdin, P. M., and Berg, D. K. (1979). Inhibition of neuronal acetylcholine sensitivity by α-toxins from *Bungarus multicinctus* venom, *Proc. Natl. Acad. Sci. USA* **76**:2072–2076.

Reynolds, J., and Karlin, A. (1978). Molecular weight in detergent solution of acetylcholine receptor from *Torpedo californica* , *Biochemistry* **17**:2035.

Riker, W. K. (1965). Effects of tetraethylammonium on synaptic transmission in the frog sympathetic ganglion, *J. Pharmacol. Exp. Ther.* **147**:161–171.

Rinken, A. A., Langel, Y. L., and Jarv, J. L. (1984). Solubilization by digitonin of muscarinic cholinoreceptor and its complex with quinuclidinyl benzylate, *Biol. Membr.* **1**:341–348.

Rinne, J. O., Laakso, K., Lönnberg, P., Mölsä, P., Paljärvi, L., and Rinne, J. K. (1985). Brain muscarinic receptors in senile dementia, *Brain Res.* **336**:19–25.

Ritchie, A. K., and Fambrough, D. M. (1975). Ionic properties of the acetylcholine receptor in cultured rat myotubes, *J. Gen. Physiol.* **65**:751–767.

Role, L. (1984). Substance P modulation of acetylcholine-induced currents in embryonic chicken sympathetic and ciliary ganglion neurons, *Proc. Natl. Acad. Sci. USA* **81**:2924–2928.

Romano, C., and Goldstein, A. (1980). Stereospecific nicotine receptors on rat brain membranes, *Science* **210**, 647–649.
Roper, S. (1976). The acetylcholine sensitivity of the surface membrane of multiply-innervated parasympathetic ganglion cells in the mudpuppy before and after partial denervation, *J. Physiol.* (*London*) **254**:455–473.
Rosengart, B. V., and Zhorov, B. S. (1983). Distance between ammonium groups in polymethylenebistrimethylammonium compouds from the data of theoretical conformational analysis, *Dokl. Akad. Nauk SSSR* **273**:505–508.
Rotter, A., Birdsall, N. J. M., Field, P. M., and Raisman, G. (1979). Muscarinic receptors in the central nervous system of the rat. II. Distribution of binding of [^{3}H]propylbenzilylcholine mustard in the midbrain and hindbrain, *Brain Res. Rev.* **1**:167–183.
Rueb, K. P., and Liefländer, M. (1979). Action of detergents on covalently labeled, membrane bound muscarinic acetylcholine-receptor of bovine nucleus caudatus, *Biochem. Biophys. Res. Commun.* **88**:627–633.
Ruff, R. L. (1977). A quantitative analysis of local anesthetic alteration of miniature end-plate currents and end-plate current fluctuations, *J. Physiol.* (*London*) **264**:89–124.
Saji, M., and Miura, M. (1982). Nicotinic and muscarinic property of ACh receptors on cultured brain stem neurons, *J. Physiol (Jpn.)* **32**:1015–1021.
Sakmann, B., Patlak, J., and Neher, E. (1980). Single acetylcholine-activated channels show burst-kinetics in presence of desensitizing concentrations of agonist, *Nature* (*London*) **286**:71–73.
Sakmann, B., Hamill, O. P., and Bormann, J. (1983a). Patch-clamp measurements of elementary chloride currents activated by the putative inhibitory transmitters GABA and glycine in mammalian spinal neurons, *J. Neural Transm. Suppl.* **18**:83–95.
Sakmann, B., Noma, A., and Trautwein, W. (1983b). Acetylcholine activation of single muscarinic K^+ channels in isolated pacemaker cells of the mammalian heart, *Nature* (*London*) **303**:250–253.
Sakmann, B., Methfessel, C., Mishina, M., Takahashi, T., Takai, T., Kurasaki, M., Fukuda, K., and Numa, S. (1985). Role of acetylcholine receptor subunits in gating of the channel, *Nature* (*London*) **318**:538–543.
Salmoiraghi, G. C., and Stefanis, C. N. (1965). Patterns of central nervous responses to suspected transmitters, *Arch. Ital. Biol.* **103**:705–724.
Salvaterra, P. M., and Moore, W. J. (1973). Binding of [^{125}I] α-bungarotoxin to particulate fractions of rat and guinea-pig brain, *Biochem. Biophys. Res. Commun.* **55**:1311–1318.
Sasaki, K. (1985). Selective blocking action of tetraethylammonium ion on three types of acetylcholine receptors in *Aplysia* ganglion cells, *Jpn. J. Physiol.* **35**:109–122.
Sastry, B. R. (1980). Excitatory postsynaptic potentials in the mammalian central nervous system associated with an increase in the membrane resistance, *Life Sci.* **27**:1403–1407.
Sato, M., Naruhashi, S., Yai, H., and Shozushima, M. (1983). Effects of various enzymes and chemical modification reagents on the Na^+- and Cl^--dependent responses of the ganglion cells to acetylcholine, *Jpn. J. Physiol.* **53**:757–778.

Sato, T., Sato, M., and Sawada, M. (1976). Effects of disulfide bond reduction on the excitatory and inhibitory postsynaptic responses of *Aplysia* ganglion cells, *Jpn. J. Physiol.* **26**:471–485.

Schmidt, J. T. (1979). Movement of optic terminals in goldfish tectum after local pre- or postsynaptic blockade of transmission, *Soc. Neurosci. Abstr.* **5**:2164.

Schulman, J., and Weight, F. (1976). Synaptic transmission: long-lasting potentiation by a postsynaptic mechanism, *Science* **194**:1437–1439.

Schwartz, R. D., McGee, R., and Kellar, K. J. (1982). Nicotinic cholinergic receptors labeled by [^{3}H]acetylcholine in rat brain, *Mol. Pharmacol.* **22**:56–62.

Schwartzkroin, P. A., and Stafstrom, C. E. (1980). Effects of EGTA on the calcium-activated afterhyperpolarization in hippocampal CA3 pyramidal cells, *Science* **210**:1125–1126.

Scott, K. A., and Mautner, H. G. (1967). Sulfur and selenium isologs related to acetylcholine and choline. IX. Further comparative studies of the pharmacological effects of acetylcholine and its thio and seleno analogs and their hydrolysis products, *Biochem. Pharmacol.* **16**:1903–1918.

Segal, M. (1982). Multiple actions of acetylcholine at a muscarinic receptor studied in the rat hippocampal slice, *Brain Res.* **246**:77–87.

Selyanko, A. A. (1979). Actions of acetylcholine and serotonin on rabbit superior cervical ganglion neurons, Dissertation, Kiev.

Selyanko, A. A. (1983). Mechanism of action of α-neurotoxins from snake venoms on nicotinic cholinoreceptors in rabbit sympathetic ganglion neurons, *Neirofiziologiya* **15**:377–383.

Selyanko, A. A. (1984). Cd^{2+} suppresses a time-dependent Cl^- current in rat sympathetic neuron, *J. Physiol.* (*London*) **350**:49P.

Selyanko, A. A., and Skok, V. I. (1979). Activation of acetylcholine receptors in mammalian sympathetic ganglion neurons, *Prog. Brain Res.* **49**:241–252.

Selyanko, A. A., Derkach, V. A., and Skok, V. I. (1979a). Fast excitatory postsynaptic currents in voltage-clamped mammalian sympathetic ganglion neurons, *J. Auton. Nerv. Syst.* **1**:127–137.

Selyanko, A. A., Skok, V. I., and Derkach, V. A. (1979b). Voltage dependence of excitatory postsynaptic current in mammalian sympathetic ganglion neurons, *Dokl. Akad. Nauk SSSR* **247**:1007–1009.

Selyanko, A. A., Derkach, V. A., and Skok, V. I. (1981a). Kinetics of ionic channels of nicotinic cholinoreceptors in sympathetic ganglion neurons, *Dokl. Akad. Nauk Ukr. SSR*, No. 10:79–83.

Selyanko, A. A., Derkach, V. A., and Skok, V. I. (1981b). Effects of some ganglion-blocking agents on fast excitatory postsynaptic currents in mammalian sympathetic ganglion neurons, in: *Physiology of Excitable Membranes* (J. Salanki, ed.), Akad. Kiado, Budapest, pp. 329–342.

Selyanko, A. A., Derkach, V. A., and Skok, V. I. (1982). Voltage-dependent actions of short-chain polymethylene bis-trimethylammonium compounds on sympathetic ganglion neurons, *J. Auton. Nerv. Syst.* **6**:13–21.

Selyanko, A. A., Derkach, V. A., and Skok, V. I. (1985a). Effect of Ca^{2+} ions on channel-blocking action of hexamethonium in sympathetic ganglion, *Dokl. Akad. Nauk SSSR* **284**:225–228.

Selyanko, A. A., Derkach, V. A., and Skok, V. I. (1985b). Action of a ganglion-blocker on single cholinoreceptor, *Dokl. Akad. Nauk Ukr. SSR*, No. 12:60–63.

Shain, W., Greene, L. A., Carpenter, D. O., Sytkowski, A. J., and Vogel, Z. (1974). *Aplysia* acetylcholine receptors: blockade by and binding of α-bungarotoxin, *Brain Res.* **72**:225–240.

Shefter, E., and Mautner, H. G. (1969). Acetylcholine and its thiolester and selenoester analogs: confirmation, electron distribution, and biological activity, *Proc. Natl. Acad. Sci. USA* **63**:1253–1260.

Shepherd, G. M. (1979). *The Synaptic Organization of the Brain*, Oxford University Press, New York–Oxford.

Shibahara, S., Kubo, T., Perski, H. J., Takahashi, H., Noda, M., and Numa, Sh. (1985). Cloning and sequence analysis of human genomic DNA encoding subunit precursor of muscle acetylcholine receptor, *Eur. J. Biochem.* **146**:15–22.

Shulz, G. E., and Schirmer, R. H. (1979). *Principles of Protein Structure*, Springer-Verlag, New York.

Siegelbaum, S. A., and Tsien, R. W. (1983). Modulation of gated ion channels as a mode of transmitter action, *Trends Neurosci.* **6**:307–313.

Siegelbaum, S. A., Camardo, J. S., and Kandel, E. R. (1982). Serotonin and cAMP close single K channels in *Aplysia* sensory neurons, *Nature (London)* **299**:413–417.

Sigworth, F. J. (1985). Open channel noise. I. Noise in acetylcholine receptor currents suggests conformational fluctuations. *Biophys. J.* **47**:709–720.

Silinsky, E. M., and Ginsborg, B. L. (1983). Inhibition of acetylcholine release from preganglionic frog nerve by ATP but not adenosine, *Nature (London)* **305**:327–328.

Simonneau, M., Tauc, L., and Baux, G. (1980). Quantal release of acetylcholine examined by current fluctuation analysis of an identified neuronal synapse of *Aplysia*, *Proc. Natl. Acad. Sci. USA* **77**:1661–1665.

Sims, S. M., Singer, J. J., and Walsh, J. V. (1985). Cholinergic agonists suppress a potassium current in freshly dissociated smooth muscle cells of the toad, *J. Physiol. (Paris)* **367**:503–529.

Sine, S. M., and Steinbach, J. H. (1984a). Activation of a nicotinic acetylcholine receptor, *Biophys. J.* **45**:175–185.

Sine, S. M., and Steinbach, J. H. (1984b). Agonists block currents through acetylcholine receptor channels, *Biophys. J.* **46**:277–284.

Sine, S. M., and Steinbach, J. H. (1986). Acetylcholine receptor activation by a site-selective ligand: nature of brief open and closed states in BC3H-1 cells, *J. Physiol. (London)* **370**:357–379.

Skok, V. I. (1973). *Physiology of Autonomic Ganglia*, Igaku Shoin, Tokyo.

Skok, V. I. (1980). Ganglionic transmission: morphology and physiology, in: *Pharmacology of Ganglionic Transmission* (D. A. Kharkevich, ed.), Springer-Verlag, Berlin–Heidelberg–New York, pp. 9–38.

Skok, V. I. (1981). Introductory remarks: new approaches in the study of transmission in autonomic ganglia, in: *Physiology of Excitable Membranes* (J. Salanki, ed.), Akad. Kiado, Budapest, pp. 295–304.

Skok, V. I. (1985). Localization and structure of binding sites for bis-ammonium compounds in nicotinic cholinoreceptor, *Biol. Membr.* **2**:245–255.

Skok, V. I. (1986). Channel-blocking mechanism ensures specific blockade of synaptic transmission, *Neuroscience* **17**:1–9.

Skok, V. I., and Heerog, S. S. (1975). Synaptic delay in superior cervical ganglion of the cat. *Brain Res.* **87**:343–353.

Skok, V. I., and Ivanov, A. Ya. (1983a). Analysis of tonic activity in rabbit sympathetic ganglion neurons, *Neirofiziologiya* **15**:295–300.

Skok, V. I., and Ivanov, A. Ya. (1983b). What is the ongoing activity of sympathetic neuron? *J. Auton. Nerv. Syst.* **7**:263–270.

Skok, V. I., and Savich, V. (1967). Electrical properties of neurons and synaptic transmission in sympathetic ganglion, *Fiziol. Zh.* **53**:377–384.

Skok, V. I., and Selyanko, A. A. (1978a). Actions of local ionophoretic application of acetylcholine and serotonin on rabbit superior cervical ganglion neurons, *Neirofiziologiya* **10**:519–524.

Skok, V. I., and Selyanko, A. A. (1978b). Ionic mechanisms of excitatory actions of transmitter, exogenous acetylcholine, and serotonin on rabbit superior cervical ganglion neurons, *Neirofiziologiya* **10**:637–644.

Skok, V. I., and Selyanko, A. A. (1979). Acetylcholine and serotonin receptors in mammalian sympathetic ganglion neurons, in: *Integrative Functions of the Autonomic Nervous System* (Ch. Mcl. Brooks, K. Koizumi, and A. Sato, eds.), University of Tokyo Press and Elsevier–North-Holland Biomedical Press, Amsterdam.

Skok, V. I., Selyanko, A. A., and Derkach, V. A. (1981). Voltage-dependent action of polymethylene bistrimethylammonium compounds on sympathetic ganglion neurons, *Dokl. Akad. Nauk Ukr. SSR,* No. 12:74–75.

Skok, V. I., Selyanko, A. A., and Derkach, V. A. (1982). Two modes of activity of nicotinic acetylcholine receptor channels in sympathetic neurons, *Brain Res.* **238**:480–483.

Skok, V. I., Selyanko, A. A., and Derkach, V. A. (1983). Channel-blocking activity is a possible mechanism for a selective ganglionic blockade, *Pflügers Arch.* **398**:169–171.

Skok, V. I., Selyanko, A. A., Derkach, V. A., Gmiro, V. E., and Lukomskaya, N. Y. (1984). Mechanism of ganglion-blocking actions of bis-ammonium compounds, *Neirofiziologiya* **16**:54–61.

Slater, N. T., Carpenter, D. O., Freedman, J. E., and Snyder, S. H. (1983). Vipoxin both activates and antagonizes three types of acetylcholine response in *Aplysia* neurons, *Brain Res.* **278**:266–270.

Slater, N. T., Carpentr, D. O., Haas, H. L., and David, J. A. (1984). Blocking kinetics at excitatory acetylcholine responses on *Aplysia* neurons, *Biophys. J.* **45**:24–25.

Slater, N. T., Carpenter, D. O., Freedman, J. E., and Snyder, S. H. (1985). Dual effects of the snake venom polypeptide vipoxin on receptors for acetylcholine and biogenic amines in *Aplysia* neurons, *Neuroscience* **14**:723–733.

Smart, L., Meyers, H., and Hilgenfeld, R. (1984). A structural model for the ligand-binding sites at the nicotinic acetylcholine receptor, *FEBS Lett.* **178**:64–68.

Smith, M. A., Margiotta, J. F., and Berg, D. K. (1983). Differential regulation of acetylcholine sensitivity and α-bungarotoxin binding sites on ciliary ganglion neurons in cell culture, *J. Neurosci.* **3**:2395–3402.

Smith, P. A., and Weight, F. F. (1977). Role of electrogenic sodium pump in slow synaptic inhibition is re-evaluated, *Nature* (*London*) **267**:68–70.

Smith, P. A., and Weight, F. F. (1986). The pathway for the slow inhibitory postsynaptic potential in bullfrog sympathetic ganglia, *J. Neurophysiol.* **56**:823–834.

Smolen, A. J. (1983). Specific binding of α-bungarotoxin to synaptic membranes in rat sympathetic ganglion: computer best-fit analysis of electron microscope radioautographs, *Brain Res.* **289**:177–188.

Snetkov, V. A. (1977). Effects of phenylalkylamines on voltage dependence of end-plate current in frog muscle fiber, *Dokl. Akad. Nauk SSSR* **234**:947–950.

Soejima, M., and Noma, A. (1984). Mode of regulation of the ACh-sensitive K-channel by the muscarinic receptor in rabbit cells, *Pflügers Arch.* **400**:424–431.

Sokolovsky, M., Gurwitz, D., and Kloog, J. (1983). Biochemical characterization of the muscarinic receptors, *Adv. Enzymol.* **55**:137–196.

Spivak, C. E., Gund, T. M., Liang, R. F., and Waters, J. A. (1986). Structural and electronic requirements for potent agonists at a nicotinic receptor, *Eur. J. Pharmacol.* **120**:127–131.

Stallcup, W. B., and Patrick, J. (1980). Substance P enhances cholinergic receptor desensitization in a clonal nerve cell line, *Proc. Natl. Acad. Sci. USA* **77**:634–638.

Steinbach, J. H. (1980). Activation of nicotinic acetylcholine receptors, in: *The Cell Surface and Neuronal Function* (C. W. Cotman et al., eds.), Elsevier–North-Holland, Amsterdam, pp. 119–156.

Stevens, C. F. (1972). Inferences about membrane properties from electric noise measurement, *Biophys. J.* **22**:1028–1047.

Stevens, C. F. (1976). Molecular basis for postjunctional conductance increases induced by acetylcholine, in: The Synapse, *Cold Spring Harbor Symp. Quant. Biol.* **40**:169–173.

Stroud, R. M. (1974). A family of protein-cutting proteins, *Sci. Am.* **231**:74–89.

Stubbins, J. F., and Hudgins, P. M. (1971). Sulfhydryl group reagents. Effect on intestinal smooth muscle, *Experientia* **27**:669–670.

Suria, A. (1976). Cyclic GMP modulates the intensity of post-tetanic potentiation in bullfrog sympathetic ganglia, *Neuropharmacology* **15**:11–16.

Szepaniak, A. C. (1974). Effect of α-bungarotoxin and *Dendroaspis* neurotoxins on acetylcholine responses of snail neurons, *J. Physiol.* (*London*) **241**:55–56P.

Takahashi, T., Kuno, M., and Numa, Sh. (1985). Location of functional regions of acetylcholine receptor α subunit by site-directed mutagenesis, *Nature* (*London*) **313**:364–369.

Takai, T., Noda, M., Furutani, Y., Takahashi, H., Notake, M., and Shimizu, Sh. (1984). Primary structure of γ-subunit precursor of calf-muscle acetylcholine receptor deduced from the cDNA sequence, *Eur. J. Biochem.* **143**:109–115.

Takeda, K., and Trautmann, A. (1984). A patch-clamp study of the partial agonist actions of tubocurarine on rat myotubes, *J. Physiol.* (*London*) **349**:353–374.

Takeda, K., Barry, P. H., and Gage, P. W. (1980). Effect of ammonium ions on endplate channels, *J. Gen. Physiol.* **75**:589–613.

Takeshige, C., and Volle, R. L. (1964a). The effects of barium and other inorganic cations on sympathetic ganglia, *J. Pharmacol. Exp. Ther.* **146**:327–334.

Takeshige, C., and Volle, R. L. (1964b). Modification of ganglionic responses to cholinomimetic drugs following preganglionic stimulation, anticholinesterase agents, and pilocarpine, *J. Pharmacol. Exp. Ther.* **146**:335–343.

Takeuchi, A., and Takeuchi, N. (1959). Active phase of frog's end-plate potential, *J. Neurophysiol.* **22**:395–411.

Tanabe, T., Noda, M., Furutani, Y., Takai, T., and Takahashi, H. (1984). Primary structure of β-subunit precursor of calf-muscle acetylcholine receptor deduced from cDNA sequence, *Eur. J. Biochem.* **144**:11–17.

Tanaka, K., Minota, S., Kuba, K., Koyano, K., and Abe, T. (1986). Differential effects of apamine on Ca^{2+}-dependent K^+ currents in bullfrog sympathetic ganglion cells, *Neurosci. Lett.* **69**:233–238.

Tashiro, N., and Nishi, S. (1972). Effects of alkali-earth cations on sympathetic ganglion cell of the rabbit, *Life Sci.* **11**:941–948.

Tauc, L., and Gerschenfeld, H. (1962). A cholinergic mechanism of inhibitory synaptic transmission in a molluscan nervous system, *J. Neurophysiol.* **25**:236–262.

Tiedt, T. M., Albuquerque, E. X., Barky, N. M., Eldefrawi, M. E., and Eldefrawi, A. T. (1979). Voltage- and time-dependent actions of piperocaine on the ion channel of acetylcholine receptor, *Mol. Pharmacol.* **16**:909–921.

Tokimasa, T. (1984). Muscarinic agonists depress calcium-dependent g_K in bullfrog sympathetic neurons, *J. Auton. Nerv. Syst.* **10**:107–116.

Tokimasa, T. (1985a). Intracellular Ca^{2+}-ions inactivate K^+-current in bullfrog sympathetic neurons, *Brain Res.* **337**:386–391.

Tokimasa, T. (1985b). Spontaneous muscarinic suppression of the Ca-activated K-current in bullfrog sympathetic neurons, *Brain Res.* **344**:134–141.

Tokimasa, T., and North, R. A. (1984). Calcium entry through acetylcholine-channels can activate potassium conductance in bullfrog sympathetic neurons, *Brain Res.* **295**:364–367.

Toldi, J., Joo, F., Adam, G., Beher, O., and Wolff, J. R. (1983). Inhibition of synaptic transmission in the rat superior cervical ganglion by intracarotid infusion of bungarotoxin, *Brain Res.* **262**:323–327.

Tosaka, T., and Libet, B. (1965). Slow postsynaptic potentials recorded intracellularly in sympathetic ganglia of frog, in: *Abstr. 23rd Intern. Congr. Physiol. Sci.*, Tokyo, p. 386.

Tosaka, T., Tosaka, J., Miyazaki, T., and Libet, B. (1983). Hyperpolarization following activation of K^+ channels by excitatory postsynaptic potentials, *Nature (London)* **305**:148–150.

Tosaka, T., Chichibu, S., and Libet, B. (1968). Intracellular analysis of slow inhibitory and excitatory postsynaptic potentials in sympathetic ganglia of the frog, *J. Neurophysiol.* **31**:396–409.

Trautmann, A. (1982). Curare can open and block ionic channels associated with cholinergic receptors. *Nature (London)* **298**:272–275.

Trautwein, W., Taniguchi, J., and Noma, A. (1982). The effect of intracellular cyclic nucleotides and calcium on the action potential and acetylcholine response of isolated cardiac cells, *Pflügers Arch.* **392**:307–314.

Trčka, V. (1980a). Relationship between chemical structure and ganglion-blocking activity: (a) Quaternary ammonium compounds, in: *Pharmacology of Ganglionic Transmission* (D. A. Kharkevich, ed.), Springer-Verlag, Berlin, pp. 123–153.

Trčka, V. (1980b). Relationship between chemical structure and ganglion-blocking activity: (b) Tertiary and secondary amines, in: *Pharmacology of Ganglionic Transmission* (D. A. Kharkevich, ed.), Springer-Verlag, Berlin, pp. 155–184.

Triggle, D. J. (1965). *Chemical Aspects of the Autonomic Nervous System*, Academic Press, London–New York.

Triggle, D. J., and Triggle, C. R. (1976). *Chemical Pharmacology of the Synapse*, Academic Press, London–New York.

Trinus, K. F., and Skok, V. I. (1979). Disulfide bonds in cholinoreceptors of frog sympathetic ganglion neurons, *Neirofiziologiya* **11**:593–600.

Tsetlin, V. I., Pluzhnikov, K. A., Karelin, A. A., and Ivanov, V. I. (1983). Acetylcholine receptor interaction with the neurotoxin, II. Photoactivable derivatives, in: *Toxins as Tools in Neurochemistry*, Walter de Gruyter, Berlin–New York, pp. 162–169.

Tsetlin, V. I., Pluzhnikov, K. A., Karelin, A. A., Karlsson, E., and Ivanov, V. T. (1984). Mutual deposition of subunits in acetylcholine receptor and neurotoxin bound to it, *Bioorg. Khim.* **10**:176–187.

Tuček, S. (1978). *Acetylcholine Synthesis in Neurons*, Chapman and Hall, London.

Turpaev, T. M. (1962). *Mediator Function of Acetylcholine and Nature of Cholinoreceptor* [in Russian], Izd. Akad. Nauk SSSR, Moscow

Turpaev, T. M., Yurchenko, O. A., and Sh.-Rozsa, K. (1983). Two types of interaction between serotonin and acetylcholine in identified neurons of snail brain, *Dokl. Akad. Nauk SSSR* **270**:1505–1508.

Turpaev, T. M., Yurchenko, O. A., and Grigorjev, N. G. (1986). Changes in *Limnea stagnalis* neuronal response to acetylcholine during intracellular and extracellular perfusion with serotonin. *Neirofiziologiya* **18**:238–243.

Uchida, S., Matsumoto, K., Takeyasu, K., Higuchi, H., and Yoshida, A. (1982). Molecular mechanism of the effects of guanine nucleotide and sulfhydryl reagent on muscarinic receptors in smooth muscle studied by radiation inactivation, *Life Sci.* **31**:201–209.

Van Helden, D., Hamill, O. P., and Gage, P. W. (1977). Permeant cations alter end-plate channel characteristics, *Nature* (*London*) **269:**711–713.

Van Rossum, J. M. (1962a). Classification and molecular pharmacology of ganglionic blocking agents. I. Mechanisms of ganglionic synaptic transmission and mode of action of ganglionic stimulants, *Int. J. Neuropharmacol.* **1**:97–110.

Van Rossum, J. M. (1962b). Classification and molecular pharmacology of ganglionic blocking agents, II. Mode of action of competitive and noncompetitive ganglion blocking agents, *Int. J. Neuropharmacol.* **1**:403–421.

Varanda, W. A., Aracava, Y., Sherby, S. M., Van Meter, W. G., Eldefrawi, M. E., and Albuquerque, E. X. (1985). The acetylcholine receptor of the neuromuscular junction recognizes mecamylamine as a noncompetitive antagonist, *Mol. Pharmacol.* **28**:128–137.

Villegas, J. (1975). Characterization of acetylcholine receptors in the Schwann cell membrane of the squid nerve fiber, *J. Physiol.* (*London*) **249**:679–689.

Vladimirova, I. A., and Shuba, M. F. (1978). Effect of strychnine, hydrastine, and apamine on synaptic transmission in smooth muscle cells, *Neirofiziologiya* **10**:295–299.

Vogel, S. M., Watanabe, S., Yeh, J. Z., Farley, J. M., and Narahashi, T. (1984). Current-dependent block of end-plate channels by guanidine derivatives, *J. Gen. Physiol.* **83**:901–918.

Volkenstein, M. V. (1975). *Molecular Biophysics* [in Russian], Nauka, Moscow.

Volle, R. L. (1962). The actions of several ganglion blocking agents on the postganglionic discharge induced by diisopropyl phosphorofluoridate (DFP) in sympathetic ganglia, *J. Pharmacol. Exp. Ther.* **135**:45–53.

Volle, R. L. (1980). Nicotinic ganglion-stimulating agents, in: *Pharmacology of Ganglionic Transmission* (D. A. Kharkevich, ed.), Springer-Verlag, Berlin, pp. 281–312.

Volle, R. L., Quenzer, L. F., Patterson, B. A., Alkadhi, K. A., and Henderson, E. G. (1981). Cyclic guanosine 3',5'-monophosphate accumulation and ^{45}Ca-uptake by rat superior cervical ganglia during preganglionic stimulation, *J. Pharmacol. Exp. Ther.* **219**:338–343.

Vulfius, C. A. (1978). Study of cholinoreceptors by the chemical modification method, in: *Cholinoreceptors* (B. N. Verprintsev, ed.) [in Russian], VINITI, Moscow.

Vulfius, C. A., Yurchenko, O. P., and Zeimal, E. V. (1969). On the mutual disposition of cholinoreceptors on the membrane of gastropoda neurons and cholinesterase activity of neurons, *Dokl. Akad. Nauk SSSR* **186**:1445–1448.

Vulfius, C. A., Yurchenko, O. P., Il'in, V. I., Bregestovski, P. D., and Verprintsev, B. N. (1979). Acetylcholine receptor of *Limnaea stagnalis* neurons. Study of the active site functional groups, in: *The Cholinergic Synapse* (S. Tucek, ed.), *Prog. Brain Res.* **49**:293–302.

Vulfius, C. A., Yurchenko, O. P., Khromov-Borisov, N. V., Alexandrova, L. N., and Andrianova, I. P. (1980), Interaction between acetylcholine and other agonists at cholinoreceptive membrane of mollusc neurons, *Brain Res.* **200**:190–194.

Wamsley, J. K. (1983). Muscarinic cholinergic receptors undergo axonal transport in the brain, *Eur. J. Pharmacol.* **86**:309–310.

Wamsley, J. K., Zarbin, M. A., and Kuhar, M. J. (1981). Muscarinic receptors flow in the sciatic nerve, *Brain Res.* **217**:155–161.

Waser, P. G. (1961). Chemistry and pharmacology of muscarine, muscarone, and some related compounds, *Pharmacol. Rev.* **13**:465–515.

Watanage, S., and Koketsu, K. (1973). 5-HT hyperpolarization of bullfrog sympathetic ganglion cell membrane, *Experientia* **29**:1370–1372.

Watson, M., Roeske, W. R., Johnson, P. C., and Yamamura, H. J. (1984). [^{3}H]pirenzepine identifies putative M_1 muscarinic receptors in human stellate ganglia, *Brain Res.* **290**:179–182.

Weight, F. F., and Padjen, A. (1973a). Slow synaptic inhibition: evidence for synaptic inactivation of sodium conductance in sympathetic ganglion cells, *Brain Res.* **55**:219–224.

Weight, F. F., and Padjen, A. (1973b). Acetylcholine and slow synaptic inhibition in frog sympathetic ganglion cells, *Brain Res.* **55**:225–228.

Weight, F. F., and Smith, P. A. (1981). IPSP reversal: evidence for increased potassium conductance combined with decreased sodium conductance, *Adv. Physiol. Sci.* **4**:351–354.

Weight, F. F., and Votava, J. (1970). Slow synaptic excitation in sympathetic ganglion cells: evidence for synaptic inactivation of potassium conductance, *Science* **170**:755–757.

Weight, F. F., Petzold, G., and Greengard, P. (1974). Guanosine-3',5'-monophosphate in sympathetic ganglia: increase associated with synaptic transmission, *Science* **186**:942–944.

Weight, F. F., Smith, P. A., and Schulman, J. A. (1978). Postsynaptic potential generation appears independent of synaptic elevation of cyclic nucleotides in sympathetic neurons, *Brain Res.* **158**:197–202.

Weitsen, H. A., and Weight, F. F. (1977). Synaptic innervation of sympathetic ganglion cells in the bullfrog, *Brain Res.* **128**:192–211.

Werman, R. (1969). An electrophysiological approach to drug-receptor mechanisms, *Comp. Biochem. Physiol.* **30**:997–1017.

White, R. L., and Gardner, D. (1983). Physostigmine prolongs the elementary event underlying decay of inhibitory postsynaptic currents in *Aplysia*, *J. Neurosci.* **3**:2607–2613.

White, A., Hendler, F., Smith, E., Hill, R. L., and Lehman, I. R. (1978). *Principles of Biochemistry*, McGraw-Hill, New York.

Whiting, P., and Lindstrom, J. (1987a). Purification and characterization of a nicotinic acetylcholine receptor from rat brain, *Neurobiology* **84**:595–599.

Whiting, P., and Lindstrom, J. (1987b). Affinity labelling of neuronal acetylcholine receptors localizes acetylcholine-binding sites to their α-subunits, *FEBS Lett.* **213**:55–60.

Wien, R., and Mason, D. F. J. (1953). The pharmacological actions of a series of phenylalkane *p*-ω-bis(trialkylammonium) compounds, *Br. J. Pharmacol.* **8**:306–314.

Wien, R., Mason, D. F. J., Edge, N. D., and Langston, G. T. (1952). The ganglion blocking properties of homologous compounds in the methonium series, *Br. J. Pharmacol.* **7**:534–541.

Wilson, S. P., Lentz, T. L., and Hawrot, E. (1985). Determination of the primary amino acid sequence specifying the α-bungarotoxin binding site on the α subunit of the acetylcholine receptor from *Torpedo californica*, *Neurobiology* **82**:8790–8794.

Wilson, S. P., and Kirshner, N. (1977). The acetylcholine receptor of the adrenal medulla, *J. Neurochem.* **28**:687–695.

Witte, O. W., Speckmann, E.-J., and Walden, J. (1985a). Acetylcholine responses of identified neurons in *Helix pomatia*, II. Pharmacological Properties of acetylcholine responses, *Comp. Biochem. Physiol.* **C80**:25–35.

Witte, O. W., Speckmann, E.-J., and Walden, J. (1985b). Acetylcholine responses of identified neurons in *Helix pomatia*, III. Ionic composition of the depolarizing currents induced by acetylcholine, *Comp. Biochem. Physiol.* **C80**:37–45.

Wood, J. D., and Mayer, C. J. (1979). Serotonergic activation of tonic type enteric neurons in guinea-pig small bowel, *J. Neurophysiol.* **42**:582–593.

Woodhull, A. M. (1973). Ionic blockade of sodium channels in nerve, *J. Gen. Physiol.* **61**:687–708.

Woodward, J. K., Bianchi, C. P., and Erulkar, S. D. (1969). Electrolyte distribution in rabbit superior cervical ganglion, *J. Neurochem.* **16**:289–299.

Yamamoto, D., Jay, Z., and Narahashi, T. (1984). Voltage-dependent calcium block of normal and tetrodotoxin-modified single sodium channels, *Biophys. J.* **45**:337–344.

Yamamura, H. I., and Snyder, S. H. (1974). Postsynaptic localization of muscarinic cholinergic receptor binding in rat hippocampus, *Brain Res.* **78**:320–326.

Yoshida, K., and Imura, H. (1979). Nicotinic cholinergic receptors in brain synaptosomes, *Brain Res.* **172**:453–459.

Young, E. F., Ralson, E., Blake, J., Ramachandran, J., Hall, Z. W., and Stroud, R. M. (1985). Topological mapping of thc acctylcholinc rcccptor: cvidcncc for a model with five transmembrane segments and a cytoplasmic COOH-terminal peptide, *Proc. Natl. Acad. Sci. USA* **82**:626–630.

Yu, L., LaPolla, R. J., and Davidson, N. (1986). Mouse muscle nicotinic acetylcholine receptor α-subunit: cDNA sequence and gene expression, *Nucleic Acids Res.* **14**:3539–3554.

Zarbin, M. A., Wamsley, J. K., and Kuhar, M. J. (1982). Axonal transport of muscarinic cholinergic receptors in rat vagus nerve: high and low affinity agonist receptors move in opposite directions and differ in nucleotide sensitivity, *J. Neurosci.* **2**:934–941.

Zeimal, E. V., and Katchman, A. N. (1983). Blocking action of furosemide on acetylcholine and γ-aminobutyric acid-induced membrane chloride conductances in molluscan neurons, *Neirofiziologiya* **15**:589–595.

Zhorov, B. S. (1986). Determination of productive conformations for ligands and topography of adrenoreceptors with the use of theoretical conformational analysis, Dissertation, Kiev.

Zingsheim, H. P., Barrantes, F. J., Hanicke, W., and Neugebauet, D.-C.(1982). Direct structural localization of two toxin-recognition sites on an ACh receptor protein, *Nature* (*London*) **299**:81–84.

Ziskind, L., and Werman, R. (1975a). Sodium ions are necessary for cholinergic desensitization in molluscan neurons, *Brain Res.* **88**:171–176.

Ziskind, L., and Werman, R. (1975b). At least three molecules of carbamylcholine are needed to activate a cholinergic receptor, *Brain Res.* **88**:177–180.

INDEX

MIX
Papier aus verantwortungsvollen Quellen
Paper from responsible sources
FSC® C105338

If you have any concerns about our products,
you can contact us on
ProductSafety@springernature.com

In case Publisher is established outside the EU,
the EU authorized representative is:
Springer Nature Customer Service Center GmbH
Europaplatz 3, 69115 Heidelberg, Germany

Printed by Libri Plureos GmbH
in Hamburg, Germany